Advances in

BOTANICAL RESEARCH

incorporating *Advances in Plant Pathology*

Advances in

BOTANICAL RESEARCH

incorporating *Advances in Plant Pathology*

VOLUME 31

PLANT TRICHOMES

Advances in

BOTANICAL RESEARCH

incorporating *Advances in Plant Pathology*

PLANT TRICHOMES

Edited by

D. L. HALLAHAN

Central Research and Development
The Dupont Company
Biochemical Sciences and Engineering
Wilmington, DE, USA

and

J. C. GRAY

Department of Plant Sciences
University of Cambridge
Cambridge, UK

Series Editor

J. A. CALLOW

School of Biological Sciences
University of Birmingham
Birmingham, UK

VOLUME 31

2000

ACADEMIC PRESS

A Harcourt Science and Technology Company
San Diego San Francisco New York Boston
London Sydney Tokyo

This book is printed on acid-free paper

Academic Press
32 Jamestown Road, London NW1 7BY, UK
http://www.academicpress.com

Academic Press
A Harcourt Science and Technology Company
525 B Street, Suite 1900, San Diego, California 92101-4495, USA
http://www.academicpress.com

ISBN 0-12-005931-2

A catalogue for this book is available from the British Library

Typeset by Keyset Composition, Colchester, Essex

Printed and bound by Antony Rowe Ltd, Eastbourne

Transferred to digital printing 2005

00 01 02 03 04 05 MP 9 8 7 6 5 4 3 2 1

CONTENTS

Monoterpenoid Biosynthesis in Glandular Trichomes of Labiate Plants

D. L. HALLAHAN

Current and Potential Exploitation of Plant Glandular Trichome Productivity

S. O. DUKE, C. CANEL, A. M. RIMANDO, M. R. TELLEZ, M. V. DUKE and R. N. PAUL

Chemotaxonomy Based on Metabolites from Glandular Trichomes

O. SPRING

Anacardic Acids in Trichomes of *Pelargonium*: Biosynthesis, Molecular Biology and Ecological Effects

D. J. SCHULTZ, J. I. MEDFORD, D. COX-FOSTER, R. A. GRAZZINI, R. CRAIG and R. O. MUMMA

Specification of Epidermal Cell Morphology

B. J. GLOVER and C. MARTIN

Trichome Initiation in *Arabidopsis*

A. R. WALKER and M. D. MARKS

Trichome Differentiation and Morphogenesis in *Arabidopsis*

M. HÜLSKAMP and V. KIRIK

CONTRIBUTORS TO VOLUME 31

C. CANEL *USDA, ARS, Natural Products Utilization Research Unit, P.O. Box 8048, University, Mississippi 38776, USA*

D. COX-FOSTER *Department of Entomology, College of Agricultural Science, The Pennsylvania State University, University Park, PA 16802, USA*

R. CRAIG *Department of Horticulture, The Pennsylvania State University, University Park, PA 16802, USA*

M. V. DUKE *USDA, ARS, Southern Weed Science Research Unit, P.O. Box 350, Stoneville, MS 38776, USA*

S. O. DUKE *USDA, ARS, Natural Products Utilization Research Unit, P.O. Box 8048, University, Mississippi 38776, USA*

A. FAHN *Department of Plant Sciences, The Hebrew University of Jerusalem, Givat Ram, 91904 Jerusalem, Israel*

B. J. GLOVER *Department of Plant Sciences, University of Cambridge, Downing Street, Cambridge, CB2 3EA, UK*

R. A. GRAZZINI *Center Analytical Laboratories, State College, 487 Nimitz Avenue State College, PA 00016, USA*

D. L. HALLAHAN *Biochemical Sciences and Engineering, Central Research and Development, The Dupont Company, P.O. Box 80328, Wilmington, DE 19880-0328, USA*

M. HÜLSKAMP *Institut für Entwicklungsgenetik, Auf der Morgenstelle 1, D-72076 Tübingen, Germany*

V. KIRIK *Institut für Entwicklungsgenetik, Auf der Morgenstelle 1, D-72076 Tübingen, Germany*

M. D. MARKS *Department of Genetics and Cell Biology, University of Minnesota, St Paul, MN 55108, USA*

C. MARTIN *Department of Genetics, John Innes Centre, Norwich Research Park, Colney Lane, Norwich, NR4 7UH, UK*

J. I. MEDFORD *Department of Biology, Colorado State University, Fort Collins, CO 80523, USA*

R. O. MUMMA *Department of Entomology, College of Agricultural Science, The Pennsylvania State University, University Park, PA 16802, USA*

R. N. PAUL *USDA, ARS, Southern Weed Science Research Unit, P.O. Box 350, Stoneville, MS 38776, USA*

A. M. RIMANDO *USDA, ARS, Natural Products Utilization Research Unit, P.O. Box 8048, University, Mississippi 38776, USA*

D. J. SCHULTZ *Department of Botany and Plant Pathology, 166 Plant Biology Laboratory, Michigan State University, East Lansing, MI 48824, USA*

O. SPRING *Institut für Botanik, Universität Hohenheim (210), Postfach 70593, Stuttgart, Germany*

M. R. TELLEZ *USDA, ARS, Natural Products Utilization Research Unit, P.O. Box 8048, University, Mississippi 38776, USA*

E. WAIGMANN *Department of Biochemistry, University of Vienna, Dr Bohrg. 9, A-1030 Austria*

A. R. WALKER *Department of Plant Sciences, University of Cambridge, Downing Street, Cambridge CB2 3EA, UK*

E. WERKER *Department of Botany, The Hebrew University of Jerusalem, Givat Ram, 91904 Jerusalem, Israel*

P. ZAMBRYSKI *Department of Plant and Microbial Biology, 111 Koshland Hall, University of California, Berkeley, CA 94720, USA*

CONTENTS OF VOLUMES 18–30

Contents of Volume 18

Contents of Volume 19

Contents of Volume 20

Contents of Volume 21

Contents of Volume 22

Contents of Volume 23

Contents of Volume 24

Contents of Volume 25

THE PLANT VACUOLE

Contents of Volume 26

Contents of Volume 27

CLASSIC PAPERS

Contents of Volume 28

Contents of Volume 29

Contents of Volume 30

PREFACE

All outgrowths from the epidermis are termed "trichomes". In plants, therefore, the term strictly encompasses all unicellular and multicellular structures emanating from the epidermis, including root hairs and larger "emergences" such as thorns. However, the term is more commonly used to refer to hairs emerging specifically from the aerial epidermis of plants. These may be simple hairs or more elaborate structures, such as glands that accumulate a variety of phytochemicals, many of which (e.g. the essential oils of the Labiate plants) are items of commerce.

Plant trichomes have interested botanists ever since the development of the microscope in the seventeenth century. Over the centuries, and in particular during the latter, this interest has expanded from an initial fascination with their structural diversity to encompass studies of their ultrastructure, biochemistry, differentiation and development, and physiological and ecological roles. In 1962, Uphof updated Netolitzky's work of 30 years earlier (*Encyclopedia of Plant Anatomy: Plant Hairs*), which provided an overview of studies of plant trichomes to that date. Over 20 years later, in 1984, Rodriguez, Healey and Mehta edited a volume (*Biology and Chemistry of Plant Trichomes*) which originated in a symposium held in 1980. Since then, significant advances in our understanding of trichome development, biochemistry and ecological significance have been made. It therefore seemed appropriate at this time to bring together some of this work in this themed volume.

The various chapters in this volume document developments in a number of areas of research involving plant trichomes and indicate opportunities for further research. An update on the diversity of structure and functions of plant trichomes is provided by Ella Werker, and this sets the stage for subsequent chapters. Glandular trichomes and their productivity is presently an area of considerable interest. The relationship between structure and secretion is described by Abraham Fahn, the biosynthesis of monoterpenoids of Labiate oils by one of the editors, and the exploitation of this productivity is discussed by Stephen O. Duke and colleagues. The use of trichome-accumulated chemicals for plant taxonomy is then discussed by Otmar Spring. The biosynthesis of anacardic acids in glandular trichomes of *Pelargonium* spp., and its ecological significance, is discussed in a chapter describing some 40 years of cross-disciplinary collaborative work. This chapter emphasizes that research on glandular trichomes has often been a meeting-point for scientists with differing areas of expertise and interests. In addition to being a "focal point of chemical and structural interactions"

(Duke, S. O. *International Journal of Plant Sciences* **155**, 617–620, 1994), trichomes are now a focus for the molecular genetic analysis of cell differentiation and development (Chasan, R. *The Plant Cell* **6**, 1029–1031, 1994). This is largely because they are both visible and (in the growth chamber at least!) non-essential. Studies of the differentiation and development of the simple trichomes of *Arabidopsis* have yielded significant advances in our understanding of these processes, and a number of genes involved have been identified. The molecular aspects of trichome differentiation and development are discussed in chapters authored by Beverley J. Glover and Cathie Martin; Mandy Walker and M. David Marks; and Martin Hülskamp and Viktor Kirik. Finally, an example of using plant trichomes to study another fundamental aspect of plant cell biology, cell-to-cell movement through plasmodesmata, is described by Elisabeth Waigmann and Patricia Zambryski.

We would like to thank all the authors, as well as the staff involved at our publisher (Sarah Stafford, Clare Nehammer) and the Editorial Board of *Advances in Botanical Research* for helping us to pull together this timely volume. Given the current pace of research in this area, we are confident that it will not be another 20 years before research on plant trichomes merits publication of a similar compendium.

David L. Hallahan
John C. Gray

Trichome Diversity and Development

E. WERKER

Department of Botany, The Hebrew University of Jerusalem, Jerusalem 91904, Israel

I. INTRODUCTION AND DEFINITIONS

A great variety of projections may be found on the surface of plant organs, and a very large number of species bear them. They vary in size, shape, number of cells, origin, location, function, etc. No method of their classification is wholly satisfactory; each has its own drawbacks, exceptions and overlaps. Different methods of classification were summarized by Uphof (1962), who added in conclusion that the final decision for classification is subjective.

Among epidermal appendages, trichomes constitute (structurally and developmentally) an intermediate group between papillae and emergences (Figs 1–3). The difference between trichomes and emergences is based on

Advances in Botanical Research Vol. 31
ISBN 0-12-005931-2

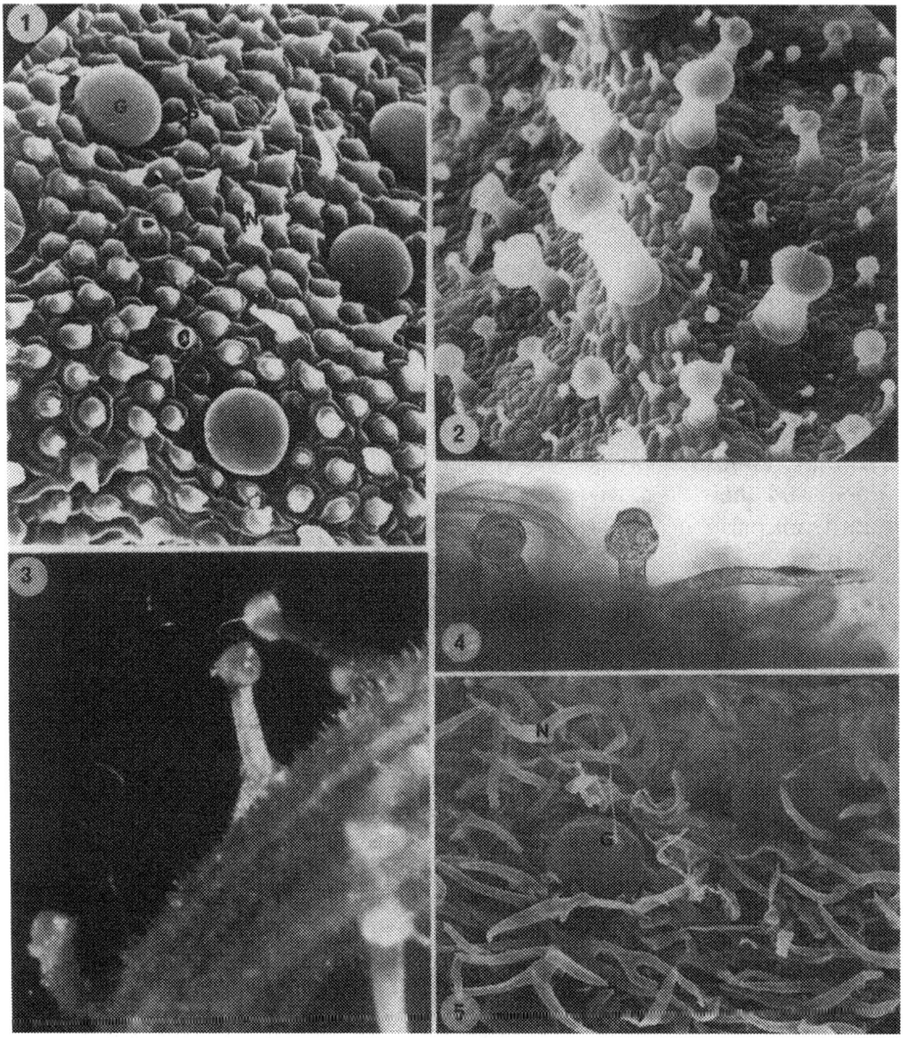

Fig. 1. *Coridothymus capitatus* leaf with papillae (P), non-glandular hairs (N) and peltate glandular trichomes (G), ×159. Fig. 2. Calyx of *Salvia sclarea* with various types of capitate glandular trichomes, ×106. Fig. 3. Calyx of *Plumbago capensis* with mucilage-secreting emergences, ×20. Figs 4 and 5. Trichomes of *Rosmarinus officinalis* leaf. Fig. 4. Branched glandular trichome, ×363. Fig. 5. Dendritic non-glandular (N) and peltate (G) trichomes, ×291. (Figs 1 and 4 reproduced, with permission from Laser Pages Publishing, from Werker *et al.*, 1985a.)

ontogenetic development, although the end result may be similar in shape and function. *Trichomes* are defined as unicellular or multicellular appendages, which originate from epidermal cells only, and develop outwards on the surface of various plant organs. Some authors restrict the definition to projections produced by a single epidermal cell (see Uphof, 1962). *Emergences* are appendages that, in addition to epidermal cells, originate also from subepidermal tissues. The distinction between trichomes and emergences is, however, not always clear owing to many intergrading forms. The number of subepidermal cells included in the construction of an emergence may be from one to all of the tissues that locally constitute the organ. There are cases in which no subepidermal cells take part in the construction of a trichome but they are differentiated locally below the trichome. A definition of the type of projection may thus be impossible without ontogenetic study. Therefore, some authors use "trichomes" as a general term for both trichomes and emergences, for practical purposes.

The trichomes are considered as such when they include both cell wall and lumen, and their height reaches above a certain critical size, to distinguish them from papillae, bladder cells, tubercules, epidermal glands, etc. Distinction between trichomes and the latter projections is also not always easy, since one type passes gradually into the other (Fig. 1). The distinction, therefore, has been based on the ratio between the width and height of the epidermal appendages, and arbitrarily deciding on the boundary between the two (Barthlott and Ehler, 1977). For capitate trichomes, it has been defined that the length of the stalk should be more than half the height of the head (Abu-Asab and Cantino, 1987). Among salt glands, for example, there is a gradual increase in height, from the epidermal glands of *Tamarix* and *Limonium*, which are more or less level with the epidermis, through salt glands of certain grasses with part of their basal cell and a small head cell protruding above the epidermis, to *Atriplex*, whose glandular hairs consist of 1–3 elongated stalk cells and a large bladder cell. During leaf development, owing to unequal growth in the leaf tissue, trichomes may become sunken to various extents (Fig. 14), up to almost complete enclosure within cavities.

In seeds and fruits the adaxial surface of the seed coat and pericarp, in contrast to leaves, is internal. In certain species this inner surface bears trichomes. In the fruit of *Acer pseudoplatanus* the endocarp bears internal hairs (Kaniewski and Wazynska, 1970) and in seeds of *Symphonia clusioides* fibre-like hairs protrude from the endotegmen towards the embryo (Netolitzky, 1926; Corner, 1976). More exceptional are the uniseriate multicellular hairs of the fruit of *Anacardium occidentale*, which are formed by the endothelium of its secretory cavities (Roth, 1977).

A gross distinction between trichomes has been made, according to shape, into hairs, scales, bristles, etc. However, the term "hair" is often much more liberally used and the other forms are coined as scaly or peltate hairs, bristle hairs, etc.

II. TRICHOME VARIABILITY AND CLASSIFICATION

Trichomes can be found on all parts of the plant: vegetative (leaves, young stems and bracts, and roots a short distance above their apical meristem); and reproductive (sepals, petals, stamens, gynoecium, seeds and fruits). The function of the trichomes may vary in accordance with the organ that bears them. A distinction is also made between all the trichomes that grow on the aerial parts of the plant and those that grow on the subterranean parts. Trichomes of the first group are very diverse in their features, while those of the second consist almost exclusively of root hairs.

Owing to the enormous diversity of trichomes in morphology, origin, size, location, surface microstructure, capability to secrete, mode of secretion, timing of activity, function, etc., classification is not easy. The use of any one of the above criteria results in overlapping of characteristics between types. Classification into types is therefore based on a combination of features, and a different combination may be needed for different groups of plants.

The major distinction between trichomes is between "glandular" and "non-glandular" trichomes. The main characteristic for distinction within non-glandular trichomes is their morphology. The primary characteristic for distinction between types of glandular trichomes is usually the secretory materials they excrete, accumulate or absorb. Morphology and function are usually characteristics for further classification within each such group.

According to their shape, trichomes have acquired descriptive terms, such as stellate, T-shaped, dendritic, scaly, shaggy, spiny, hooked, rounded, flattened, spiral, etc. (e.g. Solereder, 1908; Metcalfe and Chalk, 1950, 1979; Carlquist, 1961; Hummel and Staesche, 1962; Uphof, 1962; Fahn, 1990). A glossary of terminology in English for plant trichomes, non-glandular and glandular, was written by Payne (1978). It deals with descriptors of individual trichomes, and with terms having to do with the indumentum or masses of hair with special appearance or function.

There are, however, trichomes which cannot be defined unambiguously into one of the above two types – glandular or non-glandular. For instance, in several genera of the Lamiaceae, the leaf bears unbranched glandular hairs and branched non-glandular hairs. In addition, compound glandular hairs can be observed similar to branched non-glandular hairs but with one glandular branch (*Meriandra*, Bokhari and Hedge, 1971; *Rosmarinus officinalis*, Werker *et al.*, 1985a; *Phlomis*, Azizian and Cutler, 1982; *Hyptis*, Rudall, 1980) (Figs 4–9). In *Rosmarinus* the lowermost stalk cell in these hairs had either very thick cutinized walls as in all the non-glandular hairs of the species or relatively thin, non-cutinized walls, as in part of the glandular hairs. The question arises of when final determination of the type of hair, glandular or non-glandular, occurs during hair development.

A different example is that of *Nicotiana tabacum* (Nielsen *et al.*, 1991). The leaves of this species bear glandular trichomes consisting of a multicellular

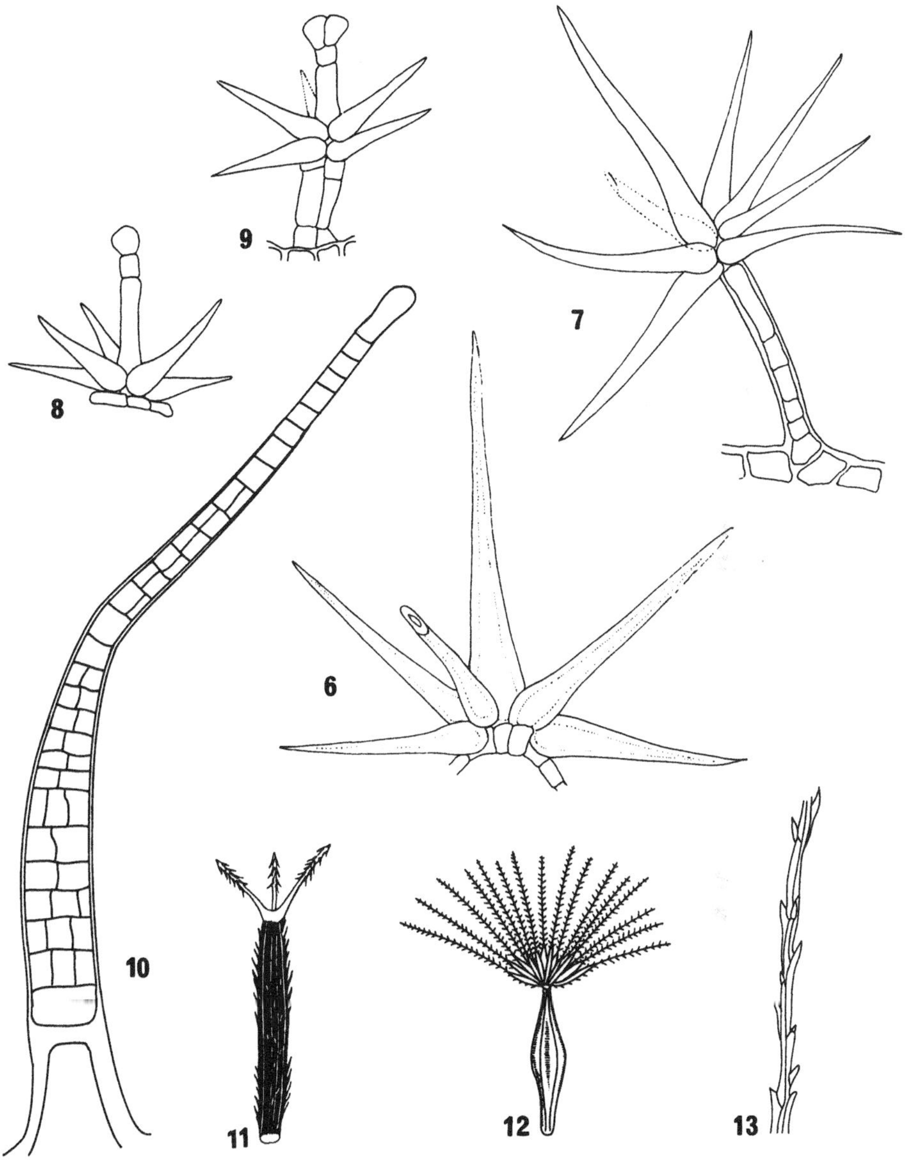

Figs 6–9. Trichomes of *Phlomis* leaf. Fig. 6. Stellate trichome. Fig. 7. Long-stalked tufted trichome. Figs 8 and 9. Glandular branched trichomes. Fig. 10. Glandular trichome from seed (rare) and fruit of *Hibiscus esculentus*. Fig. 11. Cypsela of *Bidens pilosa* showing basiscopic and acroscopic bristle hairs. Fig. 12. Cypsela of an Asteraceae showing a feathery pappus. Fig. 13. Part of a pappus ray of *Taraxacum officinale*. (Figs 6–9 reproduced, with permission, from Azizian and Cutler, 1982; Figs 11 and 12 reproduced, with permission, from Fahn and Werker, 1972.)

stalk and a 6–8-celled head. A certain genotype of the species has the same type of trichomes; however, these are non-secreting, apparently due to interrupted normal chloroplast development.

A. NON-GLANDULAR TRICHOMES

The non-glandular trichomes are diverse in morphology, anatomy and microstructure. Basically they are classified according to their morphology (Figs 6–13). They may be unicellular or multicellular, and both types can be branched or unbranched. Unbranched multicellular trichomes can be uniseriate, biseriate or multiseriate. They can be distinctly articulated between cells or cross-walls can be indistinguishable on the surface. They may differ in length, size and cell shape, and can be symmetrical or asymmetrical (Fig. 15), uniform in width or their width may change along the hair, and they can be tapering or blunt at the apex. They can vary in wall thickness and in the materials impregnated in their walls, which may render them soft, sinuous or stiff (forming bristles) (Figs 18–22). Branched multicellular trichomes may have unicellular or multicellular branches, of various lengths, equal or unequal, in one vertical or horizontal plane, alternate, opposite or in one to many whorls, acquiring a tufted appearance (Figs 5–9, 17, 18).

Often more than one type of non-glandular trichomes are found on leaves, sometimes at different locations. In *Coridothymus capitatus* (Economou-Amilli *et al.*, 1982), for example, the following types were distinguished: (1) unicellular trichomes covered with small granules, and varying in length, distributed at the greatest part of the leaf surface; (2) a bundle of two-celled trichomes at the base of the abaxial side of winter leaves; and (3) isolated two-celled trichomes scattered mainly on leaf margins, mixed with (4) long multicelled trichomes. Our understanding of why different types of hairs are produced on the same organ and whether each has a specific role is meagre.

B. GLANDULAR TRICHOMES

The term "glandular trichomes" refers to a great variety of glands. They vary in the following ways: in the chemical composition of the substances they secrete, accumulate or absorb, and in their mode of production; in their structure; in their location (similar glands but on vegetative or reproductive organs); and, according to all of the above, in their function. All these differences serve at some level, often overlapping levels, in classification.

The glandular trichomes, like the non-glandular ones, can be unicellular or multicellular, uniseriate or multiseriate, and are variously shaped. In unicellular glandular hairs, a morphological distinction can be observed between

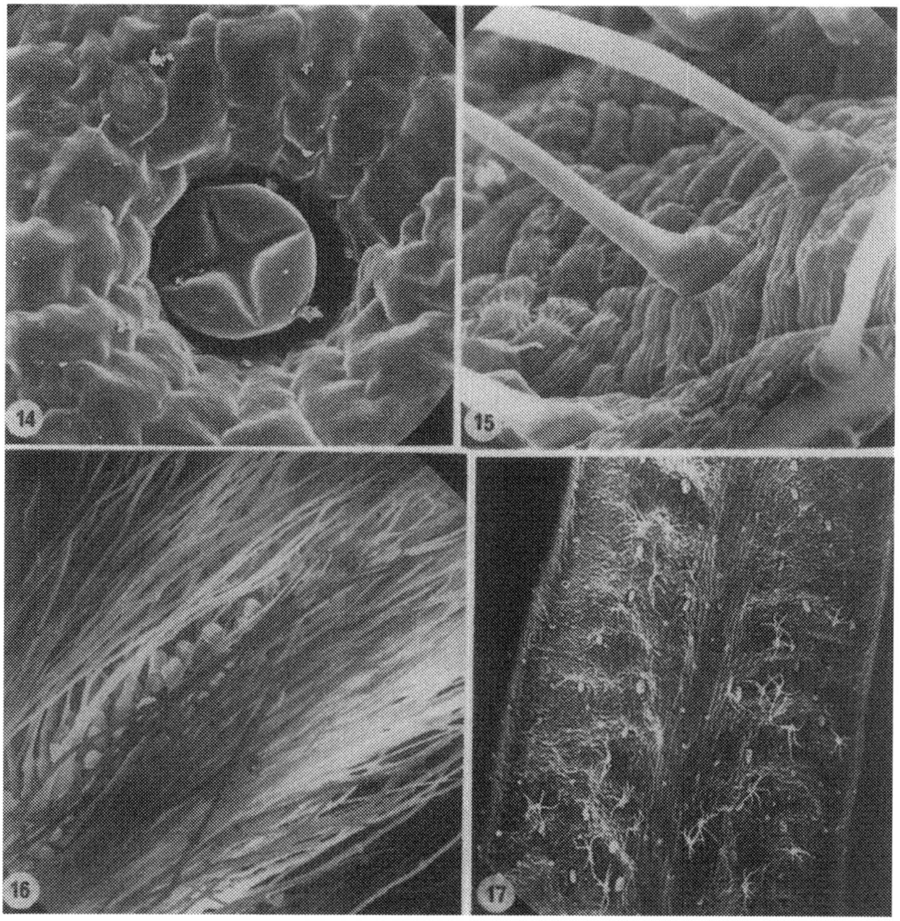

Fig. 14. Sunken glandular peltate trichome with four-celled secretory head of a leaf of *Ocimum basilicum*, ×246. Fig. 15. Non-glandular asymmetrical trichomes of a petiole of *Lupinus palaestinus*, ×315. Fig. 16. Very young leaf of *Inula graveolens* showing much elongated non-glandular hairs covering glandular hairs, ×63. Fig. 17. Leaf of *Artemisia dracunculus* with glandular and non-glandular dendroid trichomes, ×30. (Fig. 14 reproduced, with permission from Wiley & Sons, from Werker *et al.*, 1993; Fig. 17 reproduced with permission, from Werker *et al.*, 1994.)

the apical and basal part of the trichome, e.g. the stinging hairs of *Urtica* (Figs 57 and 58), but sometimes the distinction is only ultrastructural. In multicellular uniseriate trichomes there is usually a differentiation between the different cells along the hair, and in multiseriate trichomes sometimes it may also be across the trichome.

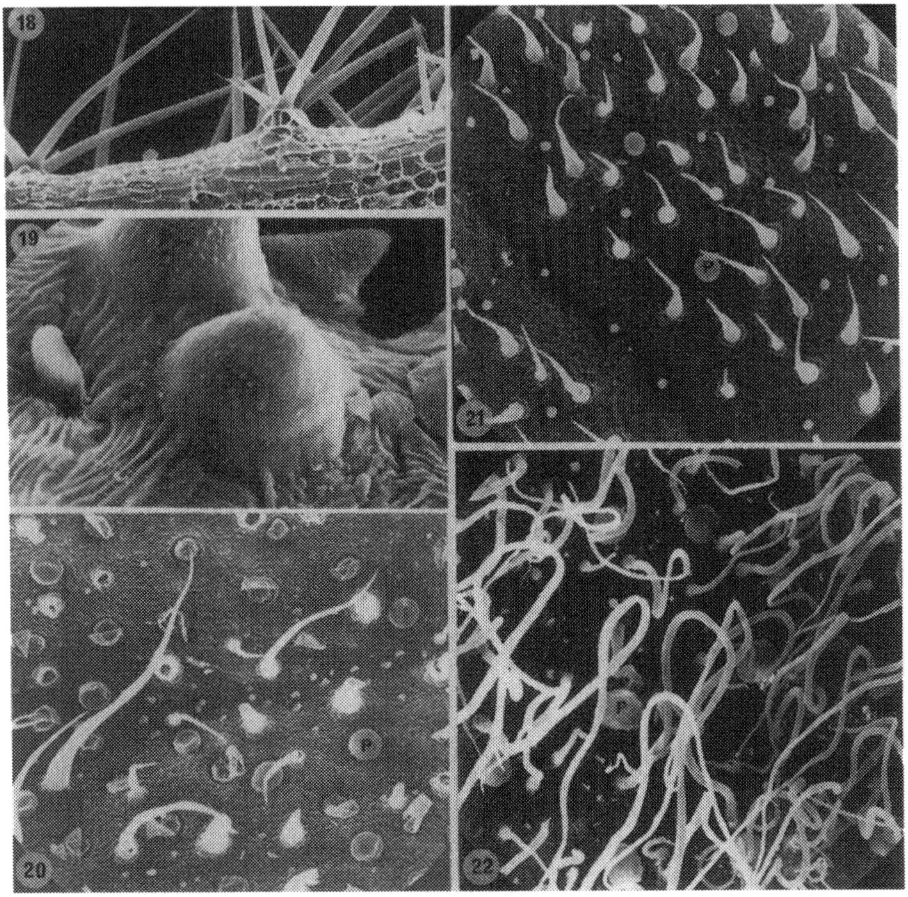

Fig. 18. Petiole of *Lavatera cretica* with stellate trichomes, ×52. Fig. 19. *Satureja thymbra* leaf with warty non-glandular trichomes, ×390. Fig. 20. *Origanum vulgare* leaf with non-glandular and glandular peltate (P) and small capitate trichomes, the latter more concentrated along the vein, ×12. Fig. 21. *Melissa officinalis* leaf with non-glandular and glandular peltate (P) and small capitate trichomes, the latter scattered and along the vein, ×60. Fig. 22. Leaf of *Salvia fruticosa* showing long filiform non-glandular trichomes and glandular peltate (P) and small capitate trichomes, ×60. (Fig. 21 reproduced, with permission from Laser Pages Publishing, from Werker *et al.*, 1985a.)

The morphology of a typical multicellular glandular hair is the following: a terminal secretory head with one to many secretory cells; a stalk of one to many cells; a base of one to a few cells; and sometimes also a neck cell(s) between the secretory head and the stalk, differing morphologically from the two (Figs 1, 2, 4, 8, 9, 26, 34, 43, 51, 54). There are many variations to this basic structure. A stalk cell may be lacking, as in certain "colleters", which develop mainly on stipules and young leaves, and contain sticky substances, lipophilic or carbohydrate, or both (Fahn, 1979). In certain salt glands the basal cells are enlarged into "collecting cells". The spouting glandular trichomes of *Dictamnus* (Rutaceae) are exceptional in that they possess a lysigenous cavity and a "lid", which breaks off when touched, spouting their essential oils that accumulate in the cavity (Amelunxen and Arbeiter, 1967; Fahn, 1979).

The distinctions between the different cell groups of trichomes are morphological and ultrastructural. However, there is not enough information regarding the differential ultrastructure and function of the different groups of cells constituting a hair, and between different cells of each group. The secretory head of the glandular hairs on the fruit of *Mallotus philipinensis*, for example, consists of 70–120 secretory cells (Roth, 1977). Whether all the cells of a secretory head secrete the same materials and are active at the same time is not known. There are some indications that there are differences between the different cells of a stalk and between those within a secretory head. Among stalk cells, the lowermost cell(s) can be highly vacuolate while the upper ones may have dense cytoplasm. Among head cells, in *Inula viscosa* different types of organelles were observed for each pair of cells of the secretory head, and with the Gomori test for acid phosphatase, reaction was observed in different head cells, even in one of a pair in different trichomes (Werker and Fahn, 1981). On *Artemisia annua* leaves there are glandular trichomes composed of 10 cells in five cell pairs, each pair apparently having different functions as deduced from ultrastructural differences (Duke and Paul, 1993). In the resin-secreting multiseriate glandular hair of *Cuphea*, the stalk consists of one type of outer cells and an inner "core". This differs not only in location but also in stainability of cell walls and protoplasts (Amarasinghe *et al.*, 1991), which apparently relates to the different functions each one fulfils.

Very often, species with glandular hairs also bear non-glandular hairs. More than one type of each of the two groups may be present on the same organ. In *Inula viscosa* (Werker and Fahn, 1981) there are two types – one stalked and one sessile. It is not known whether they secrete the same or different substances. In the latter case they may function differently. In other species trichome types may differ considerably in their structure or in their secretory substances. For instance in *Nicotiana*, one of the two capitate trichomes secretes resinous material, while the other secretes aqueous droplets in which nicotine is dissolved (Meyberg *et al.*, 1991). In most species, however, the secretory substances of each of the gland types have not been analysed separately. In *Salvia sclarea* and *S. dominica* (Werker *et al.*, 1985c), in addition to a dense indumentum of uniseriate, multicellular, non-glandular hairs, there is a great variability in structure of the capitate hairs (Fig. 2). Non-glandular hairs, in addition to providing protection to the leaf, might perhaps function (when the indumentum is dense) in providing protection to

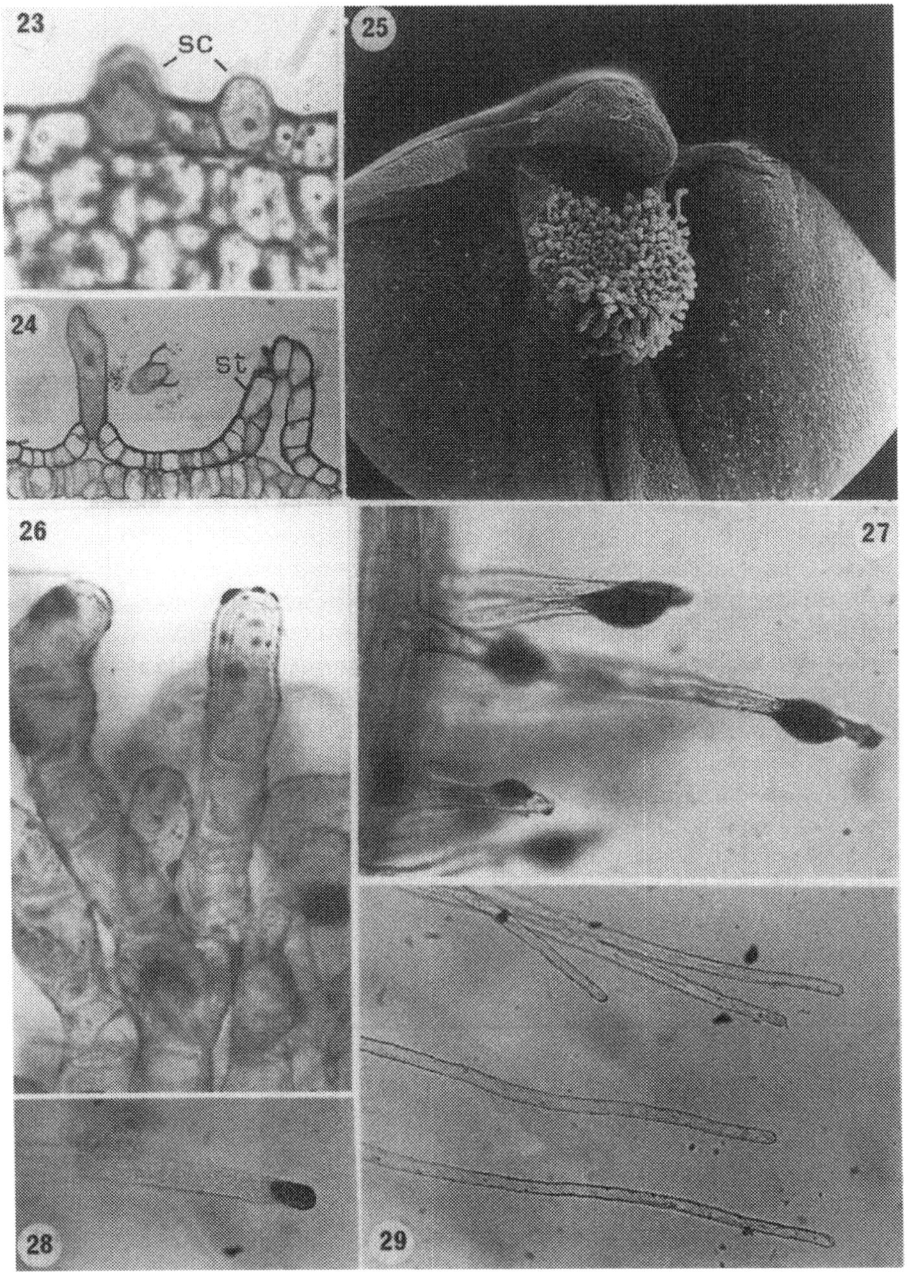
23
sc
24
st
25
26
27
28
29

the shorter, more energy-consuming glandular hairs. Glandular hairs might also be protected by being sunken or situated in cavities or furrows.

In many Lamiaceae, two main types of glandular trichomes are found, capitate and peltate, coined according to the shape of their secretory head. Descriptive terms according to morphology are not, however, always unequivocal. The head of the capitate hairs consists of 1–4 more or less rounded, usually horizontally oriented, secretory cells subtended by a stalk, one to several cells long, and a basal cell. The head cell can sometimes be very large, as in some *Salvia* species (Werker *et al.*, 1985c). The head of the peltate hairs (glandular scales) consists of 4–18 more flattened cells on a horizontal plane (Figs 14 and 46), subtended by one stalk cell and a basal cell. Thus, intermediate forms may be found.

1. Classification According to Types of Substances Secreted

There is a great variability in the secretory materials of glandular trichomes – polysaccharides, sugars, salts, lipids, essential oils, resin, proteins, etc. Meyen (1837; cited by Uphof, 1962) suggested classifying glandular hairs according to the nature of their secretion products. Fahn (1979) classified secretory substances in plants in general into two groups (restricted here to trichomes): (1) substances unmodified or only slightly so, such as the salt secreted by glandular trichomes and nectar secreted by trichomes; and (2) substances synthesized by constituent cells of the secretory tissue. This may be hydrophilic (as in mucilage-secreting trichomes, stinging hairs and digestive glandular trichomes of carnivorous plants) or lipophilic (as, for example, in the glandular trichomes of Lamiaceae, Asteraceae, Geraniaceae, Solanaceae and Cannabaceae).

The difficulty with this type of classification is that some glandular hairs secrete more than one type of substance. For instance, the glandular hairs of *Inula viscosa* secrete lipophilic substances, polysaccharides and protein, apparently at different stages during the life span of the hair and by different

---

Figs 23 and 24. Two stages of development of the glandular trichomes of *Fagonia molis*. SC, very young secretory cells; St, stalk formed by neighbouring epidermal cells. Fig. 23, ×638; Fig. 24, ×263. Figs 25 and 26. *Sternbergia clusiana*. Fig. 25. The tip of the perianth showing a cluster of glandular trichomes of a nectary, ×18. Fig. 26. Enlargement of glandular trichomes of Fig. 25; lipid droplets stained with Sudan IV on and inside the apical cell. Figs 27 and 28. Secretory root hairs of *Sorghum* showing dark mucilaginous droplets outside (Fig. 27) and inside (Fig. 28) the tip. Fig. 27, ×353; Fig. 28, ×203. Fig. 29. Ordinary root hairs of *Setaria italica*, ×203. (Figs 23 and 24 reproduced, with permission, from Fahn and Shimony, 1996; Fig. 25 reproduced, with permission, from Dafni and Werker, 1982; Figs 27 and 29 reproduced, with permission, from Werker and Kislev, 1978.)

organelles (Werker and Fahn, 1981). In some carnivorous plants the same glandular hairs produce both alluring (nectar) and digestive (enzyme) substances. In the Lamiaceae, in addition to lipophilic substances, relatively large amounts of polysaccharides, which vary in amount between species, are secreted by certain capitate hairs (Werker *et al.*, 1985a; Antunes and Sevinate-Pinto, 1991).

The same secretory substances can fulfil various functions, and each function can be fulfilled by various secretory materials. Mucilage, for example, induces the formation of bacterial nodules in the leaf of *Psychotria bacteriophila* (Lersten and Horner, 1967; Horner and Lersten, 1968), serves as a lubricant to facilitate leaf expansion (Modenesi *et al.*, 1984), or aids in seed dispersal (Werker, 1997).

Since many of the secretory materials are phytotoxic, it has been suggested that trichomes, being biosynthetically and structurally isolated, can produce large quantities of materials, which the leaf cannot (Nielsen *et al.*, 1991; Duke, 1994).

C. TRICHOME CELL WALLS

The cell wall of non-glandular trichomes, like that of other plant cells, can consist of a primary wall only or of a thick secondary wall. This may be evenly or unevenly thickened. All or part of a trichome cell wall may be impregnated by various materials like lignin, cutin, suberin, etc. (summarized by Peterson and Vermeer, 1984). In the trichomes of certain xeromorphic species, the lateral walls of the lowest cells are completely cutinized, acting like endodermal cells preventing apoplastic water flow (Fahn, 1986). In many glandular trichomes the lateral walls of the stalk cells become cutinized or suberized or both, also acting as Caspari strips in preventing apoplastic backflow of the secretory substances (Schrödter, 1926; Schnepf, 1969; Fahn, 1990). This is not the case, for example, in the peltate trichomes of *Nepeta* (Bourett *et al.*, 1994).

The trichomes are usually covered by a cuticle, continuous with that of the epidermal cells. In non-glandular trichomes, the cuticle may acquire different thicknesses. The outer surface of the trichomes can be smooth or may exhibit micro-ornamentation, which may show high diversity, causing it to appear micropapillate, warty, reticulate, striate, etc. (Fig. 19). These micromorphological structures may arise from the cell wall, the cuticle, or cuticular or subcuticular inclusions. Epicuticular wax or related chemical substances are rare on trichomes.

Certain non-glandular hairs, dead or living, are capable of movement in response to external conditions, such as the hairs aiding seed dispersal, trigger hairs and water-collecting scales. The mechanisms of movement will not be dealt with here.

D. FUNCTIONS OF TRICHOMES

The functions of many trichomes, glandular or non-glandular, are only either guessed at or totally unknown. Many of the hypotheses concerning trichome functions have not been experimentally tested. The functions of non-glandular trichomes depend on their morphology, the organ on which they are situated and their direction or orientation. Non-glandular dead hairs play a different role on leaves than on a dispersal unit. Basiscopic bristles on dispersal units, like those on the awns of the cypsela of *Bidens* (Carlquist, 1961), for example, aid in epizoochory (Fig. 11). Non-glandular pointed, ridged, trichomes, downwardly directed towards a trap, serve as trapping trichomes in *Sarracenia*. Acroscopic trichomes on the awns of certain grasses and at the base of the mericarp of *Erodium* aid in penetration of the diaspore into the soil and prevent its movement upwards.

Glandular trichomes, in accordance with their location, the substances they secrete and timing of activity, also may serve different functions. Structurally, similar trichomes can produce different materials, but even when similar materials are produced, when the trichomes are differently located, they may have different functions. For instance, the functions of mucilage trichomes on seeds differ considerably from those on leaves.

Different types of glandular trichomes on an organ may differ in the period of time they are active in secretion relative to the stage of development of the organ. Within the Lamiaceae, from a functional viewpoint and in accordance with their mode of secretion, the glandular hairs can be classified into two main types, capitate and peltate. The secretory materials of certain types of the capitate glandular hairs are extruded to the outside soon after their production, when the leaves are still young and developing, either through a porous cuticle or by its rupture. The lipophilic fraction apparently acts as a repellent on the young leaves even before being touched, and the entire process of secretion is usually soon terminated. In the peltate glandular hairs, on the other hand, the materials secreted are gradually secreted from the head cells into a space formed by elevation of the cuticle and part of the wall (Fig. 45). The compound cuticle will rupture if touched by a predator, releasing the secreted materials, which may then act as repellents. From a functional viewpoint, and in accordance with the mode and timing of secretion, the glandular hairs can be classified into "short-term" glandular hairs corresponding with the first group of capitate glandular hairs and "long-term" glandular hairs corresponding with the peltate trichomes (Werker, 1993).

Some of the suggested functions for trichomes on leaves and stems are given in the following paragraphs.

(a) Protection. The plant may need protection from various external factors: herbivores and pathogens, extensive light, extreme temperatures,

excessive water loss, allelopathy against competitor plants, etc. When non-glandular trichomes are densely spread (sometimes forming a mat thicker than the leaf itself, as found in many plants which grow under arid conditions), they may serve as a mechanical barrier against most of the above factors. A correlation between trichome density and insect pest resistance has been shown by Levin (1973). Glandular hairs, which secrete lipophilic substances, can serve in chemical protection against various types of herbivores and pathogens by deterring or poisoning them. Secreted mucilage may trap insects by stickiness. Secreted resin in *Betula pendula* (Lapinjoki *et al.*, 1991) in northern latitudes serves as a glue and moisture eliminator in protecting the developing tissues from the harsh winter and spring weather, and in deterring birds and insects early in the spring. In some individual seedlings of this species, resin production was found to be very poor, but their twigs were usually covered by long non-glandular hairs.

(b) Water absorption. Root hairs actively absorb water and minerals from the soil. They are usually short-lived, unicellular, unbranched, straight and much elongated (Fig. 29), although curved, tortuous, dichotomously ramified, hooked, spiral, tendril-like with 2–3 entwined hairs, or with one or more swellings, also sometimes occur (Uphof, 1962). However, in a few species, part of the root hairs act as secretory hairs. The simple, unicellular, root hairs of apple (Head, 1964) and certain grasses like *Sorghum* (Kislev *et al.*, 1978; Werker and Kislev, 1978) may function as secretory hairs, secreting mucilaginous droplets. In *Sorghum*, secretion was found to be correlated with short root hairs and was absent in long ones (Figs 27 and 28). A suggested function for these short root hairs was the establishment of the plant in soil.

Epiphytic plants, which have no contact with the soil, as well as plants that grow under xeric conditions, sometimes have other types of hair for collecting water from fog and dew. In certain Bromeliaceae like *Tillandsia* (Uphof, 1962; Tomlinson, 1969; Benzing, 1990; Fahn and Cutler, 1992) and *Ananas comosus* (Sakai and Sanford, 1980), they consist of a large multi-celled, single-layered shield of dead cells with thick walls of cellulose and pectin, a stalk of 2–5 living cells, the upper one of which is level with the epidermis and the rest embedded in the leaf tissue, and two basal cells. Dead cells of the central disc of the shield are active in involution of the shield when dry and its straightening when wet. The living stalk cells absorb and conduct the accumulated water. An ultrastructural study of *Ananas* (Sakai and Sanford, 1980) revealed polysaccharide material within vacuoles and as an inner layer to the cell wall in the stalk cells of dry trichomes. After wetting for 12 h, vacuoles were enlarged and appeared to contain little material and the inner wall layer disappeared. In both states numerous mitochondria were observed. In *Eria ornata* (Orchidaceae) (Uphof, 1962) the trichome consists of an apical tuft of 1–7 mucilaginous, water-absorbing cells above the level of the epidermis and a few-celled, funnel-like, water-conducting basal part, deeply sunk in a depression formed by the epidermis.

An example for hairs suggested to absorb water in plants under xeric conditions is *Spartocytisus filipes* (Lyshede, 1976). The leaf is densely covered by three-celled T-shaped hairs. Swelling upon hydration of the horizontal cell, which has very thick pectic walls, suggests an ability to collect water that is absorbed by the two lower cells.

(c) Salt secretion. Salt-secreting glandular trichomes (or, as they were previously termed, "hydathode-trichomes") consist of living cells that actively secrete a solution of different minerals. Salt glandular trichomes may differ in morphology, in the salts they secrete and in the manner of secretion. For instance, among salt-secreting trichomes in *Atriplex*, the apical large bladder cell accumulates the salt solution in its expanding vacuole until the cell ruptures (Thomson and Platt-Aloia, 1979; Thomson and Healey, 1984). *Phillyrea latifolia* has a collecting cell, one stalk cell and a multicelled secretory head (Gravano *et al.*, 1998), which excretes the salt solution to the outside. One of the hair types in *Monarda fistulosa* (Lamiaceae) is a uniseriate salt glandular trichome (Heinrich *et al.*, 1983). Trichomes secreting other types of salts are found, for instance, in certain Leguminosae, such as *Cicer* (where they consist of a multicelled oval head and a uniseriate stalk), *Phaseolus* and *Lathyrus*, and in certain Malvaceae (Uphof, 1962; Fahn, 1990).

(d) Alluring trichomes. To this type belong the secretory hairs of certain nectaries, like the unicellular hairs of *Tropaeolum* and *Lonicera*, and the multicellular hairs of *Abutilon* (Fahn and Rachmilevitz, 1970; Fahn, 1990).

Carnivorous plants have special requirements particular to the group and to its individual species. In order to catch their prey, they need alluring, trapping, sensory (tactile, trigger), digestive and absorbing devices. These functions are mainly fulfilled by trichomes (Fahn, 1979; Juniper *et al.*, 1989; Owen and Thomson, 1991).

If the function of an organ changes during its ontogeny, the function of its trichomes may change as well. On the bracts of *Cannabis sativa* (Mahlberg *et al.*, 1984) there is a decrease in density of one type of glandular trichome and an increase of another type. These changes are suggested to relate to protection of floral organs from desiccation for young bracts, and protection of maturing seeds against herbivory as well as desiccation for mature bracts.

1. Location-dependent Functions

(a) Flower trichomes. Most of the above-mentioned functions referred to trichomes on leaves and young stems. Not many flowers have been thoroughly examined for trichomes (glandular or non-glandular) on all their

parts. There is still much to be learned about the structure, ultrastructure, secretory substances and function of each type of these trichomes. They may have different functions on each flower part, even between the abaxial and adaxial sides of the flattened parts. These functions can also be different in different species. Polyploidy has been reported in glandular trichomes in flowers, further indicating their specificity.

The conflicting functions of both protection (against undesirable animals or their visits at an undesirable stage of flower development) and the need (for those plants pollinated by animals) to attract them, have to be taken into consideration. In *Cannabis sativa* (Mahlberg *et al.*, 1984) capitate stalked trichomes are exclusive to floral-related organs, bracts and adjacent leaves of pistillate flowers, while the rest of the vegetative organs bear various other types.

In the young flower, the calyx encloses all the inner flower parts. The calyx in Lamiaceae, for example, is densely covered on its outer abaxial side, by all the types of glandular hairs typical of the vegetative organs of the species, as well as with non-glandular hairs (Figs 2, 30, 31). The obvious role of these trichomes appears to be protection. However, it is not necessarily the glandular trichomes' only function since, in certain species, differences have been found in the relative amount of the various types of glandular hairs and in the proportions of their chemical constituents, in comparison with those of the leaves (Werker *et al.*, 1985c; Dudai *et al.*, 1988; Werker, 1993).

Lamiaceae species, such as *Majorana syriaca*, *Origanum vulgare*, certain *Salvia* species (Werker *et al.*, 1985b,c; Dudai *et al.*, 1988) and *Leonotis leonurus* (Ascensão *et al.*, 1995), possess peltate glandular trichomes on the outer, abaxial side of the corolla lobes, as the only or main type. The inner, adaxial side shows great versatility of trichomes and their location on the corolla within the family: glandular trichomes may be sparse (*Origanum vulgare*, *Leonotis leonurus*), or may possess mainly one specific type of capitate trichomes as in *Salvia dominica* (a species very rich in various types of glandular trichomes elsewhere on the plant) (Figs 32, 33). *Lavandula officinalis* has a rather peculiar type of glandular hairs on the inside of the

---

Figs 30–35. Flower trichomes. Fig. 30. Calyx of *Origanum vulgare* with peltate, capitate and non-glandular trichomes, ×17. Fig. 31. Young calyx of *Salvia sclarea* with non-glandular hairs on veins and glandular trichomes on the entire surface, ×21. Fig. 32. A few peltate trichomes on the adaxial side of the corolla lobes of *Origanum vulgare*, ×49. Fig. 33. Peltate and non-glandular trichomes on the abaxial side of the corolla of *Origanum vulgare*, ×41. Fig. 34. Long glandular hairs on the inner surface of the corolla tube of *Lavandula officinalis*, ×7. Fig. 35. Large peltate trichomes on the underside of the connective of a stamen of *Majorana syriaca*, ×7. P, peltate trichomes. (Fig. 30 reproduced, with permission, from Werker *et al.*, 1985b; Fig. 34 reproduced, with permission from John Wiley & Sons, from Werker, 1993.)

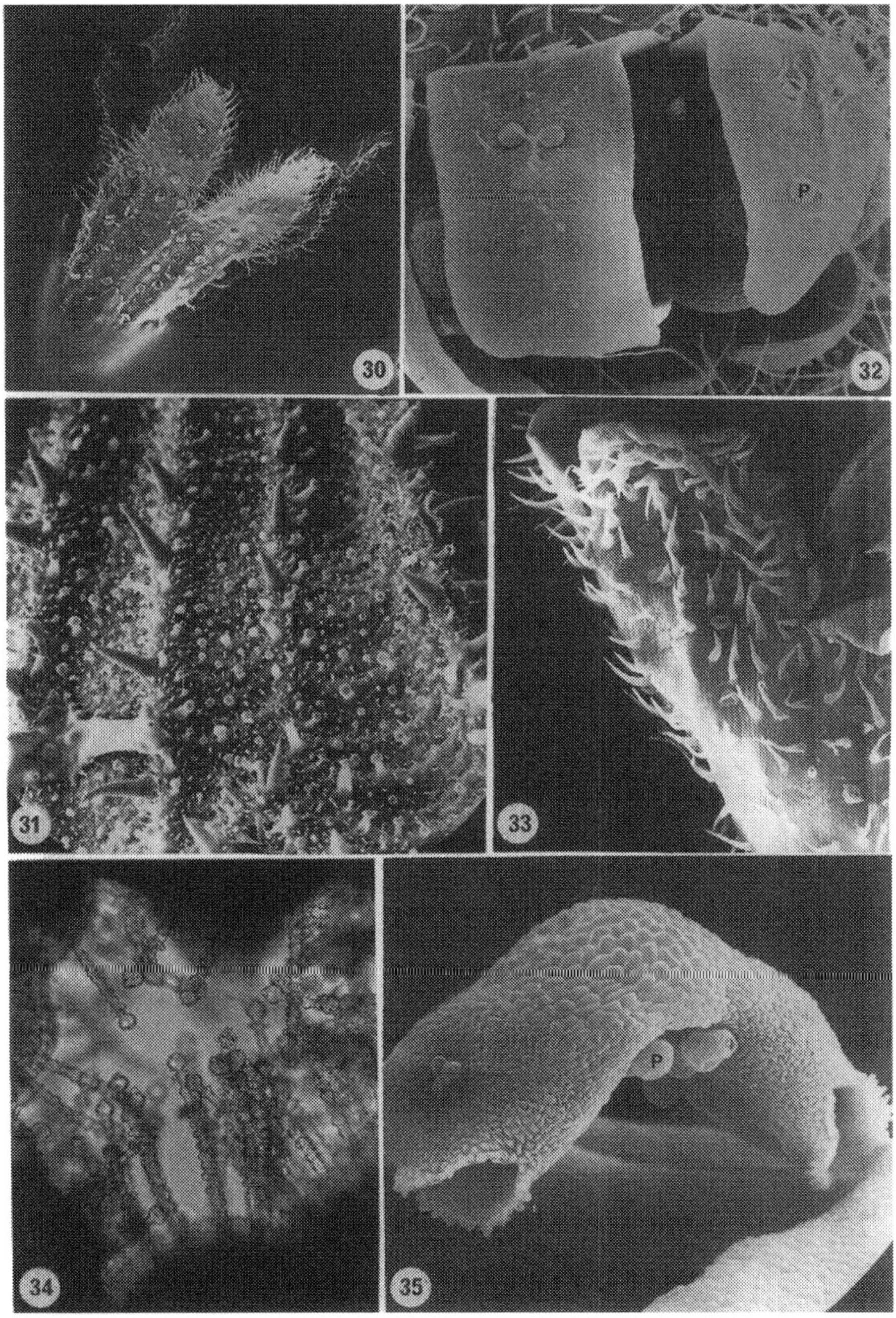
30
32
P
31
33
34
35
P

corolla tube, with one head cell and many swellings on the stalk (Fig. 34). On the two sides of the corolla, therefore, two totally different types of trichomes may be present, which suggests that the inner ones may act as attractants. Separate chemical analysis of the composition of essential oils on each side of the corolla is needed. In *Leonotis leonurus* early in development, uniseriate multicellular orange-coloured non-glandular trichomes occur on the abaxial side of the posterior corolla lobe forming on a mature flower a villous indumentum (Ascensão *et al.*, 1995).

Sternbergia clusiana (Amaryllidaceae) flowers possess an appendix on the adaxial side, close to the tip of the perianth segments, covered with a cluster of multicellular, mostly uniseriate, sometimes branched, glandular hairs (Dafni and Werker, 1982) (Figs 25, 26). Cytohistological staining showed sugars, polysaccharidic and lipophilic substances in the hairs. The plant is pollinated by two main types of pollinators: the honey bees, which collect nectar and pollen from the shorter inner whorl of stamens and the lowermost part of the funnel-shaped perianth tube; and hover flies, which visit mainly the appendages on the perianth segments. The lipophilic secreted substances apparently serve as attractants. Here there is, apparently, a specialization of two types of nectaries for different pollinators. The perianth nectary acts both in alluring flies and in supplying them with nectar.

In various species, large glandular trichomes were observed on the stamens, for instance, on the underside of the anther lobes in *Cannabis sativa* (Mahlberg *et al.*, 1984), *Majorana syriaca* (Dudai *et al.*, 1988) (Fig. 35) and *Leonotis leonurus* (Ascensão *et al.*, 1995). Whether these serve as attractants or, on the contrary, as deterrents to crawling insects (or both, but aimed at different animals) is not known.

On the gynoecium, various types of trichomes were observed, for instance, small glandular trichomes on the ovarial lobes of *Satureja thymbra* (Dafni *et al.*, 1988) and peltate ones in *Leonotis leonurus* (Ascensão *et al.*, 1995). In some families the stigma may possess "brush hairs", which serve for collecting pollen grains.

(b) Fruit and seed trichomes. Fruits, similarly to leaves and stems, need protection during their growth period. Fruit trichomes have been reviewed by Roth (1977); variously shaped hairs, non-glandular and glandular, may be found on developing fruits. In some, hairs may be shed on the mature fruit, as for instance the non-glandular unicellular hairs of Pomaceae. Scale hairs that include raphides are found on certain young fruits, such as *Cocos nucifera*. In *Galbulimima* (Himantandraceae) the fruit bears numerous peltate scales (Doweld and Shevyryova, 1998) with a basal part of 4–8 thick-walled cells bearing the scaly part of 25–30 long, radially elongated, thin-walled cells. Glandular trichomes are also found on some species, such as *Theobroma cacao* and *Hibiscus esculentus* (Fig. 10).

Seeds of the angiosperms develop inside the ovary, which protects them from unfavourable external conditions. Upon maturity many seeds become exposed and need protection; part of them also develop certain means for dispersal. These two requirements are sometimes achieved by papillae, trichomes and emergences. Seed trichomes are reviewed by Werker (1997).

In mature fruits and seeds, which constitute the dispersal unit, trichomes may aid in the more specific functions of fastening to a substrate, dispersal by wind or clinging with bristles and hooks to the fur and feathers of animals (Figs 11–13). For instance, the multicellular trichomes present on young pods of certain Leguminosae transform into burs. Empty air-filled hairs form a greatly increased surface area and therefore may aid in anemochory and hydrochory, as, for example, in the seeds of *Nymphoides*, and *Trigonia* (Boesewinkel, 1987). When hairs are arranged in a parachute-like tuft, as in certain achenes of the Asteraceae and seeds of *Tamarix* and *Salix* or, in rare cases, in the form of a wing, as in *Vochysia* (Boessewinkel and Venturelli, 1987), the diaspore may be carried by air currents or water.

Glandular trichomes on seeds are often mucilaginous, and are found in various plant families. They may be spread over the entire seed or restricted to certain zones. They can be unicellular or multicellular. Their mucilage can be "pectic" or "fibrillar" containing elementary cellulose microfibrils. Elementary microfibrils may also form massive spirals or rings. The functions of mucilage-secreting trichomes may be manyfold, and sometimes various functions are fulfilled by the same trichome in succession or in accordance with external conditions. Some of the functions of mucilaginous trichomes are: adherence of wetted mucilaginous seeds to animals (epizoochory); water retention during imbibition, thereby decreasing desiccation during germination; glueing the seed to the soil after it becomes wet and dry again and when the next rain comes providing a counterforce and lubrication to the radicle (myxospermy); hindering the passage of oxygen in an excess of water, or serving in hydrochory by reduction of specific weight.

E. TRICHOME DISTRIBUTION

There is variability in coverage, by both non-glandular and glandular trichomes and different proportions of these types, between species and even between the two sides of organs, such as leaves, bracts, sepals and tepals. They may appear on one side only or on both sides, equally or unequally. For instance, in *Origanum* × *intercedens* peltate trichomes differ between the two sides of the leaf, in number, size of the subcuticular space and composition of the secretory substances (Bosabalidis and Skoula, 1998). Trichomes may be spread more or less evenly on the surface, or be more localized. Non-glandular trichomes may be present only along the margins and/or along the veins (Figs 63–65), or be restricted to intercostal regions, in clefts, etc. In *Leonotis* (Ascensão *et al.*, 1995) both capitate and peltate trichomes are found on each side of the leaf, being predominant on the abaxial surface. On this side, the glandular trichomes are particularly concentrated on the intervein areas, while the non-glandular trichomes are abundant on the veins and leaf margins. In some other species of the Lamiaceae, a correlation was found between the density of capitate hairs, which secrete both lipophilic and polysaccharidic substances and their location above veins (Figs 20, 21). In the xeromorphic leaves of *Nerium oleander* and *Xanthorrhoea gracilis*, non-glandular hairs are concentrated around the stomata covering the cavities in which these are sunken (Fahn, 1990). Non-glandular trichomes are often present at the base of the leaf.

III. DEVELOPMENT

Trichome development commences at an early stage of leaf development, often prior to stomatal development, and sometimes even before the leaf primordium can be distinguished. The non-glandular hairs on the leaves of *Inula viscosa*, for example, reach nearly their full size when the leaf is about 2 mm long. At this stage they are orientated towards the leaf tip, covering the young leaf. The young leaves and shoot apex are thus physically protected by these long hairs (Fig. 16). With further leaf development these hairs become almost erect.

For glandular trichomes, three phases of development have been distinguished (Ascensão and Pais, 1998): presecretory, secretory and postsecretory. Here mainly the presecretory phase will be dealt with. For the secretory phase, see Fahn (this volume).

The various types of trichome development that can be distinguished are described below.

(a) One epidermal cell gives rise to a unicellular or multicellular, uniseriate or multiseriate trichome. A protodermal cell that gives rise to a trichome becomes conspicuous by its denser cytoplasm and larger nucleus (Bosabalidis and Tsekos, 1982a,b, 1984; Karousou *et al.*, 1992) (Figs 36, 47). As seen for the capitate trichomes in *Leonotis leonurus* (Ascensão and Pais, 1998), this initial contains small vacuoles, occasional Golgi stacks and a few cisternae of rough endoplasmic reticulum and plastids with a few starch grains. In uniseriate trichomes, the protodermal cell enlarges and protrudes as a papilla, which shows polarization into a lower vacuolate part and an upper part dense with cytoplasm. On organs that bear non-glandular and glandular hairs, distinction between the two at the first stage of ontogeny may be impossible. In the Lamiaceae, both the capitate, peltate and non-glandular hairs start by enlargement of a protodermal cell. Uphof (1962) used the plane of division of the initial protodermal cell for classification of the modes of trichome development. The development of uniseriate multicellular trichomes is achieved by a series of periclinal divisions (Figs 49, 52, 53). After the first one, the lower or upper cell (or both) continues to divide periclinally. These divisions may be acropetal – when the uppermost cell is the one undergoing divisions, or basipetal – when divisions of the lowermost cell take place (Uphof, 1962). Sometimes, however, the order of divisions is not consistent with this classification.

The uniseriate five-celled glandular hairs of *Pelargonium scabrum* (Oosthuizen and Coetzee, 1983) originate by a periclinal first division forming a basal cell, which undergoes no further divisions, and an apical cell which undergoes a periclinal division into a head cell and a stalk cell. This stalk cell divides periclinally to form two stalk cells, the upper one of which divides periclinally again to form the final number of three stalk cells. In all the examined peltate trichomes of the Lamiaceae, after the first periclinal division, a second one occurs in the upper cell. The uppermost of the three forms the initial of the secretory head and the middle one forms the stalk (e.g. Amelunxen, 1964; Heinrich, 1973; Bosabalidis and Tsekos, 1982a,b, 1984; Karousou *et al.*, 1992; Ascensão *et al.*, 1995) (Figs 36–42). In *Nicotiana* (Meyberg *et al.*, 1991) there are long capitate trichomes consisting of 4–6 stalk cells and 1–6 head cells. The authors suggest that the number of head cells is one in young trichomes and increases with maturation of the trichomes.

In biseriate and multiseriate trichomes, the first division is in many cases anticlinal or oblique, as in *Chrysanthemum morifolium* (Vermeer and Peterson, 1979). However, it may also occur after one to several periclinal divisions. In *Inula viscosa* (Werker and Fahn, 1981) (Figs 47–54), distinction between the uniseriate non-glandular hairs and the biseriate glandular hairs can be made with certainty after the first anticlinal division in any of the cells above the basal cell takes place. In this species, which possesses both stalked and sessile glandular hairs, the head cells in the stalked ones enlarge before cell divisions in the stalk are completed. Elongation of the stalk cells starts when the head cells have almost reached their final size. In the stalk, one of the cells of a pair sometimes undergoes a further transverse division, resulting in a different number and different length of the cells in the two tiers of the stalk cells (Fig. 54). At the end of the presecretory phase, the basal cell and the lowermost of the stalk cells develop a central vacuole. The head cell cytoplasm is dense and rich with those organelles that will produce the secretory substances. In various species, chloroplasts develop in the stalk cells. Crystals may develop in various cells of a trichome.

(b) One epidermal cell gives rise to a uniseriate or multiseriate trichome, but neighbouring epidermal cells give rise to a pedestal of the trichome. In many species, the epidermal cells surrounding a hair, non-glandular or glandular, may differ in shape and/or orientation around the trichomes, often

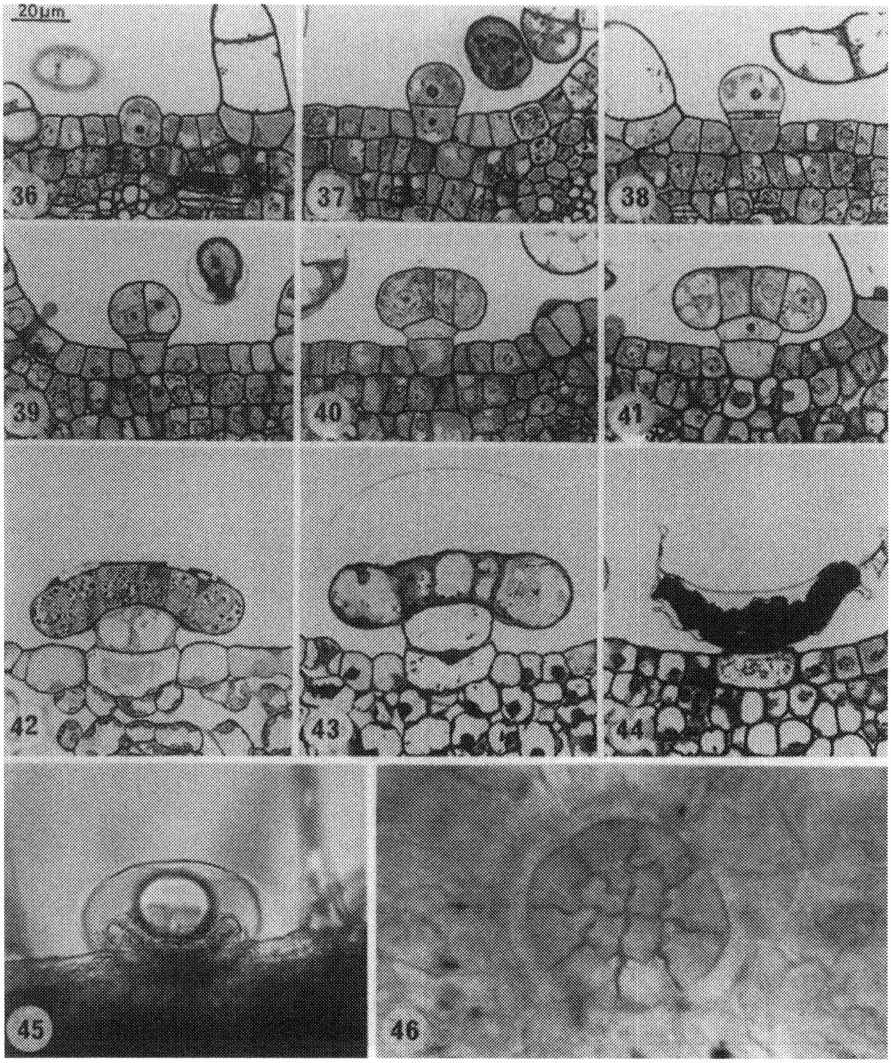

Figs 36–46. Successive stages of development of glandular peltate trichomes of leaves of *Origanum × intercedens*, ×424; Figs 42 and 43. Slight and advanced detachment of the cuticle from the lipid-collecting chamber. Fig. 44. An aged trichome. Figs 45 and 46. *Origanum vulgare*. Fig. 45. Intact peltate trichome with a large drop of secreted material in the subcuticular chamber, ×30. Fig. 46. Surface view of the secretory head cells of a peltate trichome after removal of the elevated cuticle, ×408. (Figs 36–44 reproduced, with permission from the University of Chicago Press, from Bosabalidis and Exarchou, 1995; Fig. 45 reproduced, with permission, from Werker *et al.*, 1985b.)

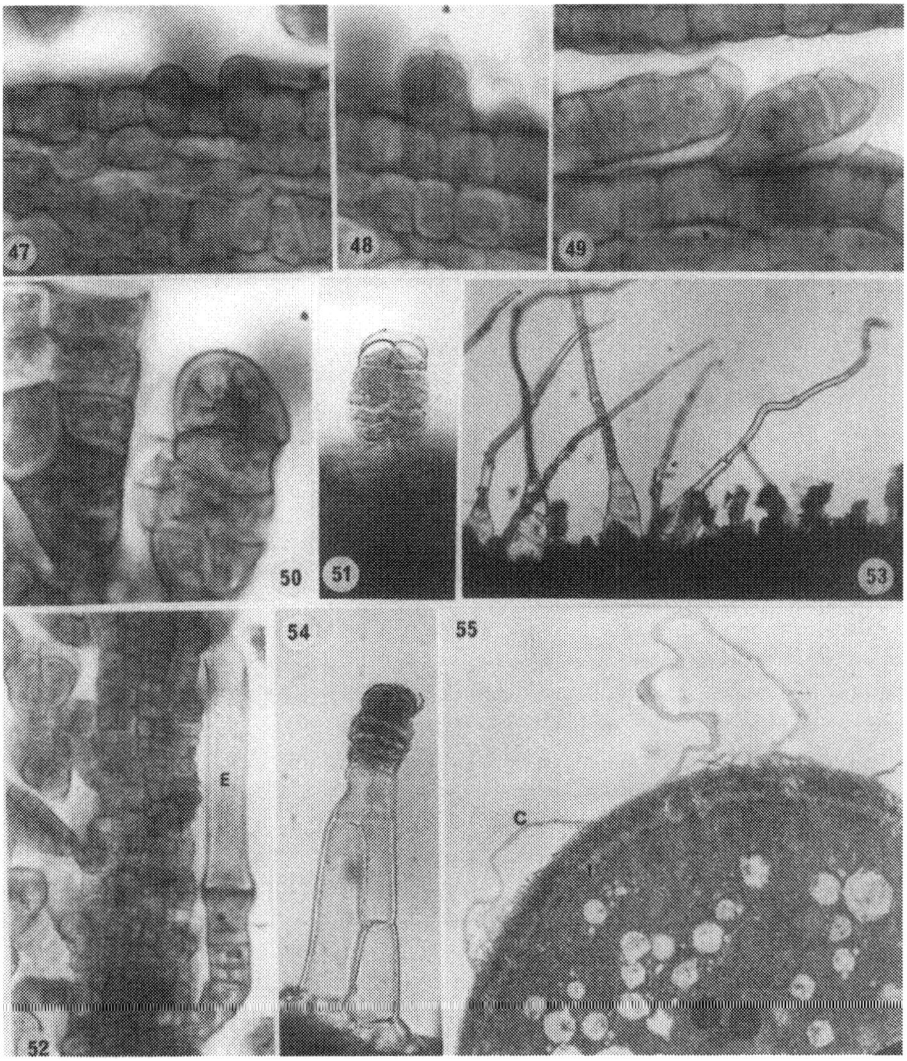

Figs 47–55. Successive stages of development of the glandular biseriate and non-glandular uniseriate trichomes of *Inula viscosa*. Fig. 47. Enlarged and protruding initials of trichomes. Fig. 48. Anticlinal division of the trichome initial. Fig. 49. Two young trichomes with periclinal divisions and an enlarged apical cell. Fig. 50. Two glandular trichomes periclinally and partly anticlinally divided. Figs 47–50. ×671; Fig. 51. Mature sessile glandular trichome with slightly elevated cuticle. ×299; Fig. 52. Young non-glandular trichome with elongated apical cell (E), ×299. Fig. 53. Mature non-glandular trichomes, ×77. Fig. 54. Mature stalked glandular trichome with an uneven number of stalk cells and a torn cuticle, ×171. Fig. 55. Transmission electron micrograph of a head cell with invaginations (I) of the wall and the cuticle (C) partly elevated, ×4. (Figs 51 and 54 reproduced, with permission from the University of Chicago Press, from Werker and Fahn, 1981.)

forming a circle. Their initials, as that of the trichome proper, also may show a denser cytoplasm than that of neighbouring epidermal cells. In certain species the surrounding epidermal cells grow in such a way that they protrude above the organ's surface, elevating the trichome. The pedestal cells may also form a sheath of various heights around the hair.

Pelargonium scabrum leaves bear unicellular spiny hairs, often warty, raised on a pedestal of swollen epidermal cells, between which the enlarged base of the trichome cell is embedded (Oosthuizen and Coetzee, 1983). Similarly, in *Leonotis leonurus* (Ascensão *et al.*, 1995; Ascensão and Pais, 1998), the leaves, stems and sepals bear uniseriate multicellular, pointed non-glandular trichomes and one type of capitate glandular hair, both supported by a cellular pedestal. In *Fagonia* (Fahn and Shimony, 1996) a single epidermal cell develops into an elongated unicellular secretory cell. When this secretory cell has almost reached its full length, anticlinal division and elongation start in the surrounding epidermal cells, causing their elevation and thereby elevating the differentiating secretory cell above an intercellular space formed by detachment of these epidermal cells from the mesophyll. In mature trichomes these cells form a kind of stalk in which the lowermost part of the secretory cell is embedded (Figs 23 and 24).

The stinging trichomes of *Urtica* appear to be situated and partly sheathed by a "pedestal" (Figs 57 and 58). In fact the unicellular hair develops on an emergence and the surrounding epidermal cells of the emergence form a sheath around the base of the hair (Uphof, 1962).

(c) "Twin hairs" are typical to the pericarp of many Asteraceae. They were classified by Metcalfe and Chalk (1979) as two armed hairs, and regarded by others (Roth, 1977; and literature cited therein) as a special type of hair. Twin hairs develop from a single epidermal cell, which divides anticlinally, producing a basiscopic cell and an acroscopic cell, with respect to the longitudinal symmetry of the fruit. The basiscopic cell usually subdivides by an oblique wall and the acroscopic one by a periclinal wall. With further periclinal divisions, two cell rows can originate. The cells of the two rows may differ considerably in structure. The hair cells are either united along their

---

Fig. 56. Compound aseptate hair (C) of *Elaeocarpus edulis*. Figs 57 and 58. Stinging hair of *Urtica dioica*. E, emergence; Sh, sheath. Fig. 57. Surface view. Fig. 58. Longitudinal section. Figs 59–61. Twin hairs. Fig. 59a–g. Successive stages of development; Fig. 60. Mature hair of *Calendula officinalis*. Fig. 61. Mature hair of *Senecio umbrosus*. Fig. 62. Schematic drawing of the apical portions of a tip-growing root hair (left) and a diffuse growing cotton lint hair (right). (Fig. 56 reproduced, with permission from Cambridge University Press, from Corner, 1976; Figs 57 and 58 reproduced, with permission, from Fahn, 1990; Figs 59–61 reproduced, with permission, from Roth, 1977; Fig. 62 reproduced, with permission from NRC Research Press, from Tiwari and Wilkins, 1995.)

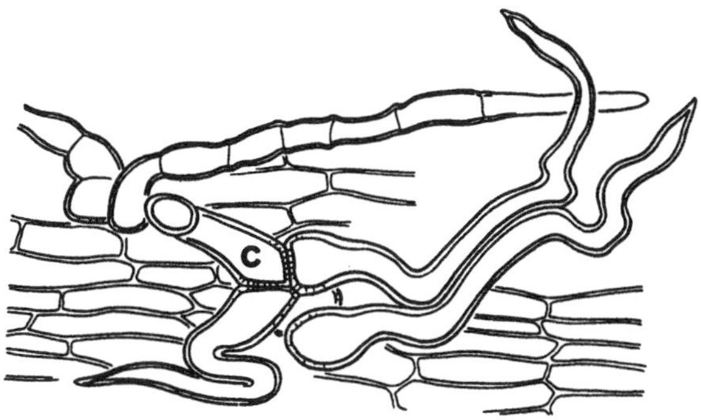

56

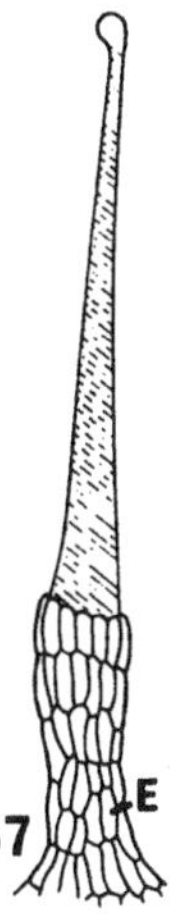

57

58

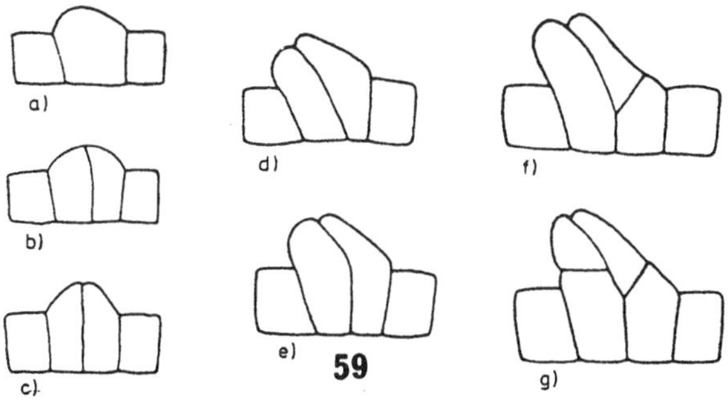

59

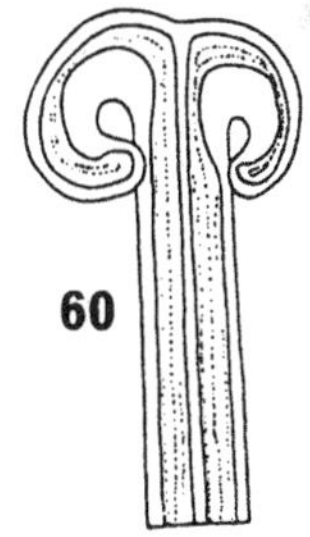

60

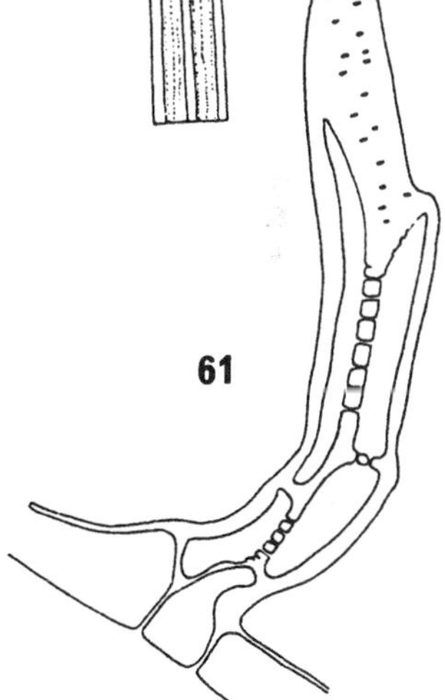

61

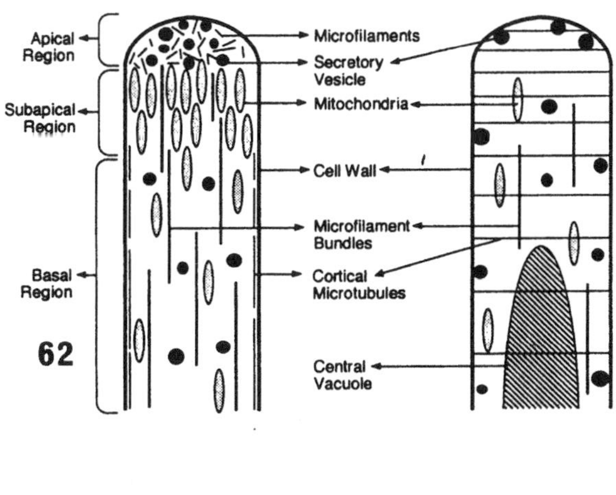

62

entire length or separated from each other at their apex, where the cells may become inflated (Figs 59–61).

(d) More than one epidermal cell gives rise to a multiseriate trichome. In the spouting glandular trichomes of *Dictamnus* the initial first splits anticlinally producing four cells, which then divide periclinally. The upper four cells form the lid, the four lower ones, the centre, whereas neighbouring epidermal cells form the mantle.

Several neighbouring epidermal cells can produce together a "compound" or "fascicled" hair. For example, in *Helicteres hirsuta*, unicellular, thick-walled, pointed hairs on the leaves appear in groups of 3–8, touching at the base but with the rest as far as the tip free, at a slight angle to one another (Uphof, 1962). The seeds of *Elaeocarpus edulis* bear two types of hair: (1) closely appressed unicellular acute hairs, thick-walled and lignified, which may appear in clusters of 2–8; and (2) erect, uniseriate multicellular hairs, slightly thick-walled and lignified (Corner, 1976) (Fig. 56).

(e) Hair development by splitting of cells. Such hairs develop not by the protrusion of a protodermal cell but by splitting of cells of an anticlinally elongated epidermis. Such hairs are found, for instance, on the seeds of the monocotyledon *Cephalostemon riedelianus* (Venturelli and Bouman, 1988). During early seed development, the epidermal cells of the seed coat at the chalazal end elongate much more than those elsewhere around the seed, reaching a length of up to 1500 μm, the longest found on the centre of the appendage. The individual cells gradually split apart into unicellular hairs, which remain mutually attached only at their bases. A similar process occurs on the seed of certain Bromeliaceae, where a pappus is formed by splitting of epidermal cells (Netolitzky, 1926). The internal hairs of the pod of *Vicia faba* (Kaniewski, 1963) start to develop by periclinal divisions in the inner epidermis, forming a multilayered epidermis in anticlinal cell rows, which later separate from each other producing multicellular hairs.

(f) Subepidermal cells, in which one to all the local tissues subtending the epidermis may take part in the formation of a trichome/emergence. Examples are the glandular emergences on the calyx of *Plumbago capensis* (Fig. 3) and those on *Drosera* leaves (Fahn, 1979).

A combination of types (b) and (f) is found in the stalked capitate hairs of *Cannabis sativa* (Mahlberg *et al.*, 1984). At the stage in which the multicellular head approaches its full size, it becomes elevated on a multicellular "stalk" formed by surrounding epidermal as well as subepidermal cells, which elongate vertically to raise the base of the trichome proper.

(g) "False hairs" develop by partial disintegration of cell walls. In the seeds of *Solanum* and *Lycopersicum*, the anticlinal walls of the radially elongated epidermal cells are thickened more or less along their angles, gradually thinning distally. The thin parts of the wall become mucilaginous and disperse, leaving the anticlinal angles as "hairs" (Souèges, 1907; Netolitzky, 1926; Hayward, 1938; Czaja, 1963; Fahn, 1990; Werker, 1997).

A. CELL WALL GROWTH

Various aspects of wall development in trichomes, during cell elongation and during secretion, have been investigated. The process of rapid elongation of unicellular, long, simple hairs, like root hairs and the lint hairs of *Gossypium hirsutum* seeds, serves as a model for investigation of the building of new cell walls. In root hairs the primary and secondary cell walls are deposited simultaneously in the extreme tip and the tubular part of the root hair, respectively while there is a temporal separation between primary and secondary wall deposition in cotton hairs. It has been found that root hairs are exclusively tip-growing hairs (Lloyd and Wells, 1985; Emons *et al.*, 1990), while those of cotton are of the diffuse growing type (review by Ryser, 1985; Tiwari and Wilkins, 1995). Tip-growing cells display several ultrastructural features: organelle zonation, i.e. dictyosome-derived vesicles at the apical part, mitochondria in the subapical part, cellulosic microfibrils at random orientation at the tip, and longitudinal or helical (steeply pitched) in the subapical or basal regions, axial orientation of microtubules, and actin microfilament meshwork at the cell apex (Fig. 62). In addition, disruption of actin filaments stops trichome growth. Cells growing by the diffuse mechanism do not show any of the above characteristics: organelles are dispersed throughout the cytoplasm; the addition of new wall material is not restricted to the apical part of the hair, since its cellulose microfibrils occur in transverse orientation throughout the fibre length including the tip region; and cortical microtubules are oriented perpendicularly or obliquely to the axis. In addition, disruption of actin filaments does not stop trichome growth, but disruption of microtubules prevents polarized cell expansion (Tiwari and Wilkins, 1995).

In many excreting glandular trichomes, development of the cell walls and cuticle of the secretory cells does not cease when the trichome reaches its full size. Two processes, which will be only briefly mentioned here, may take place in the wall during the secretory phase: (1) the formation of short ingrowths on the inner side of the tangential wall of the secreting cells (e.g. Fahn and Rachmilevitz, 1970; Schnepf, 1974; Vermeer and Peterson, 1979) (Fig. 55); and (2) splitting of the cuticle from the primary wall or together with a cutinized part of the primary wall occurs, forming a space where secretory materials accumulate for short or long periods of time (Figs 42–45, 51, 55).

Repeated secretion of essential oils, and apparently the development of a new cuticle as the old one ruptures, has been recorded in *Pelargonium scabrum* (Oosthuizen and Coetzee, 1983).

B. TIMING OF TRICHOME DEVELOPMENT

The question has often arisen as to whether the number of trichomes is fixed during the life span of the leaf or whether it increases with age. For Lamiaceae, some authors have suggested that the number of glandular

Figs 63–66. Leaves of *Ocimum basilicum* at different stages of development, showing distribution of glandular trichomes. Fig. 63. Leaf primordium, 0.5 mm long, ×82. Fig. 64. The tip of a 10-mm-long leaf, showing glandular trichomes far dispersed, ×14. Fig. 65. The basal part of an 8.5-mm-long leaf, showing very dense glandular trichomes, ×25. Fig. 66. The lower part of a 30-mm-long leaf with far-dispersed glandular trichomes, slightly more dense at the base; non-glandular hairs on the large veins, ×7. (Figs 63 and 66 reproduced, with permission, from Werker *et al.*, 1993.)

trichomes is established early during leaf differentiation (Ascensão and Pais, 1987), while others have concluded that it increases throughout all stages of leaf development (Unzelman and Healey, 1974; Turner *et al.*, 1980; Croteau *et al.*, 1981; Maffei *et al.*, 1986, 1989). In *Ocimum basilicum* (Werker *et al.*, 1993) glandular hairs start to be produced at very early stages of leaf development (Fig. 63) and the rate of their differentiation is rapid. The leaf itself has its own pattern of differentiation, at different rates in various regions, thus both mature and meristematic regions may be found in a developing leaf. As long as cell divisions occur in the leaf, new glandular hairs may be produced. Young regions of the leaf are, therefore, densely covered with glandular hairs, but once its cells start to expand the glandular hairs gradually become more widely spaced (Figs 64–66). In contrast, the salt glandular hairs of *Atriplex* (Black, 1954), which consist of short- and long-stalked hairs densely packed together, show a continual development of new trichomes. They start to be formed on leaf primordia and this continues up to leaf senescence; old hairs die while new ones are formed.

C. LIFE SPAN OF TRICHOMES

There are trichomes, non-glandular and glandular, which start functioning very early and senesce before the organ reaches its full size, and there are others which are still viable and functioning when it is mature. Examples of early shed trichomes are the branched non-glandular hairs of *Artemisia dracunculus* (Lassanyi and Stieber, 1976) (Fig. 17) and the glandular, peltate, resin-secreting trichomes of *Betula pendula* (Lapinjoki *et al.*, 1991).

In glandular trichomes the end of secretion is correlated with senescence of cells (Fig. 44), except for the basal cell, which becomes ultrastructurally more active and is suggested to act in terpene catabolism (Croteau, 1988). Non-glandular trichomes may remain on an organ and be functional also when dead, for example, on leaves as a protection, and on fruits and seeds aiding in dispersal. There are cases in which only part of the trichome cells stay alive. In the non-glandular hairs of *Inula viscosa* (Werker and Fahn, 1981), the lowest 5–6 cells are short, thick-walled and their protoplast is retained even when the leaf is fully mature, while in the terminal, much-elongated 2–4 cells, the protoplast disintegrates very early during leaf development (Fig. 53). The apical cell is the first to elongate during hair development and the first to die (Fig. 52). Some of the water-absorbing trichomes mentioned above consist of both dead and living cells.

IV. CONCLUSIONS

A vast number of trichomes of various types, of many plant species, was investigated from various aspects. This chapter gives a short review of classification, structure, some aspects of ultrastructure, function and ontogeny. These fields of research raise many yet unsolved questions, the answers to which need interdisciplinary research work.

Trichomes are immensely diverse in shape, size, structure, location, capability to secrete, function, etc. Each of these features serves, at some level, for classification purposes. Since there are many transitional features and much overlapping between groups, no system of classification is as yet fully satisfactory.

Intergrading forms between the glandular and non-glandular trichomes of a species may appear on an organ. What are the factors which control the differentiation of a trichome into one type or another and at what stage of development?

The ultrastructure of the head cells of multicellular glandular trichomes has often been investigated; much less so was that of the other morphologically differing cells of the trichome. Morphological differences may also be found between the cells of a multicellular secretory head. Further investigation of all the groups of cells in a trichome and the interaction between them, may provide new interesting information.

More than one type of both non-glandular and glandular trichomes are often present on an organ, sometimes differently located. On the other hand, similar types of trichomes may appear on different organs, vegetative and reproductive. Such trichomes which differ in type and/or location probably differ in other, not always investigated, properties and (accordingly) also in function. The greatest part of the research has been done on trichomes of leaves, much less so on those of flowers with their various parts and at different stages of development. The types of trichomes on flowers may be similar or very different from those on the vegetative organs. The functions of all these trichomes are often guessed or speculated rather than experimentally investigated.

REFERENCES

Abu-Asab, M. S. and Cantino, P. D. (1987). Phylogenetic implications of leaf anatomy in subtribe Melittidinae (Labiatae) and related taxa. *Journal of the Arnold Arboretum* **68**, 1–34.

Amarasinghe, V., Graham, S. A. and Graham, A. (1991). Trichome morphology in the genus *Cuphea* (Lythraceae). *Botanical Gazette* **152**, 77–90.

Amelunxen, F. (1964). Elektronenmikroskopische Untersuchungen an den Drüsenhaaren von *Mentha piperita* L. *Planta Medica* **12**, 121–139.

Amelunxen, F. and Arbeiter, H. (1967). Untersuchungen an den Spritzdrüsen von *Dictamnus albus* L. *Zeitschrift für Pflanzenphysiologie* **58**, 49–69.

Antunes, T. and Sevinate-Pinto, I. (1991). Glandular trichomes of *Teucrium scorodonia* L. morphology and histochemistry. *Flora* **185**, 65–70.

Ascensão, L. and Pais, M. S. (1987). Glandular trichomes of *Artemisia campestris* (ssp. *maritima*): ontogeny and histochemistry of the secretory product. *Botanical*

Gazette **148**, 221–227.

Ascensão, L. and Pais, M. S. (1998). The leaf capitate trichomes of *Leonotis leonurus*: histochemistry, ultrastructure and secretion. *Annals of Botany* **81**, 263–271.

Ascensão, L., Marques, N. and Pais, M. S. (1995). Glandular trichomes on vegetative and reproductive organs of *Leonotis leonurus* (Lamiaceae). *Annals of Botany* **75**, 619–626.

Azizian, D. and Cutler, D. F. (1982). Anatomical, cytological and phytochemical studies on *Phlomis* L. and *Eremostachys* Bunge (Labiatae). *Botanical Journal of the Linnean Society* **85**, 249–281.

Barthlott, W. and Ehler, N. (1977). Rasterelektronenmikroskopie der Epidermis-Oberflächen von Spermatophyten. *Tropische und Subtropische Pflanzenwelt* **19**, 367–467.

Benzing, D. A. (1990). "Vascular Epiphytes." Cambridge University Press, Cambridge.

Black, R. F. (1954). The leaf anatomy of Australian members of the genus *Atriplex*. *Australian Journal of Botany* **2**, 269–286.

Boesewinkel, F. D. (1987). Ovules and seeds of Trigoniaceae. *Acta Botanica Neerlandica* **36**, 81–91.

Boesewinkel, F. D. and Venturelli, M. (1987). Ovule and seed structure in Vochysiaceae. *Botanische Jahrbuecher für Systematik* **108**, 547–566.

Bokhari, M. H. and Hedge, I. C. (1971). Observations on the tribe Meriandreae of the Labiatae. *Notes from the Royal Botanic Garden, Edinburgh* **31**, 53–67.

Bosabalidis, A. M. and Exarchou, F. (1995). Effect of NAA and GA on leaves and glandular trichomes of *Origanum* × *intercedens* Rech.: morphological and antomical features. *International Journal of Plant Sciences* **156**, 488–495.

Bosabalidis, A. M. and Skoula, M. (1998). A comparative study of the glandular trichomes on the upper and lower leaf surfaces of *Origanum* × *intercenens* Rech. *Journal of Essential Oil Research* **10**, 277–286.

Bosabalidis, A. M. and Tsekos, I. (1982a). Ultrastructure of the essential secretion in glandular scales of *Origanum dictamnus* L. leaves. *In* "Aromatic Plants: Basic and Applied Aspects" (N. Margaris, A. Koedam and D. Vokou, eds), pp. 3–12. Martinus Nijhoff Publishers, The Hague.

Bosabalidis, A. M. and Tsekos, I. (1982b). Glandular scale development and essential oil secretion in *Origanum dictamnus* L. *Planta* **156**, 496–504.

Bosabalidis, A. M. and Tsekos, I. (1984). Glandular hair formation in *Origanum* species. *Annals of Botany* **53**, 559–563.

Bourett, T. M., Howard, R. J., O'Keefe, D. P. and Hallahan, D. L. (1994). Gland development on leaf surfaces of *Nepeta racemosa*. *International Journal of Plant Sciences* **155**, 623–632.

Carlquist, S. (1961). "Comparative Plant Anatomy." Holt, Rinehart, Winston, New York.

Carlquist, S. (1966). The biota of long-distance dispersal. II. Loss of dispersibility in Pacific Compositae. *Evolution* **20**, 30–48.

Corner, E. J. H. (1976). "The Seeds of Dicotyledons." Cambridge University Press, Cambridge.

Croteau, R. (1988). Catabolism of monoterpenes in essential oil plants. *In* "Flavors and Fragrances: A World Perspective" (B. M. Lawrence, B. D. Mookherjee and B. J. Willis, eds). Proceedings of the 10th International Congress of Essential Oils, 1986, pp. 65–83, Elsevier, Amsterdam.

Croteau, R., Felton, M., Karp, F. and Kjonaas, R. (1981). Relationship of camphor biosynthesis to leaf development in sage (*Salvia officinalis*). *Plant Physiology* **67**, 820–824.

Czaja, A. Th. (1963). Neue Untersuchungen an der Testa der Tomatensamen. *Planta (Berlin)* **59**, 262–279.

Dafni, A. and Werker, E. (1982). Pollination ecology of *Sternbergia clusiana* (Ker-Gawler) Spreng. (Amaryllidaceae). *New Phytologist* **91**, 571–577.

Dafni, H., Lensky, Y. and Fahn, A. (1988). Flower and nectar characteristics of nine species of Labiatae and their influence on honeybee visits. *Journal of Apicultural Research* **27**, 103–114.

Doweld, A. B. and Shevyryova, N. A. (1998). Carpology, seed anatomy and taxonomic relationships of *Galbulimima* (Himantandraceae). *Annals of Botany* **81**, 337–347.

Dudai, N., Werker, E., Putievsky, E., Ravid, U., Palevitch, D. and Halevy, A. H. (1988). Glandular hairs and essential oils in the leaves and flowers of *Majorana syriaca. Israel Journal of Botany* **37**, 11–18.

Duke, S. O. (1994). Commentary, glandular trichomes – a focal point of chemical and structural interactions. *International Journal of Plant Sciences* **155**, 617–620.

Duke, S. O. and Paul, R. N. (1993). Development and fine structure of the glandular trichomes of *Artemisia annua* L. *International Journal of Plant Sciences* **154**, 107–118.

Economou-Amilli, A., Vokou, D., Anagnostidis, K. and Margaris, N. S. (1982). Leaf morphology of *Thymus capitatus* (Labiatae) by scanning electron microscopy. *In* "Aromatic Plants: Basic and Applied Aspects" (N. Margaris, A. Koedam and D. Vokou, eds), pp. 13–24. Martinus Nijhoff Publishers, The Hague.

Emons, A. M. C., Wolters-Arts, A. M. C., Traas, J. A. and Derksen, J. (1990). The effect of colchicine on microtubules and microfibrils in root hairs. *Acta Botanica Neerlandica* **39**, 19–27.

Fahn, A. (1979). "Secretory Tissues in Plants." Academic Press, London, New York, San Francisco.

Fahn, A. (1986). Structural and functional properties of trichomes of xeromorphous leaves. *Annals of Botany* **57**, 631–637.

Fahn, A. (1990). "Plant Anatomy", 4th edn. Pergamon Press, Oxford.

Fahn, A. and Cutler, D. F. (1992). "Xerophytes." Encyclopedia of Plant Anatomy. Gebrüder Borntraeger, Berlin, Stuttgart.

Fahn, A. and Rachmilevitz, T. T. (1970). Ultrastructure and nectar secretion in *Lonicera japonica. In* "New Research in Plant Anatomy" (N. K. B. Robson, D. F. Cutler and M. Gregory, eds). Academic Press, London. *Botanical Journal of the Linnean Society* **63** (Suppl. 1), 51–56.

Fahn, A. and Shimony, C. (1996). Glandular trichomes of *Fagonia* L. (Zygophyllaceae) species: structure, development and secreted materials. *Annals of Botany* **77**, 25–34.

Fahn, A. and Werker, E. (1972). Anatomical mechanisms of seed dispersal. *In* "Seed Biology" (T. T. Kozlowski, ed.), pp. 151–221. Academic Press, New York, London.

Gravano, E., Tani, C., Bennici, A. and Gucci, R. (1998). The ultrastructure of glandular trichomes of *Phillyrea latifolia* L. (Oleaceae) leaves. *Annals of Botany* **81**, 327–335.

Hanlidou, E., Kokkini, S., Bosabalidis, A. M. and Bessière, J.-M. (1991). Glandular trichomes and essential oil constituents of *Calamintha menthifolia* (Lamiaceae). *Plant Systematics and Evolution* **177**, 17–26.

Hayward, H. E. (1938). "The Structure of Economic Plants." Macmillan, New York.

Head, G. C. (1964). A study of "exudation" from the root hairs of apple roots by time-lapse cine-photomicrography. *Annals of Botany* **28**, 495–498.

Heinrich, G. (1973). Entwicklung, Feinbau und Ölgehalt der Drüsenschuppen von

Monarda fistulosa. *Planta Medica* **23**, 154–166.

Heinrich, G., Schultze, W., Pfab, I. and Böttger, M. (1983). The site of essential oil biosynthesis in *Poncirus trifoliata* and *Monarda fistulosa*. *Physiologie Vègètale* **21**, 257–268.

Horner, H. T. and Lersten, N. R. (1968). Development structure and function of secretory trichomes in *Psychotria bacteriophylla* (Rubiaceae). *American Journal of Botany* **55**, 1089–1099.

Hummel, K. and Staesche, K. (1962). Die Verbreitung der Haartypen in den natürlichen Verwandtschaftsgruppen. *In* "Encyclopedia of Plant Anatomy" IV (W. Zimmermann and P. G. Ozenda, eds), Vol. 5, 209–250. Gebr. Borntrager, Berlin, Nikolassee.

Juniper, B. E., Robins, R. J. and Joel, D. M. (1989). "The Carnivorous Plants." Academic Press, London.

Kaniewski, K. (1963). Hairs in the locules of the broad-bean (*Vicia faba* L.) fruit. *Bulletin d'Academie Polonaise des Sciences*, Cl. V. *Serie des Sciences Biologiques* **16**, 585–594.

Kaniewski, K. and Wazynska, Z. (1970). Sclerenchymatous endocarp with hairs in the fruit of *Acer pseudoplatanus* L. *Bulletin de l'Academie Polonaise des Sciences* **18**, 413–420.

Karousou, R., Bosabalidis, A. M. and Kokkini, S. (1992). *Sideritis syriaca* ssp. *syriaca*: glandular trichome structure and development in relation to systematics. *Nordic Journal of Botany* **12**, 31–37.

Keene, C. K. and Wagner, J. G. (1985). Direct demonstration of duvatrienediol biosynthesis in glandular heads of tobacco trichomes. *Plant Physiology* **79**, 1026–1032.

Kislev, M., Korach, E. and Negbi, M. (1978). Mechanisms of root penetration of seeds germinating on the soil surface. *Annals of Botany* **43**, 87–92.

Koller, D. (1956). Germination regulating mechanisms in some desert seeds – III. *Calligonum comosum* L'Her. *Ecology* **37**, 430–433.

Lapinjoki, S. P., Elo, H. A. and Taipale, H. T. (1991). Development and structure of resin glands on tissues of *Betula pendula* Roth. during growth. *New Phytologist* **117**, 219–223.

Lassányi, Z. and Stieber, G. (1976). The volatile oil secretory system of the tarragon (*Artemisia dracunculus* L.). *Acta Agronomica Academiae Scientiarum Hungaricae* **25**, 269–280.

Lersten, N. R. and Horner, H. T. Jr (1967). Development and structure of bacterial leaf nodules in *Psychotria bacteriophila* Val. (Rubiaceae). *Journal of Bacteriology* **94**, 2027–2036.

Levin, D. A. (1973). The role of trichomes in plant defense. *Quarterly Review of Biology* **48**, 3–15.

Lloyd, C. W. and Wells, B. (1985). Microtubules are at the tips of root hairs and form helical patterns corresponding to inner wall fibrils. *Journal of Cell Science* **75**, 225–238.

Lyshede, O. B. (1976). Structure and function of trichomes in *Spartocytisus filipes*. *Botaniska Notisser* **129**, 395–404.

Maffei, M., Chialva, F. and Sacco, T. (1989). Glandular trichomes and essential oils in developing peppermint leaves. I. Variation of peltate number and terpene distribution within leaves. *New Phytologist* **11**, 707–716.

Maffei, M., Gallino, M. and Sacco, T. (1986). Glandular trichomes and essential oils of developing leaves in *Mentha viridis lavanduliodora*. *Planta Medica* **52**, 187–193.

Mahlberg, P. G., Hammond, C. T., Turner, J. C. and Hemphill, J. K. (1984). Structure, development and composition of glandular trichomes of *Cannabis sativa* L. *In* "Biology and Chemistry of Plant Trichomes" (E. Rodriguez, P. L. Healey and I. Mehta, eds), pp. 23–51. Plenum Press, New York, London.

Metcalfe, C. R. and Chalk, L. (1950). "Anatomy of the Dicotyledons", Vols I and II. Clarendon Press, Oxford.

Metcalfe, C. R. and Chalk, L. (1979). "Anatomy of the Dicotyledons", Vol. 1, 2nd edn. Clarendon Press, Oxford.

Meyberg, M., Krohn, S., Brümmer, B. and Kristen, U. (1991). Ultrastructure and secretion of glandular trichomes of tobacco leaves. *Flora* **185**, 357–363.

Modenesi, P., Serato-Valenti, G. and Bruni, A. (1984). Development and secretion of clubbed trichomes in *Thymus vulgaris* L. *Flora* **175**, 211–219.

Netolitzky, F. (1926). "Anatomie der Angiospermen-Samen." Handbuch der Pflanzenanatomie Bd. X. Borntraeger, Berlin.

Nielsen, M. T., Akers, C. P., Järlfors, U. E., Wagner, G. J. and Berger, S. (1991). Comparative ultrastructural features of secreting and nonsecreting glandular trichomes of two genotypes of *Nicotiana tabacum* L. *Botanical Gazette* **152**, 13–22.

Oosthuizen, L. and Coetzee, J. (1983). Morphogenesis of trichomes of *Pelargonium scabrum. Suid-Afrikaans Tydskrift vir Plantkunde* **2**, 305–310.

Owen, T. P. and Thomson, W. W. (1991). Structure and function of a specialized cell wall in the trichomes of the carnivorous bromeliad *Brocchinia reducta. Canadian Journal of Botany* **69**, 1700–1706.

Payne, W. (1978). A glossary of plant hair terminology. *Brittonia* **30**, 239–255.

Peterson, R. L. and Vermeer, J. (1984). Histochemistry of trichomes. *In* "Biology and Chemistry of Plant Trichomes" (E. Rodriguez, P. L. Healey and I. Mehta, eds), pp. 71–94. Plenum Press, New York, London.

Roth, I. (1977). "Fruits of Angiosperms." Encyclopedia of Plant Anatomy. Gebrüder Borntraeger, Berlin, Stuttgart.

Rudall, P. J. (1980). Leaf anatomy of the subtribe Hyptidinae (Labiatae). *Botanical Journal of the Linnean Society* **80**, 319–340.

Ryser, U. (1985). Cell wall biosynthesis in differentiating cotton fibers. *European Journal of Cell Biology* **39**, 236–256.

Sakai, W. S. and Sanford, W. G. (1980). Ultrastructure of the water-absorbing trichomes of pineapple (*Ananas comosus*, Bromeliaceae). *Annals of Botany* **46**, 7–11.

Schnepf, E. (1969). Sekretion und Exkretion bei Pflanzen. *In* "Protoplasmatologia", Bd. 8. Springer-Verlag, Vienna.

Schnepf, E. (1974). Gland cells. *In* "Dynamic aspects of plant ultrastructure" (A. W. Robards, ed.), pp. 331–357. McGraw-Hill, London.

Schrödter, K. (1926). Zur physiologischen Anatomie der Mittelzellen drüseger Gebilde. *Flora* **120**, 19–86.

Solereder, H. (1908). "Systematisches Anatomy der Dikotylen." Enke, Stuttgart.

Souèges, E. C. R. (1907). Developpement et structure du tègument séminal chez le Solanacées. *Annales des Sciences Naturelles, Botanique, Ser. 9* **6**, 1–24.

Thomson, W. W. and Healey, P. L. (1984). Cellular basis of trichome secretion. *In* "Biology and Chemistry of Plant Trichomes" (E. Rodriguez, P. L. Healey and I. Mehta, eds), pp. 95–111. Plenum Press, New York, London.

Thomson, W. W. and Platt-Aloia, K. (1979). Ultrastructural transitions associated with the development of the bladder cells of the trichomes of *Atriplex. Cytobios* **25**, 105–114.

Tiwari, S. C. and Wilkins, T. A. (1995). Cotton (*Gossypium hirsutum*) seed trichomes expand via diffuse growing mechanism. *Canadian Journal of Botany* **73**, 746–757.

Tomlinson, P. B. (1969). Commelinales – Zingiberales. *In* "Anatomy of the Monocotyledons" (C. R. Metcalfe, ed.), Vol. III. Clarendon Press, Oxford.
Turner, J. C., Hemphill, J. K. and Mahlberg, P. G. (1980). Trichomes and cannabinoid content of developing leaves and bracts of *Cannabis sativa* L. (Cannabaceae). *American Journal of Botany* **67**, 1397–1406.
Unzelman, J. M. and Healey, P. L. (1974). Development, structure and occurrence of secretory trichomes of *Pharbitis*. *Protoplasma* **80**, 285–303.
Uphof, J. C. T. (1962). "Plant Hairs." Encyclopedia of Plant Anatomy IV, Vol. 5, pp. 1–206. Gebrüder Borntraeger, Berlin, Nikolassee.
Venturelli, M. and Bouman, F. (1988). Development of ovule and seed in Rapateaceae. *Botanical Journal of the Linnean Society* **97**, 267–294.
Vermeer, J. and Peterson, R. L. (1979). Glandular trichomes on the inflorescence of *Chrysanthemum morifolium* cv. *Dramatic* (Compositae). II. Ultrastructure and histochemistry. *Canadian Journal of Botany* **57**, 714–729.
Werker, E. (1993). Function of essential oil-secreting glandular hairs in aromatic plants of the Lamiaceae – a review. *Flavour and Fragrance Journal* **8**, 249–255.
Werker, E. (1997). "Seed Anatomy." Encyclopedia of Plant Anatomy. Gebrüder Borntraeger, Berlin, Stuttgart.
Werker, E. and Fahn, A. (1981). Secretory hairs of *Inula viscosa* (L.) Ait. – development, ultrastructure and secretion. *Botanical Gazette* **142**, 461–476.
Werker, E. and Kislev, M. (1978). Mucilage on the root surface and root hairs of Sorghum: heterogeneity in structure, manner of production and site of accumulation. *Annals of Botany* **42**, 809–816.
Werker, E., Putievsky, E., Ravid, U., Dudai, N. and Katzir, I. (1993). Glandular hairs and essential oil in developing leaves of *Ocimum basilicum* L. (Lamiaceae). *Annals of Botany* **71**, 43–50.
Werker, E., Putievsky, E., Ravid, U., Dudai, N. and Katzir, I. (1994). Glandular hairs, secretory cavities, and the essential oil in leaves of tarragon (*Artemisia dracunculus* L.). *Journal of Herbs, Spices and Medicinal Plants* **2**, 19–32.
Werker, E., Ravid, U. and Putievsky, E. (1985a). Structure of glandular hairs and identification of the main components of their secreted material in some species of the Labiatae. *Israel Journal of Botany* **34**, 31–45.
Werker, E., Putievsky, E. and Ravid, U. (1985b). The essential oils and glandular hairs in different chemotypes of *Origanum vulgare* L. *Annals of Botany* **55**, 793–801.
Werker, E., Ravid, U. and Putievsky, E. (1985c). Glandular hairs and their secretions in the vegetative and reproductive organs of *Salvia sclarea* and *S. dominica*. *Israel Journal of Botany* **34**, 239–252.

Structure and Function of Secretory Cells

A. FAHN

Department of Plant Sciences, The Hebrew University of Jerusalem, 91904 Jerusalem, Israel

I. INTRODUCTION

A great variety of important natural chemical compounds, which have been used by man through the ages, are produced by specialized secretory cells occurring in plants. These compounds have various functions in the plants themselves. Some of the secreted substances, for instance, serve to attract pollinators and vectors in seed dispersal; others deter phytophagous animals. Salt-secreting glands are important in plants growing in particular environments. The secretory cells occur singly or in multicellular structures and are

Advances in Botanical Research Vol. 31
ISBN 0-12-005931-2

histologically regarded as *secretory tissues*. The secretory tissues of plants are usually classified according to the substances they produce. However, the same tissue often secretes a number of different substances. The study of the large spectrum of secretory substances and their production by a range of morphological structures requires a variety of research approaches.

The period of intense research interest in secretory tissues, towards the end of the nineteenth and the early part of the twentieth century, was marked by numerous publications on subjects ranging from their anatomy and morphology to their usefulness in systematics, the nature of the secretory products and their economic value to physiology and in relation to ecology (Müller, 1866–1867; Bonnier, 1879; Tschirch, 1906; Haberlandt, 1918; Zimmermann, 1932; Frey-Wyssling, 1935; Sperlich, 1939; Kisser, 1958; Thurston and Lersten, 1969). More recently, secretory tissues have also figured in considerations of the co-evolution of plants and animals (Levin, 1973; Baker and Baker, 1975; Bentley, 1976; Cruden *et al.*, 1983). The majority of studies over the last 30 years have been biophysical, ultrastructural and cytochemical (Schnepf, 1969a; Hill and Hill, 1976; Lüttge and Schnepf, 1976; Vassilyev, 1977, 1994; Fahn, 1979a, 1988; Thomson and Healey, 1984; Kronestedt-Robards and Robards, 1991).

The cells involved in the process of secretion are usually characterized by small vacuoles and relatively dense cytoplasm containing numerous mitochondria. The frequency of other cell compartments and organelles varies according to the particular compound secreted. A general characteristic, common to most secretory trichomes, is the complete cutinization of the side walls of stalk cells, similar to that occurring in the walls of the cells of the root endodermis. The presence of these "endodermal cells" in secretory trichomes ("Mittelzelle" of Schrödter, 1926) indicates that the flow of the secreted substances back into the plant through the apoplast is prevented. The views on the manner of secretion by the secretory cells are based largely on the study of ultrastructural changes that occur during the development of the secretory cells. In addition to conventional electron microscopy, freeze substitution, cytochemistry, autoradiography and sometimes stereological methods have been applied.

The elimination of the secretory material from the cytoplasm may be carried out in two ways (Fig. 1). The secreted substance passes the plasmalemma (or tonoplast): (1) by an active molecular or ionic process (*eccrine secretion*), or (2) the secreted substance is collected in membrane-bound vesicles (*granulocrine secretion*). The membranes of the vesicles fuse with the plasmalemma (or tonoplast), or the vesicles containing the secretory material are eliminated from the cytoplasm by invaginations of the plasmalemma, which surround and detach them from the protoplast. The present views on the processes of secretion, based mainly on ultrastructural studies, will be discussed in this chapter in relation to the three most common secretory tissues.

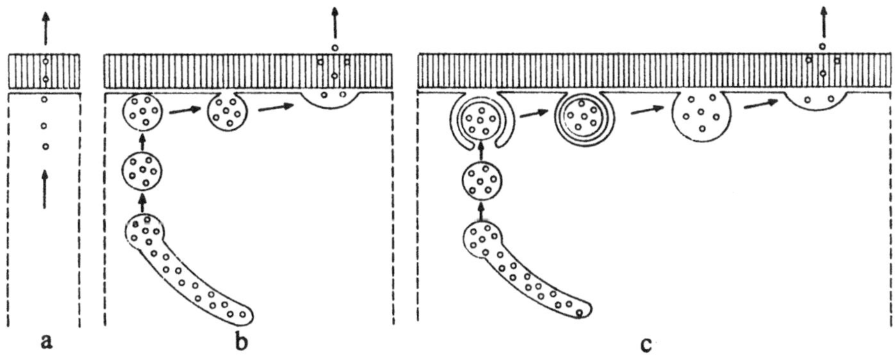

Fig. 1. Different pathways of secretion from secretory cells. (a) Membrane transport – *eccrine secretion.* (b,c) Transport by vesicles – *granulocrine secretion.* The membrane of the vesicles fuses with the plasmalemma (b). The vesicle with the secreted substance is eliminated by invaginations of the plasmalemma (c). (Reprinted, with permission, from Fahn, 1979a.)

II. SALT GLANDS

Salt glands are specialized secretory trichomes that secrete solutions of mineral salts, although their secretions often also contain organic substances. Ions reported as occurring in the secreted solutions of salt glands include Na^+, K^+, Li^+, Mg^{2+}, Ca^{2+}, Cl^-, SO_4^{2-}, NO_3^-, PO_4^{3-} and HCO_3^- (Waisel, 1972; Hagermeyer and Waisel, 1988). There is no direct connection between the salt glands and the vascular bundles, and salt may be secreted into vacuoles or to the outside of the glands.

Salt glands occur in several families of dicotyledons and in the Gramineae. In dicotyledons, they are characteristic of the Frankeniaceae, Plumbaginaceae and Tamaricaceae, but also occur in some genera of other families, such as Acanthaceae, Chenopodiaceae, Convolvulaceae and Primulaceae. Salt glands are present in many mangrove plants, e.g. *Aegialitis* (Plumbaginaceae), *Aegiceras* (Myrsinaceae), *Avicennia* (Avicenniaceae) and *Laguncularia* (Combretaceae).

A. GLANDS SECRETING SALTS INTO THE VACUOLE

In *Atriplex* spp., salt is secreted into a central vacuole of the bladder cell of the leaf trichomes. These cells are situated on top of narrow 1–3-celled stalks. In the early stages of the trichome development the bladder cell lacks vacuoles, and growth of this cell is accompanied by the formation and expansion of a central vacuole (Fig. 2a). At the same time, salt accumulates within it. Eventually, the bladder cell ruptures, releasing the salt onto the

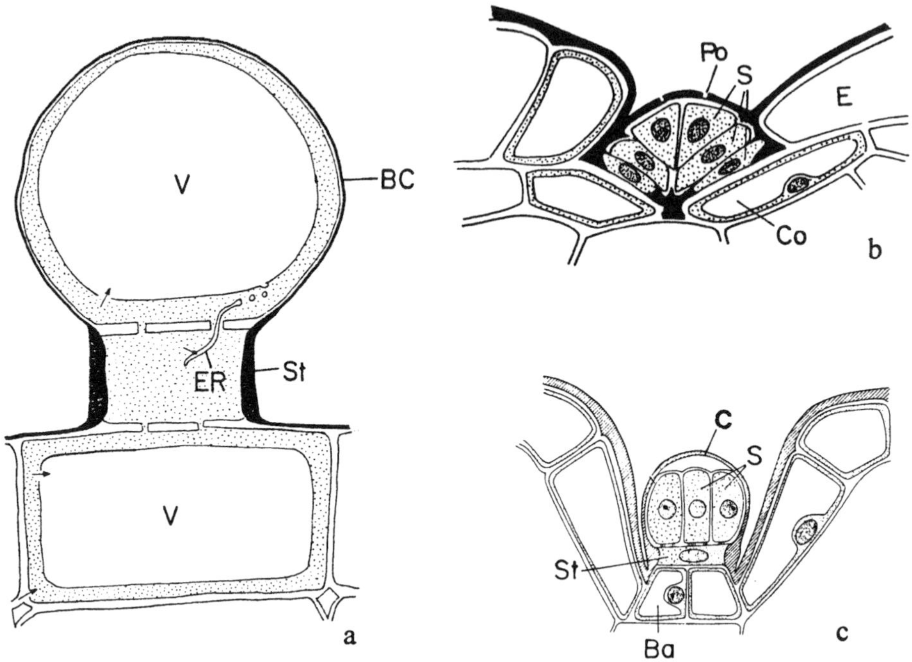

Fig. 2. Salt glands. (a) Diagram of a bladder hair of *Atriplex*, showing possible ways of chloride transfer to the bladder cell and its vacuole. Arrows indicate membrane transport. One of the vesicles occurring in the cytoplasm is seen fusing with the bladder tonoplast. BC, bladder cell; ER, endoplasmic reticulum; St, stalk cell; V, vacuole. (b) *Tamarix aphylla*. Co, collecting cell; E, epidermal cell; Po, pore in the cuticle; S, secretory cells. (c) *Avicennia marina*. Ba, basal cell; C, cuticle; S, secretory cells; St, stalk cell. [(a) Adapted from Osmond *et al.*, 1969; (b) and (c) reprinted, with permission, from Fahn, 1979a.]

surface of the leaf. The side walls of the stalk cell (or, if the stalk consists of more than one cell, those of the lowest cell), become completely cutinized (Osmond *et al.*, 1969; Thomson and Platt-Aloia, 1979).

B. GLANDS SECRETING SALTS TO THE OUTSIDE OF THE CELLS

1. Glands of Gramineae

Among the glands secreting salt to the outside of the cells, those of the Gramineae exhibit the simplest anatomy. They consist of two cells, a basal and a cap cell. The position, size and shape of the glands vary between species (Liphshitz and Waisel, 1974; Liphshitz *et al.*, 1974). The salt gland of *Spartina* possesses a cylindrical depression, the "well", between the cap cell and the four epidermal cells surrounding it. The cuticle over the upper part of the cap cells is distended and becomes detached from the wall forming a cavity between the two. In the elevated cuticle there are numerous narrow pores filled with electron-dense material. Plasmodesmata occur between the basal cell and the cap cell, and between the basal and neighbouring mesophyll cells (Skelding and Winterbotham, 1939; Levering and Thomson, 1971).

On the basis of ultrastructural studies of *Spartina*, *Cynodon* and *Distichlis*, Oross, Leonard and Thomson proposed in 1985 a model of the structure and function of a gland of *Cynodon* (Fig. 3). Secretion from the cap-cell protoplast to the permeable cap-cell wall and adjoining collecting chamber is eccrine, according to these authors.

2. Glands of Dicotyledons

The salt glands of *Limonium*, *Tamarix*, *Avicennia* and many other dicotyledonous plants are multicellular. The salt glands of *Tamarix aphylla* (L.) Karst, whose ultrastructure was studied by Thomson and Liu (1967) and by Shimony and Fahn (1968), consist of eight cells (Fig. 2b). The two vacuolated basal cells are termed *collecting cells*. The six upper cells with dense cytoplasm are the *secretory cells*, which are enclosed by a cuticle, except for portions of the walls between the two lowest secretory cells and the collecting cells (Fig. 2b). These wall portions, which are penetrated by numerous plasmodesmata, are termed *transfusion areas*. The side walls of the lowest secretory cells are completely cutinized. These walls are lined by a complex extracytoplasmic structure known as the "intrafacial apparatus", which contains a variety of vesicles and lamellae (Bosabalidis and Thomson, 1984). In the upper secretory cells, wall protuberances are present (Fig. 4a). At the top of the gland, several distinct pores traverse the cuticle (Fig. 4a). Numerous mitochondria and plastids occur in the uppermost pair of the secretory cells. The plastids are electron-dense and contain osmiophilic droplets. Vesicles interconnected by lamellae form several concentric series in the peripheral region of the plastids. Along the walls and their protuberances, there are many vesicles and small vacuoles containing electron-dense material (Fig. 4a). Multivesicular structures are also visible in the cytoplasm. Ribosomes and polysomes are more abundant in secretory cells of plants grown in high NaCl concentration than those grown in low concentrations of NaCl. The protoplasts of the collecting cells resemble those of the mesophyll cells. Electron-dense material, apparently of a pectic nature, occurs above the cuticle at the top of the gland, between this portion of the cuticle and the cell walls and in the cuticular pores (Fig. 4a). This material forms a continuous channel system with the cell walls and their protuberances, and would appear to be the route by which the secreted salt solution reaches the outside of the plant.

The salt gland of the mangrove *Avicennia* (Fig. 2c) consists of 2–4 collecting cells, one disc-like stalk cell and usually eight, but sometimes up to

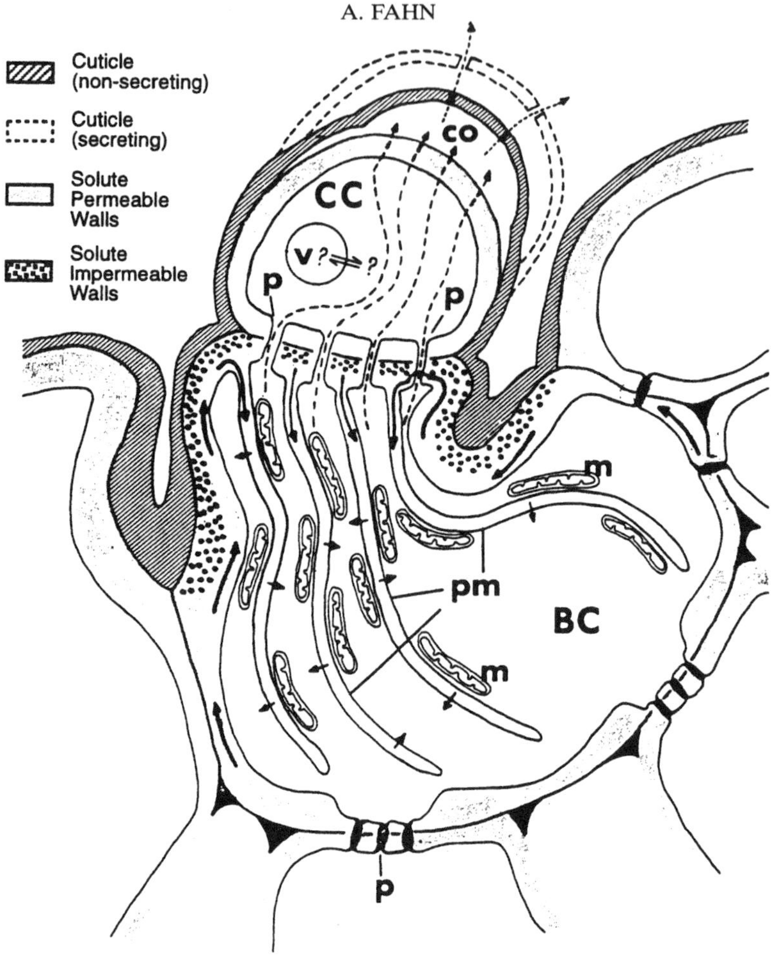

Fig. 3. A model of the structure and function of a salt gland of Bermuda grass (*Cynodon*) (Oross *et al.*, 1985). The larger, basal outer portion of the basal cell (BC) is constricted into a neck that protrudes slightly from the leaf surface and bears the cap cell (CC) as its outer terminus. Plasmodesmatal connections (p) occur between the basal cell and all adjoining cells. The raised portion of the cuticle over the cap cell forms the outer boundary of a subcuticular space known as the collecting chamber (co). The thick, highly lignified external wall layer present in the neck region of the basal cell is the only part of the gland apoplast that is impermeable to solutes (see legend on the figure). The basal cell protoplast is characterized by numerous, long invaginations of the plasmalemma (pm) that originate only near the juncture of the two gland cells. These partitioning membranes are closely associated with many mitochondria (m) and microtubules (not shown in the figure) on their intracellular side. The cap cell is relatively unspecialized in comparison to the basal cell and contains a normal complement of organelles that includes vacuoles (v) of varying size. Small arrows, energy-requiring transmembrane flux from the lumina of the partitioning membranes to the basal cell cytoplasm; heavy lined arrows, pathway of passive transport to the partitioning membranes; long dashes, pathway of diffusive flow through the gland symplast; short dashes, pressurized flow from the collecting chamber. Redrawn from Oross *et al.* (1985).

12, radially arranged secretory cells (Chapman, 1944). The electron microscopical study carried out by Shimony *et al.* (1973) on the salt glands of *Avicennia marina* (Forssk) Vierh. revealed many similarities between their ultrastructure and that of the glands of *Tamarix*. At the top of the gland, the cuticle is traversed by pores, which are numerous and very narrow (Fig. 4b). The side walls of the stalk cell are completely cutinized and the cytoplasm adheres strongly to them (Fig. 4d). The transverse walls of this cell contain numerous plasmodesmata. Between the cuticle and the upper walls of the secretory cells, moderately electron-dense homogeneous material, probably of a pectic nature, is present. It is often also seen on the inner surface of the side walls of the secretory cells (Fig. 4c). This material may function as a flow channel for the secreted salt solution in the same manner as the electron-dense material present in the glands of *Tamarix*. Inside the secretory cells, the cytoplasm occupies most of the cell volume. The nucleus is large and mitochondria and ribosomes numerous. Vesicles (Fig. 4c), some with dense contents, frequently form aggregates in the peripheral cytoplasm. Multivesicular structures often occur between the plasmalemma and the cell wall.

Since in the salt glands studied, numerous vesicles and small vacuoles occur in their secretory cells, many in the peripheral cytoplasm, several authors have suggested that they are involved in the process of secretion (Thomson and Liu, 1967; Shimony and Fahn, 1968; Osmond *et al.*, 1969; Thomson *et al.*, 1969; Shimony *et al.*, 1973). However, later, Thomson and Healey (1984) and Oross *et al.* (1985) hypothesized that the primary mechanism involved in the transport of salt across membranes into the vacuole of *Atriplex* (to the outside of the secretory cell and through the plasmalemmal invaginations in graminaceous glands, and into the vesicles and/or across the plasmalemma of the secretory cells of multicellular glands) is an eccrine process. Recent experimental work on the salt glands of *Avicennia germinans* also leads to the suggestion that membrane transport may be the primary mechanism underlying secretion in these glands (Dschida *et al.*, 1992; Balsamo *et al.*, 1995).

On the basis of a study of the localization of calcium in the various compartments of the different trichome cells in *Cicer arietinum*, which secrete organic acid, Lazzaro and Thomson (1992) concluded that secretion from the head cell of the trichome occurs predominantly through an exocytic, vesicular pathway.

The transport of the salt solution, after it has been eliminated from the protoplast of the two-celled and the multicellular glands, takes place in the wall. The cell wall of the secretory cells seems to be particularly adapted to this purpose. In many cases (e.g. *Tamarix*, *Spartina*) wall protuberances extend into the cell. The wall protuberances (and to some extent the wall itself) appear to be rich in pectic substances, thus having favourable properties for the absorption of solutions. In addition, these protuberances,

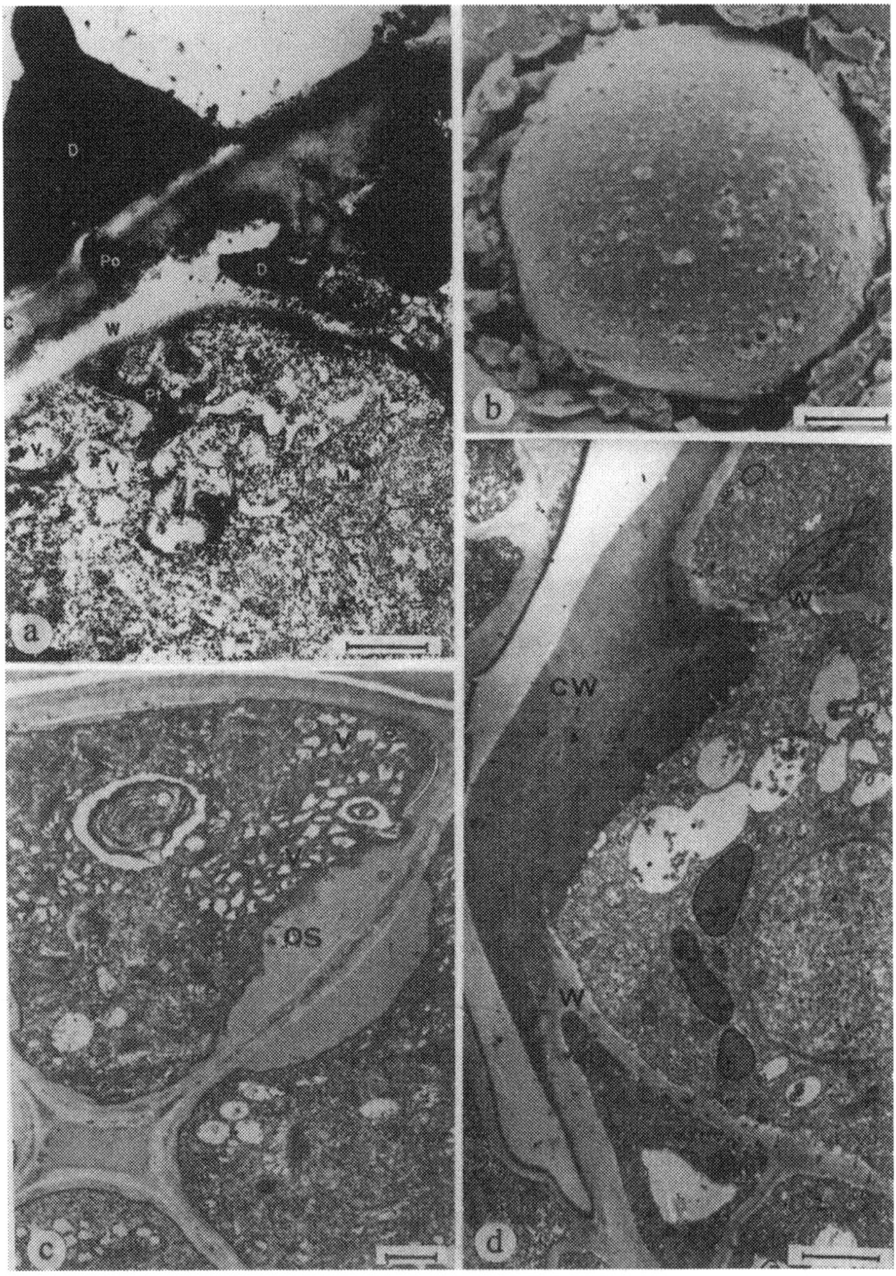
D
Po
D
W
Pt
V
V
M
a
b
V
V
OS
c
W
CW
W
d

which are lined by the plasmalemma, greatly enlarge the surface of the protoplast, as in many other cells in which short-distance transport takes place. Such transfer cell morphology (Pate and Gunning, 1972) would clearly appear to facilitate the transport of solutes between the symplast and its extracellular environment. The backflow of the secreted salt solution through the apoplast into the neighbouring tissues is blocked, as in almost all other secretory trichomes, by the endodermal cells (Schrödter, 1926; Schnepf, 1969a; Fahn, 1979a).

Gravano *et al.* (1998), who have recently studied the multicellular glandular trichomes of *Phillyrea latifolia* (Oleaceae), stated that they may function as specialized salt glands under conditions of high salinity. However, as sodium and chloride secretion is not the primary role of these trichomes, the authors suggested that their secretory activity may also contribute to the maintenance of ion concentrations within physiological limits under conditions of severe drought.

3. *Uptake of Ions from the Xylem and their Elimination from the Protoplast*

Lüttge (1969, 1971) reviewed the hypotheses about the uptake of ions from the xylem and their translocation through the mesophyll to the salt glands. According to Shimony *et al.* (1973), a downhill salt gradient appears to exist in leaves of *Avicennia* from near the xylem to the gland, and is continued through the gland itself, the secretory cells having the lowest salt content. Campbell and Thomson (1976a,b) and Van Steveninck *et al.* (1976) discussed the question of apoplastic versus symplastic transport of salt to the gland. However, this problem has not yet been resolved. As already mentioned, the cytoplasm of the secretory cells of salt glands is rich in ribosomes and contains a large nucleus and numerous organelles, particularly mitochondria. This supports the view that the elimination of ions from the cytoplasm requires energy.

Fig. 4. Electron micrographs of salt glands. (a) An upper part of a gland of *Tamarix aphylla*. C, cuticle; D, electron-dense pectic material; M, mitochondrion; Po, pore in the cuticle filled with electron-dense material; Pt, wall protuberance; V, vacuoles; W, cell wall. Scale bar = 1 μm. (b–d) *Avicennia marina*. Scanning electron micrograph of a gland on the upper leaf surface, showing numerous pores in the cuticle (b). Scale bar = 1 μm. Cross-section of a young gland, showing numerous vesicles and an opaque substance (OS) (apparently of pectic nature) between the protoplast and the cell wall (c). Scale bar = 1 μm. Longitudinal section, showing a completely cutinized side wall (CW) of the stalk cell (d). Scale bar = 1 μm. W, cutin-free cell wall. [(a) Reprinted, with permission, from Shimony and Fahn, 1968; (b–d) reprinted, with permission, from Fahn (1979b) and Shimony *et al.* (1973).]

III. NECTARIES

A. PLACES OF OCCURRENCE AND GENERAL STRUCTURE OF NECTARIES

Caspary (1848) distinguished between *floral* and *extrafloral* nectaries, the former occurring in flowers, the latter on vegetative organs. Delpino (1868–1875), who considered the function of the nectaries, proposed the terms *nuptial* and *extranuptial* nectaries; the first for nectaries occurring within the flower, which are directly associated with pollination, and the second for nectaries occurring on the outside of the outer floral parts and on vegetative organs, which are not directly associated with pollination. The terms *floral* and *extrafloral* nectaries are now commonly used in the same sense as Delpino used the terms *nuptial* and *extranuptial* nectaries (Elias and Gelband, 1976; Fahn, 1979a,b; Vogel, 1997, 1998a–c).

Nectaries occur on plant surfaces. They may be either flush with the surface of the plant or may form outgrowths or be sunken, in some cases very deeply, as for instance are the septal nectaries of many monocotyledons. The nectary consists of an epidermis, with or without trichomes, and specialized parenchyma (Fig. 5). The tissue that constitutes the nectary is called *nectariferous tissue*. The cells of this tissue, especially of the parenchyma, are usually small and contain a dense granular cytoplasm and a relatively large nucleus (Caspary, 1848; Behrens, 1879; Fahn, 1952). The nectaries either abut the ordinary vascular bundles of the organs on which they occur, or contain special bundles, which are connected with the ordinary vasculature. The special vascular bundles may consist of both phloem and xylem or of phloem alone (Frei, 1955; Kartashova, 1965; Fahn, 1979a; Said, 1982; Davis *et al.*, 1986; Dafni *et al.*, 1988; Belmonte *et al.*, 1994; Nepi *et al.*, 1996b; Gaffal *et al.*, 1998). Nectar is exuded from the nectary either by modified epidermal cells or by trichomes (Figs 5, 6), via modified stomata (Rachmilevitz and Fahn, 1973, 1975; Zer and Fahn, 1992; Davis and Gunning, 1991, 1992), or by lysigenous cavities, as for instance in the extrafloral nectaries of *Sambucus* (Fahn, 1987a).

B. NECTAR AND ITS SOURCE

Secreted nectar consists mainly of sugars but with small amounts of other substances. The most common sugars are sucrose, glucose and fructose, with smaller amounts of maltose, raffinose, galactose and melibiose. Based on quantitative relationships between sucrose and glucose/fructose, three main types of nectar have been distinguished: (1) sucrose-dominant, (2) glucose-dominant and (3) those with an equal ratio of sucrose to glucose/fructose (Fahn, 1949; Percival, 1961; Bahadur *et al.*, 1986). The ratio between the

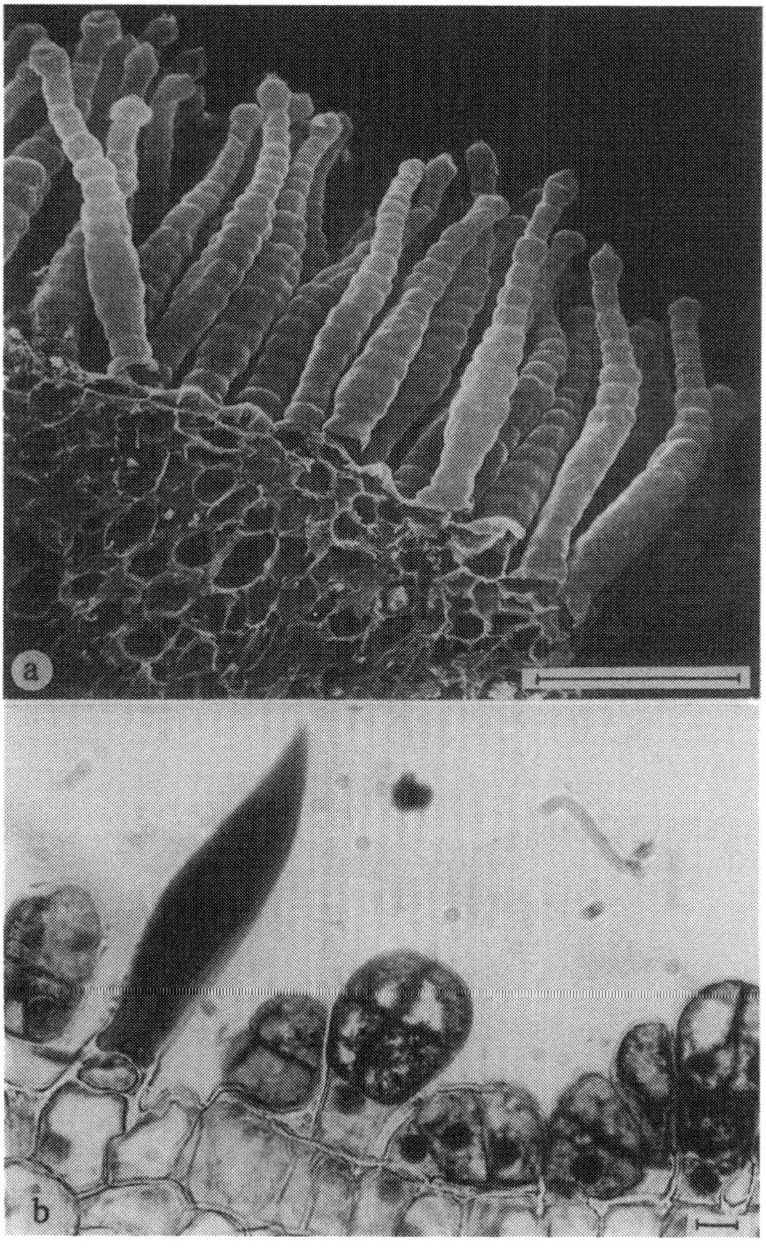

Fig. 5. Nectar-secreting trichomes. (a) Scanning electron micrograph of *Abutilon pictum*. Scale bar = 10 μm. (b) Stipular nectary of *Vicia faba*. Wall protuberances can be seen inside the secretory cells of the trichomes. Scale bar = 0.5 μm.

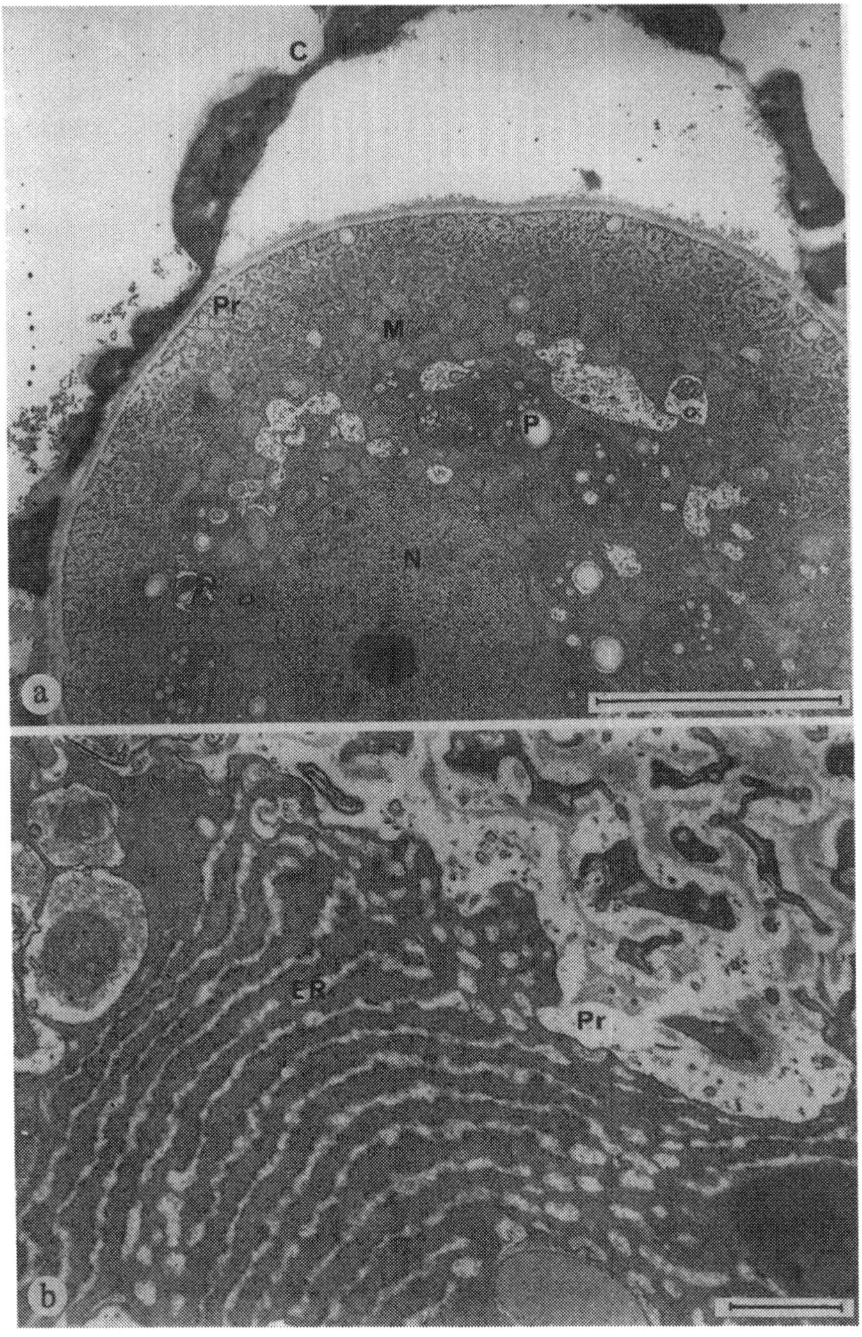
C
Pr
M
P
N
a
ER
Pr
b

different sugars is generally constant within a species but may differ markedly between species, sometimes even between closely related ones (Baker and Baker, 1983a,b). The other substances reported in nectar are polysaccharidic mucilage, proteins (mainly enzymes, among them transglucosidase and transfructosidase), amino acids, lipids, organic acids, ascorbic acid (functioning apparently as an antioxidant), mineral ions, phosphates and alkaloids (Lüttge, 1961; Baker and Baker, 1975, 1983a,b; Fahn, 1979a). Baker and Baker (1973a,b) found amino acids in a minor but significant amount in nectar from 260 of 266 plants examined. These authors concluded that the amino acids significantly affect the flower-visiting behaviour of potential pollinators. Butterfly-pollinated flowers, for instance, appear to produce nectar containing high concentrations of amino acids more consistently than bee-pollinated flowers. This difference may be related to the fact that bees have an alternative source of amino acids in pollen.

The origin of the secreted nectar is the phloem sap (Frey-Wyssling and Agthe, 1950; Frey-Wyssling, 1955). The pre-nectar moves symplastically via plamodesmata from the sieve elements to the secretory cells (Sawidis, 1991). In the nectariferous tissue, the pre-nectar may become modified by enzymatic activity and by the process of reabsorption (Lüttge, 1961; Ziegler, 1965). The changes occurring in the pre-nectar of extrafloral nectaries of *Ricinus communis* L. are recorded by Baker *et al.* (1978) (Table I).

The factors governing the preferential flow of the pre-nectar towards the secretory cells rather than to neighbouring cells are not yet clear. It has been suggested that hydrolysis of sucrose into its monomers may play a role in forming a sink for sugars in the nectariferous tissue. As a result, a sucrose concentration gradient could be maintained causing a passive flow of sucrose from the sieve elements to the nectar-secreting cells (see Lüttge and Schnepf, 1976; Meyberg and Kristen, 1981). However, this cannot be the case in nectaries in which sucrose is the dominant sugar secreted. Lüttge (1977) considers active transmembrane transport to play the main role in sugar movement, but exocytosis and endocytosis may also take place. Elimination of nectar from the secretory cells of the nectary may occur via either eccrine or granulocrine pathways (Fig. 7) (see later).

The amount of secreted nectar may differ between male and female flowers of the same plant (Fahn, 1949). In *Cucurbita pepo*, for instance, this difference is correlated with the nectary surface area and the number of

---

Fig. 6. Electron micrographs of secretory papillae of the floral nectary of *Lonicera japonica*. (a) The cuticle (C) is partly detached from the cell wall. The latter possesses a thick layer of wall protuberances (Pr); M, mitochondria; N, nucleus; P, plastids. Scale bar = 10 μm. (b) Vesicles are present at the edges of the parallel endoplasmic reticulum (ER) cisternae. Pr, wall protuberances. Scale bar = 1 μm.

TABLE I

The composition of phloem exudate and of nectar secreted by the extrafloral nectaries of Ricinus communis *plants. The concentrations are expressed in mg/ml.* (Data from Baker *et al.*, 1978.)

	Phloem exudate	Secreted nectar
Dry matter	100–125	900
Sucrose	80–106	280
Glucose α	0	107
Glucose β	0	190
Fructose α and β	0	295
Galactose	0	1.0
Protein	1.45–2.2	0.12–0.20
Amino acids (as glutamic acid)	5.2	1.7
Organic acids (as malic acid)	2.0–3.2	Trace
Nitrate	0	0.04
Phosphate	0.35–0.55	0.17
Potassium	2.3–4.4	7.0–9.4
Sodium	0.05–0.28	0.092–0.21
Calcium	0.02–0.09	0.48–0.80
Magnesium	0.11–0.12	0.15–0.22
pH	8–8.2	6.8–7.2

stomata (Nepi *et al.*, 1996a). Gaffal *et al.* (1998), who measured the volume flux via individual stomata apertures of the floral nectary of *Digitalis purpurea*, concluded that only a fraction of the total number of stomata per nectary would be sufficient to discharge the amount of nectar produced by the nectary.

Bieleski and Redgwell (1980) suggested that there is a bidirectional movement of sugars, from the nectary to the outside and from the secreted nectar back into the nectary. According to these authors, the two-way transfer of sugars plays a significant part in the final sugar composition of the nectar. Reabsorption of nectar in *Cucurbita pepo* not gathered by insects was observed by Nepi *et al.* (1996a,b). Back flow of secreted nectar into the nectariferous tissue seems, to the author of this chapter, to be possible only in nectaries secreting through stomata.

C. ULTRASTRUCTURE OF NECTAR-SECRETING CELLS

The walls of the secretory cells of many nectaries possess protuberances on their inner surface, which may be developed to different extents depending on the species. They may be very small and inconspicuous, as in the epithelial cells of the septal nectaries of *Tillandsia* (Cecchi *et al.*, 1982), somewhat

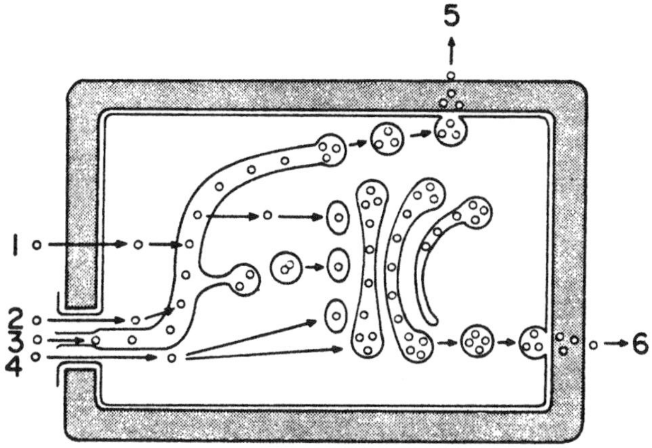

Fig. 7. Diagram presenting various possible ways of pre-nectar transport into a secretory cell and of nectar secretion from the cell. (1) Molecular transport through plasmalemma and into the endoplasmic reticulum (ER); (2) transport through the cytoplasmic annulus of a plasmodesma into the protoplast and entrance into the ER by membrane passage; (3) entering the ER of the secretory cell through the desmo-tubule of a plasmodesma; (4) as in (2) but entering a Golgi body; (5) exudation of nectar by fusion of vesicles that originated from ER with the plasmalemma; (6) exudation of nectar by fusion of Golgi vesicles with the plasmalemma. (Reprinted, with permission, from Fahn, 1979a.)

larger, as in the septal nectaries of the banana (Fahn and Benouaiche, 1979), or they may form a thick labyrinthine layer as in the septal nectaries of *Gasteria trigona* Haw. (Schnepf, 1969a), in the secretory trichomes of the floral nectaries of *Lonicera japonica* (Fig. 6) (Fahn and Rachmilevitz, 1970) and in the secretory trichomes of the extrafloral nectaries of *Vicia faba* L. (Tarkowska *et al.*, 1981). In the wall of the apical dome of the epithelial cells of the *Cynanchum* nectary, Christ and Schnepf (1985) found channel-like structures of low electron density. The free surface of the nectary may be covered by either a thin or a thick cuticle. According to Figueiredo and Pais (1992), the exudation of nectar into the spur cavity of the orchid *Limodorum abortivum* occurs by disruption of the cuticle. In nectaries that secrete through papillae or multicellular trichomes, the side walls of the lower part of the papillae or those of the lowest cells of the multicellular trichomes (the stalk cells) are completely cutinized. The cell walls at the stalk base of the stipular nectary trichomes of *Vicia faba* are thick and, as suggested by Davis *et al.* (1988), are lignified.

The volume of the cytoplasm in the nectariferous cells is relatively large. With the growth of the cells towards the stage of secretion, the volume of the vacuoles may increase. The cytoplasm is usually dense and rich in ribosomes. The plasmalemma is often highly convoluted and in many nectaries multivesicular structures occur between the infolded regions of the plasmalemma and the cell walls (Eymé, 1967; Fahn and Rachmilevitz, 1970; Rachmilevitz and Fahn, 1973; Kalman and Gulyás, 1974; Belin-Depoux and Clair-Maczulajtys, 1975; Fahn and Benouaiche, 1979; Mohan and Inamdar, 1986). In many plants, vesicles occurring in the cytoplasm can also be seen to be in contact with the plasmalemma (Fahn, 1979b; Eleftheriou and Hall, 1983). In most nectaries the endoplasmic reticulum (ER) is highly developed and its cisternae are often arranged in stacks. The ER is mainly rough but at the stage of secretion it may become in part smooth. At this stage, the ER cisternae are dilated and associated with vesicles (Fig. 6b) (Fahn and Rachmilevitz, 1970; Fahn and Benouaiche, 1979; Schnepf and Christ, 1980; Eleftheriou and Hall, 1983; Mohan and Inamdar, 1986). In many nectary cells, in addition to ER, numerous active dictyosomes are present (Eymé, 1966; Figier, 1968; Tacina, 1973; Benner and Schnepf, 1975; Fahn and Benouaiche, 1979; Christ and Schnepf, 1985; Marginson *et al.*, 1985b). At the stage of secretion, the nectary cells contain numerous mitochondria, the cisternae of which are usually well developed (Fahn and Rachmilevitz, 1970; Erikson, 1977; Schnepf and Christ, 1980; Mohan and Inamdar, 1986). In *Cynanchum vincetoxicum* Pers., most of the mitochondria are aggregated into complexes (Christ and Schnepf, 1985). Plastids occur in varying numbers, with very few thylakoids and most contain starch grains. In some plant species the amount of starch is very large at the stage preceding secretion (Durkee *et al.*, 1981; Durkee, 1982). An increase in the amount of starch grains towards the stage of secretion and its subsequent decline has recently been observed in the nectaries of *Rosmarinus officinalis* (Zer and Fahn, 1992) and in the floral nectaries of *Cucurbita pepo* (Nepi *et al.*, 1996b). Free osmiophilic globuli are often present in the cytoplasm.

D. TRANSPORT OF PRE-NECTAR

Several possible pathways have been suggested for the pre-nectar flow from the phloem endings, through the parenchymatous cells of the nectary and into the secretory cells (Fig. 7). These include: (1) via the apoplast (Vassilyev, 1969, 1971; Schnepf and Deichgraber, 1984); (2) via exocytosis and endocytosis (tentatively suggested by Eymé (1966) and Findlay and Mercer (1971)); in this pathway multivesicular structures may take part (Kalman and Gulyás, 1974); (3) molecular transport across the plasmalemma of successive cells and passage through cell walls; and (4) via plasmodesmata. The first of these possible pathways can be discarded, at least in the case of nectaries, which secrete from trichomes, because of the presence of the endodermal stalk cells. The second pathway would involve repeated secretion and absorption of pre-nectar as it passes from cell to cell through the cell wall.

Many studies lead to the conclusion that the transport of the pre-nectar is mainly in the symplast. Numerous plasmodesmata traverse the walls of nectariferous cells (Fahn and Rachmilevitz, 1970; Findlay and Mercer, 1971; Wergin *et al.*, 1975; Eleftheriou and Hall, 1983; Kronestedt *et al.*, 1986; Sawidis, 1991). Gunning and Hughes (1976) studied the nectary trichomes of *Abutilon*, and found a high frequency of plasmodesmata in the periclinal cell walls. These authors have calculated that the required sugar flux through the endodermal stalk cell of the trichome is 3–4 orders of magnitude greater than quoted in the literature for flux through cell membranes. On the other hand, the number and dimensions of the plasmodesmata in the distal periclinal wall of the endodermal stalk cell provide a low-resistance pathway for bulk flow of sugar solution. Their study thus supports the view that the pre-nectar flows through the plasmodesmata.

E. TRANSPORT OF NECTAR OUTWARD FROM THE SECRETORY CELLS.

Two main modes of transport of nectar to the outside of the protoplast of secretory cells have been suggested. There are theories of active molecular transport through membranes, eccrine secretion and of transport via vesicles whose membranes fuse with the plasmalemma, granulocrine secretion (Fig. 1).

In 1964, Schnepf expressed the view that, in the septal nectaries of *Gasteria* and some other Liliaceae, nectar transport is a molecular process at the plasmalemma; amplification of the plasmalemma, resulting from the development of wall protuberances, was associated with the onset of nectar secretion. The possibility of molecular transport of sugars within the nectary of *Abutilon* was discussed by Reed *et al.* (1971).

Fahn and Rachmilevitz (1970) and Rachmilevitz and Fahn (1973) studied the changes in the ultrastructure of the secretory cells of the nectaries of *Lonicera*, *Vinca* and *Citrus*, at various developmental stages. They observed that, towards the phase of secretion, there is a pronounced increase in the amount of the ER, with the cisternae becoming dilated and associated with numerous vesicles (Fig. 6). In some nectaries, ER cisternae and vesicles lay adjacent to the plasmalemma. On the basis of these observations, it was suggested that vesicle membranes fuse with the plasmalemma. Presumably the nectar accumulates in the ER and is transported to the plasmalemma in the vesicles budded off from its cisternae. The membranes of the two fuse and the nectar is eliminated from the protoplast by reverse pinocytosis. This view had been supported by a number of other investigations (Belin-Depoux and Clair-Maczulajtys, 1975; Eleftheriou and Hall, 1983; O'Brien *et al.*, 1996; Stpiczynska, 1997).

Suggestions that Golgi vesicles, rather than vesicles originating from the ER, are involved in outward transport of the nectar have been made by a number of authors (Eymé, 1966; Figier, 1968, 1971; Tacina, 1973; Benner and

Schnepf, 1975; Juniper *et al.*, 1989). In the nectary of the calyx spur of *Tropaeolum majus* L. (Rachmilevitz and Fahn, 1975) and in the septal nectary of *Musa paradisiaca* L. var. *sapientum* (Fahn and Benouaiche, 1979), numerous dictyosomes bud off vesicles at the stage of secretion. Dilated ER cisternae and ER vesicles were also numerous. It was thus suggested that in the nectaries of these plants both the endoplasmic reticulum and the Golgi apparatus are involved in nectar secretion. A similar suggestion was made by researchers who worked on nectaries of some other plants (Schnepf and Christ, 1980; Christ and Schnepf, 1985; Marginson *et al.*, 1985a) (see Fig. 7).

In addition to developmental ultrastructural studies, labelling and histochemical techniques have also been used in this study of secretory cells. Heinrich (1975a) studied the localization of phosphatases in the secretory epithelium of the septal nectaries of *Aloe*. He found high activity at ATPase, nucleoside diphosphatase and glucose-6-phosphatase in the ER, and generally an absence of activity of these phosphatases in the plasmalemma. Heinrich thus assumed that the endoplasmic reticulum plays an important role in nectar transport. Figier (1972), who tried to localize acid phosphatase in petiolar nectaries of *Impatiens*, also observed only limited reactions on the plasmalemma and the cell walls, except of the sieve tubes and the transfer cells, where he found heavy deposits of the reaction product. Attempts to follow the path of sugar in the nectary by means of autoradiography were made by some researchers. After feeding floral nectaries of *Lonicera* with tritiated sucrose, the ER and vesicles in the secretory cells became labelled. In the nectariferous tissue below the secretory cells, labelling was frequently observed in the plasmodesmata (Fahn and Rachmilevitz, 1975). Heinrich (1975b), on the basis of labelling the septal nectaries of *Aloe* with tritiated glucose, concluded that the nectar is secreted by vesicles, but could not decide whether they originated from ER or from dictyosomes.

A new model for the manner of nectar secretion in the nectary of *Abutilon*, as presented in Fig. 8, has been suggested by Robards and Stark (1988), who based it on evidence obtained from freeze-substitution, freeze-fracture and physiological experiments. The model takes account of all the currently available information concerning secretion in *Abutilon*, but does not predict excessively high rates of either eccrine or usual granulocrine

---

Fig. 8. Diagram illustrating the model for nectar secretion in *Abutilon*, suggested by Robards and Stark (1988), based on evidence from freeze-substitution, freeze-fracture and physiological experiments. It is suggested that pre-nectar moves into the symplast of the trichome via the numerous plasmodesmata in the transverse walls of the stalk cell. In each of the trichome cells some of the pre-nectar is loaded from the cytoplasm into the secretory reticulum. At this stage a filtration effect, which defines the chemical composition of the secreted product, takes place. The sucrose is partially

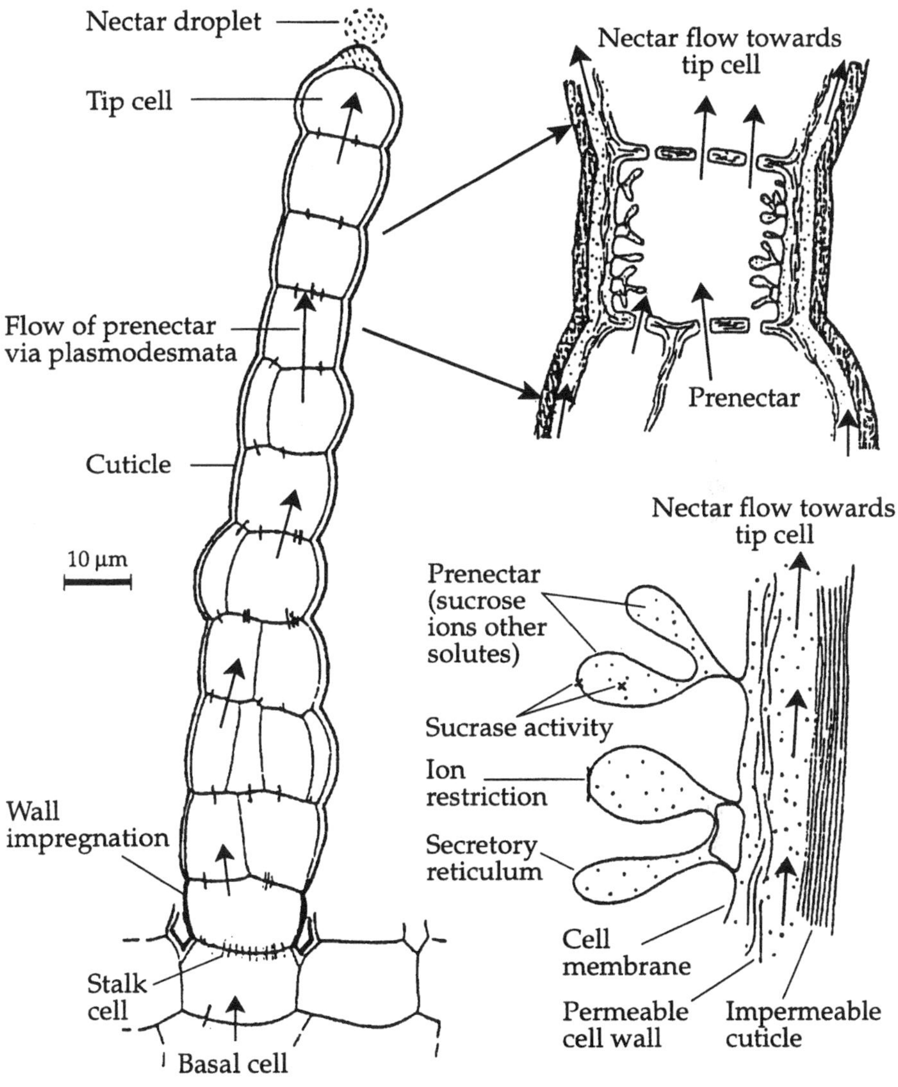

hydrolyzed to glucose and fructose, but evidence is not yet available as to whether this takes place at the membrane or within the cisternal cavity of the secretory reticulum. As loading into the secretory reticulum continues, a hydrostatic pressure builds up until a minute pulse of nectar is forced into the apoplastic space between the plasmalemma and the cuticle. The continuing build-up of pressure within this compartment ultimately reaches the level where pores in the cuticle over the tip cell become open and release a pulse of nectar to the exterior. (Redrawn from Robards and Stark, 1988.)

secretion. It also overcomes the osmotic difficulties that would otherwise arise from the high concentrations of solutes that are transported in this system.

The nectaries of some plants, such as those of the banana (Fahn and Benouaiche, 1979), secrete nectar, which contains polysaccharides and protein in addition to sugar. It has been suggested that, in these nectaries, the secretion of polysaccharides and protein also involves the ER and the Golgi apparatus.

In *Tropaeolum majus*, in addition to secretion by stomata, droplets, which presumably contain polysaccharides and a certain amount of sugar, are also secreted from hairs of the inner epidermis of the sepal spur situated above the nectariferous tissue. Results obtained using the Gomori test indicated that high acid phosphatase activity occurs, at the stage of secretion, in the nectariferous parenchyma cells between the epidermis and the region of vascular bundles. Enzyme activity was also observed in the wall thickenings at the tops of the epidermal hairs (Fig. 9) (Rachmilevitz and Fahn, 1975).

IV. CELLS SECRETING LIPOPHILIC MATERIAL

A. SUBSTANCES SECRETED AND TYPES OF SECRETORY STRUCTURES

Among the lipophilic materials secreted by plants are terpenes, fats, waxes and flavonoid aglycones. Essential oils and resins contain a great variety of terpenes (Goodwin and Mercer, 1983). Essential oils contain predominantly volatile low-molecular-weight terpenes, whereas the resins are a mixture of volatile and non-volatile terpenes. Such lipophilic materials are secreted by a variety of anatomical structures: solitary cells or idioblasts (e.g. oil cells of *Laurus*), areas of epidermal cells (e.g. epidermal spots on stipules in buds of *Populus*, which secrete flavonoid aglycones mixed with terpenes), trichomes (e.g. glandular trichomes of the Labiatae), cavities (e.g. essential-oil cavities of the Rutaceae), and ducts (e.g. resin ducts of *Pinus*). Many tissues secreting lipophilic material may also produce additional substances. The secretory trichomes of *Inula viscosa* Ait. and *Fagonia* species, for instance, secrete polysaccharides and protein, in addition to lipophilic material (Werker and Fahn, 1981; Fahn and Shimony, 1998).

Fragrance in flowers results from volatile low-molecular-weight terpenes, which occur in the form of minute droplets in the cytoplasm of the epidermal and neighbouring mesophyll cells of the sepals (Kisser, 1958). In some plants, e.g. *Ceropegia*, *Aristolochia* and species of the Orchidaceae and Araceae, fragrance production is restricted to certain areas of the floral parts known as *osmophores*, whose cells usually differ structurally from their neighbours. Osmophores may also secrete amines and ammonia in addition to terpenes

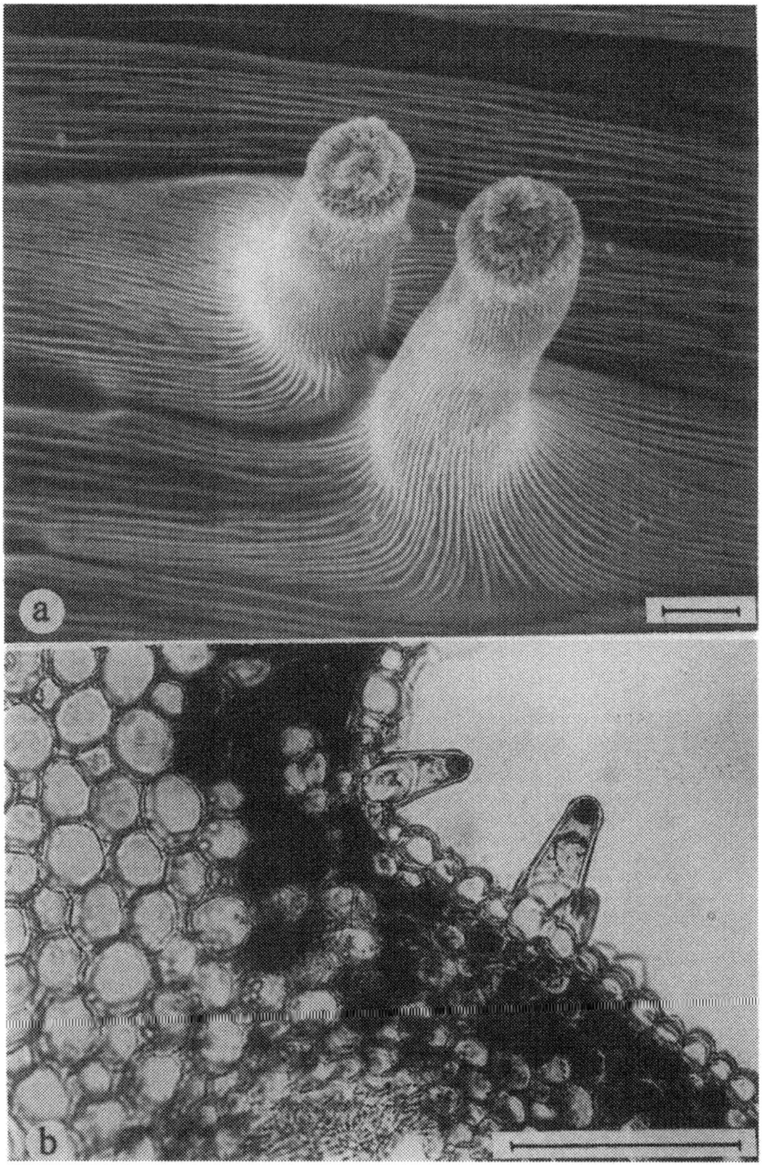

Fig. 9. *Tropaeolum majus*. (a) Scanning electron micrograph showing secretory hairs of the inner epidermis of the flower spur. Scale bar = 10 μm. (b) Part of a cross-section of the lowest portion of the spur. Distinct positive reaction to the Gomori test occurs in the secretory subepidermal secretory parenchyma cells and in the hair tips, where the wall thickening occurs. Scale bar = 100 μm. (Reprinted, with permission, from Rachmilevitz and Fahn, 1975.)

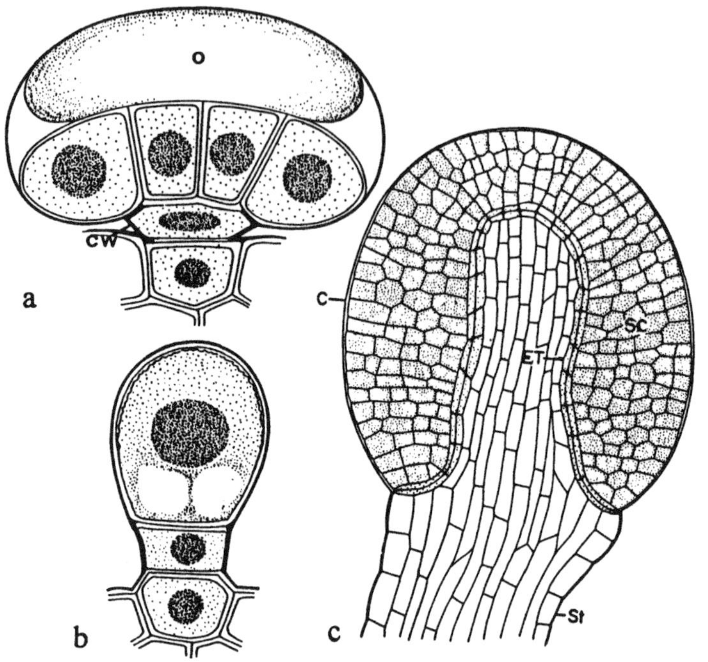

Fig. 10. Trichomes secreting lipophilic material. (a,b) Oil-secreting trichomes of *Mentha piperita*. Peltate hair (a); capitate hair (b). CW, cutinized cell wall; O, oil drop. (c) Longitudinal section of a glandular trichome of the calyx of *Plumbago capensis*, secreting a sticky material containing terpenes. C, cuticle; ET, endodermal tissue; SC, secretory cells; St, stalk. (Reproduced, with permission, from Fahn, 1979a.)

(Vogel, 1962, 1966; Pridgeon and Stern, 1983). *Oil cells* containing oil drops, commonly attached to a wall protuberance, occur in a number of families, such as Araceae, Aristolochiaceae, Calycanthaceae, Lauraceae, Magnoliaceae, Piperaceae and Saururaceae (Paech, 1952; Ziegler, 1960; Maron and Fahn, 1979).

Glandular trichomes secreting lipophilic substances occur in many families, e.g. Labiatae, Compositae, Geraniaceae, Solanaceae and Cannabinaceae. The trichomes vary in shape and structure. In the Labiateae, there are two main types of glandular trichomes: peltate and capitate. Peltate trichomes comprise a basal cell, a short stalk cell and a broad head consisting of many secretory cells arranged in one layer. The capitate trichomes consist of a basal cell, a one- to several-celled stalk and a head of one or two cells (Fig. 10a,b) (Fahn, 1979a; Werker *et al.*, 1985a,b). The glandular trichomes possess a cuticle proper and a cuticular layer. The side walls of the stalk cells, at least in peltate trichomes, are usually completely cutinized. With maturation of the trichomes, the two layers of the cuticle become detached from the pecto-cellulosic wall and the cellulosic wall, and the essential oil accumulates in the subcuticular space. The latter is particularly large in the peltate trichomes. Some of the volatile essential-oil fraction may pass through the cuticle, but most of the secreted material is discharged to the outside only after the rupture of the cuticle (Schnepf, 1969a). In some capitate trichomes, pores occur in the cuticle (Amelunxen, 1964; Ascensão and Pais, 1998). According to Ascensão and Pais (1998), the capitate trichomes of *Leonotis leonurus* secrete proteins, acid and neutral polysaccharides in addition to essential oils and flavonoid aglycones. The cellular components apparently associated with the synthesis and transport of this mucilaginous material are a highly developed ER and numerous Golgi stacks. These authors suggested that the protein–polysaccharide component reaches the plasmalemma via Golgi-derived vesicles, while the lipophilic constituents may be transported directly by the ER. The glandular trichomes of *Cannabis sativa* L., which secrete the narcotic cannabinoids, occur in several forms ranging from few-celled bulbous structures to many-celled capitate glands (Mahlberg *et al.*, 1984). The secretory cells of the capitate trichomes of *Nepeta racemosa* (Labiatae) possess extensive wall protuberances and plasmalemma invaginations similar to many nectaries (Bourett *et al.*, 1994).

The secretory trichomes of many Compositae are multicellular and biseriate. They may be stalked or sessile. The head consists of one or two pairs of chlorophyll-less summit cells and 2–3 pairs of chlorophyll-containing cells. All head cells secrete other compounds in addition to lipophilic substances. In *Inula viscosa* it has been reported that polysaccharides occur in addition to lipophilic compounds (Werker and Fahn, 1981). In *Artemisia campestris* L. ssp. *maritima*, fatty acids, terpenes, sterols and flavonol aglycones have been detected (Ascensão and Pais, 1985). The secreted material, as seen in *Inula*, is either exuded through minute pores in the cuticle, or accumulates below the cuticle and is discharged to the outside after the cuticle is ruptured.

In the large resin-secreting trichomes occurring on the calyx of *Plumbago capensis*, a 2–3-layered zone of endodermal cells is present between the mantle of secretory cells and the core of the trichome head. The core is part of the stalk (Fig. 10c) (Rachmilevitz and Joel, 1976). Flavonoid-secreting tissues are dealt with by Wollenweber (1984).

In numerous plant families, secretory cavities and ducts (canals) are present (Metcalfe and Chalk, 1983). The lumen of these structures is usually lined by several layers of secretory cells. The innermost layer is the most active in the process of secretion and is called the epithelium. During expansion of the cavity, cells of the outer layers may become incorporated into the epithelium (Fahn and Evert, 1974; Joel and Fahn, 1980a; Bosabalidis and Tsekos, 1982a; Pereira da Costa, 1985). The outermost cell layers of the secretory complex may be thick-walled and form a protective sheath, as for instance in *Citrus* (Bosabalidis and Tsekos, 1982a).

B. CELL COMPARTMENTS INVOLVED IN THE PROCESS OF SECRETION

The most common ultrastructural feature of the cells secreting a lipophilic substance is the occurrence of osmiophilic material in plastids (Fahn, 1988; Duke and Paul, 1993; Sacchetti *et al.*, 1996; Monteiro *et al.*, 1998). In most cases, these plastids are partly or completely surrounded by endoplasmic reticulum (periplastidal ER) (Fig. 13b,c) (Schnepf, 1969b; Fahn and Evert, 1974; Fahn and Benayoun, 1976; Joel and Fahn, 1980b; Bosabalidis and Tsekos, 1982a; Heinrich and Schultze, 1985; Pedro *et al.*, 1991; Ascensão *et al.*, 1997; Fahn and Shimony, 1998). In some plants, the plastids contain a peripheral reticulum, which may function in the transport of lipophilic material to the envelope (Vermeer and Peterson, 1979; Werker and Fahn, 1981; Ascensão and Pais, 1985). Plastids with a peripheral reticulum occur also in floral chromoplasts of some plants (Whatley and Whatley, 1987). The membranes of the ER in contact with osmiophilic droplets are usually smooth. Smooth tubular ER often occurs in the cytoplasm of cells secreting lipophilic material (Amelunxen and Arbeiter, 1967; Schnepf, 1969b, 1972; Joel and Fahn, 1980b; Danilova and Kashina, 1987; Pedro *et al.*, 1991). In the secretory cells of the peltate trichomes of *Nepeta racemosa*, Bourett *et al.* (1994) observed that, at the stage of secretion into the subcuticular space, tubular endomembrane elements, apparently smooth ER, occur in proximity to plastids. Osmiophilic material has also been observed in mitochondria, dictyosomes and the nuclear envelope (Vassilyev, 1970; Fahn and Evert, 1974; Fahn and Benayoun, 1976; Joel and Fahn, 1980b; Fahn and Shimony, 1998). Production of lipophilic material in the ground cytoplasm has been suggested to occur in the calyx glands of *Plumbago capensis* Thunb. (Rachmilevitz and Joel, 1976) and in oil glands of *Origanum dictamnus* L. (Bosabalidis and Tsekos, 1982b).

The occurrence of osmiophilic material in several different cell compartments in tissues that secrete essential oils or resins raises several questions regarding the possible site of synthesis of these materials. Three possible interpretations can be suggested: (1) each organelle in which osmiophilic droplets occur is capable of synthesizing all the oil or resin components independently; (2) there are several steps in essential-oil or resin synthesis, each taking place in a particular organelle; and (3) different components are synthesized by different organelles. On the basis of the fact that, in the resin ducts of *Mangifera indica* L., the lipophilic droplets associated with the dictyosomes appeared less osmiophilic than those in the plastids and the ER, and that both kinds of droplets were also found outside the plasmalemma, Joel and Fahn (1980b) suggested that the third interpretation was the most likely in this case. This view is strengthened by the results of a histochemical study that revealed droplets of different kinds in the shoot apical region of *Pinus halepenis* (Werker and Fahn, 1968).

Cheniclet and Carde (1985) suggested that there is a relationship between

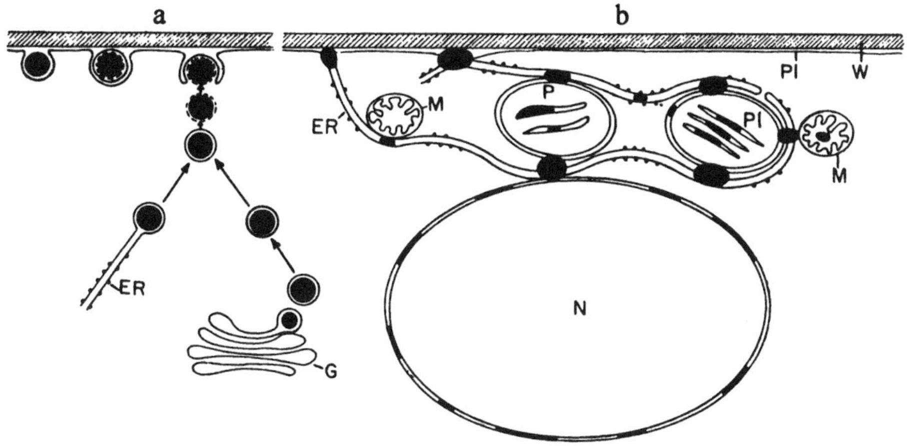

Fig. 11. Diagram showing the possible mode of intracellular transport of resin and its elimination from the protoplasts of epithelial duct cells of *Pinus halepensis*. ER, endoplasmic reticulum; G, Golgi body and its vesicles; M, mitochondrion; N, nucleus; P, plastid; Pl, plasmalemma; W, cell wall. Solid black indicates resin droplets, and interrupted lines around a resin droplet indicate that a membrane may or may not be present. (Reprinted, with permission, from Benayoun and Fahn, 1978.)

the structure of secretory cell plastids and their involvement in monoterpene synthesis. When the essential oil contains significant amounts of monoterpenes, the secretory cells contain typical leucoplasts devoid of ribosomes and thylakoids. In contrast, when the volatile extract contains little or no monoterpenes, the plastids display various structural characteristics, such as thylakoids, ribosomes and tubular networks.

The ER, in addition to taking part in the synthesis of lipophilic substances, may also take part in intracellular transport of this material (Fig. 11) (Benayoun and Fahn, 1979; Bosabalidis and Tsekos, 1982a). Benayoun and Fahn (1979), Mikulska and Zolnierowicz (1976) and Pedro *et al.* (1991) observed connections between membranes of the plastid envelope and the periplastidic ER. Benayoun and Fahn (1979) suggested that, in the epithelial cells of the resin ducts of *Pinus halepensis*, the ER is transporting the resin components from the plastids, mitochondria and nuclear envelope to the plasmalemma. By fusing with the plasmalemma, the ER releases the resin to the outside of the protoplast. The resin produced in the cytoplasm and by the Golgi apparatus may be eliminated by plasmalemma invaginations, which surround the resin droplets and detach them from the protoplast (Figs 11a, 11c and 12d). Fusion of membranes of the ER, which carries oil droplets, with the plasmalemma has also been suggested by Bosabalidis and Tsekos (1982a) for the secretory cells of the essential oil cavities of *Citrus deliciosa* Tenore. Transport of fatty oils through the plasmalemma via granulocrine

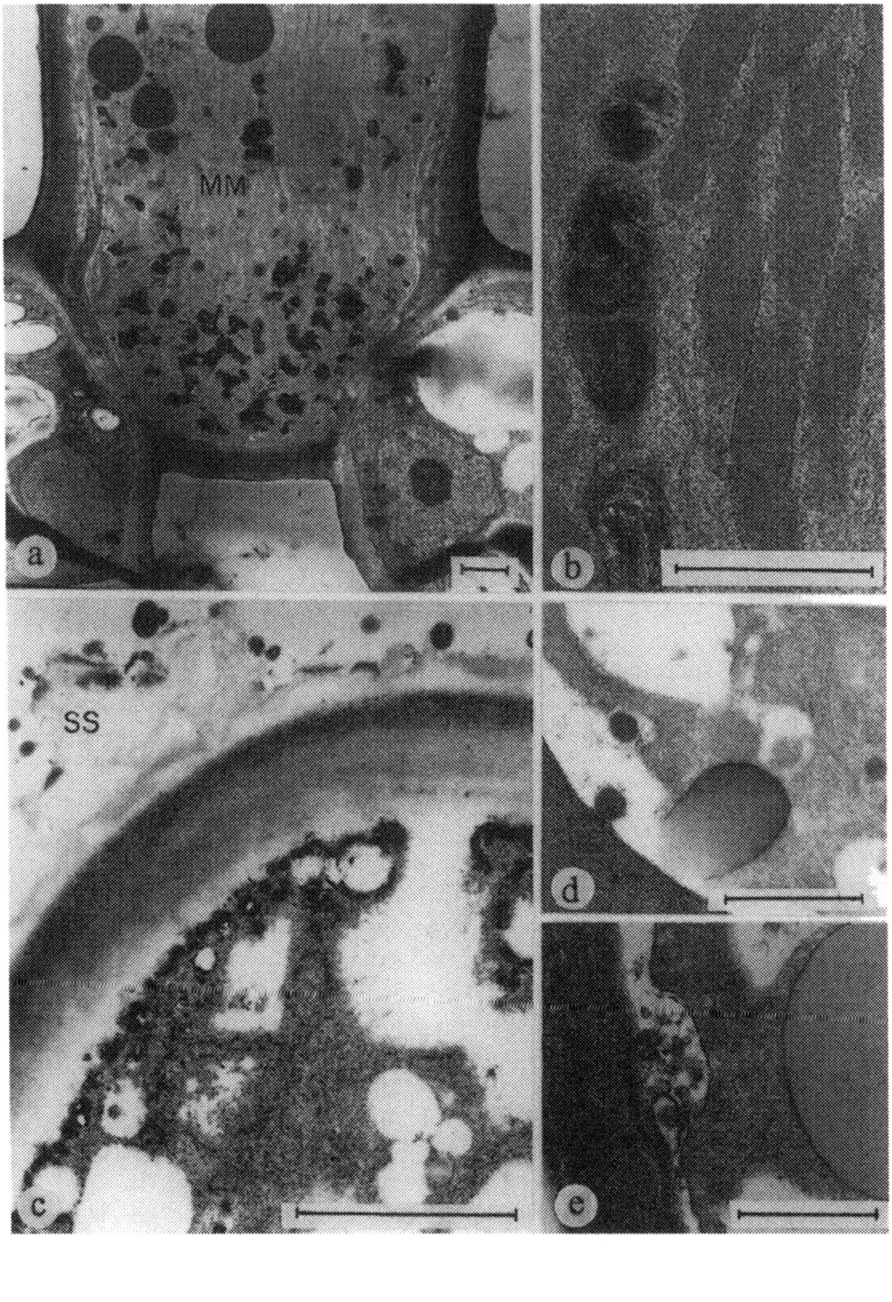
MM
SS
a
b
c
d
e

elimination has also been postulated in the osmophores of the orchid *Restrepia* (Pridgeon and Stern, 1983). Ascensão *et al.* (1997) stated that in the peltate glandular trichomes of *Leonotis leonorus* granulocrine secretion mechanism may operate alone or concurrently with an eccrine process. Pedro *et al.* (1991), who studied the procumbent trichomes of *Geranium robertianum*, suggested that in these trichomes the plastids may represent the main site of monoterpene and flavonoid biosynthesis, while the ER is the main site of sesquiterpene biosynthesis, and that the close association of periplastidic ER with plastids may indicate that the ER is also involved in the transport of the secretory material. Trichomes of the desert plant *Fagonia* secrete a sticky substance that contains mainly lipophilic compounds and polysaccharides (Fig. 12c) (Fahn and Shimony, 1996). Very interesting results were obtained by an ultrastructural study of the secretory cells of two different species of *Fagonia* (Fahn and Shimony, 1998). In *F. mollis*, the lipophilic material was found to be produced by mitochondria, which, after releasing the osmiophilic material, are modified, becoming elongated and losing the inner membranous structures (Fig. 12a,b). In the other species, *F. glutinosa*, the lipophilic material was observed to appear first as plastoglobuli in plastids with inner tubular membranes (Fig. 13). The polysaccharides in both species appear to be produced by dictyosomes. As many vesicles and multivesicular structures have been seen to occur outside plasmalemma invaginations, it was suggested that the manner of secretion of the secretory substance to the outside of the protoplast is granulocrine (Fig. 12d,e). The involvement of tubular structures appearing inside the plastids in the secretory process of lipophilic material was also suggested by Figueiredo and Pais (1994) to occur in glandular trichomes of *Achillea millefolium* ssp. *millefolium*.

A different mechanism of secretion of lipophilic material was suggested to occur in the secretory cells of the thermogenic appendices of *Sauromatum guttatum*. According to Skubatz *et al.* (1995), the rough endoplasmic reticulum (RER) plays a main role in the biosynthesis, accumulation and secretion of sesquiterpenes in the cell. The osmiophilic material accumulates in pocket-like structures of the RER, which fuse with the plasmalemma, but

---

Fig. 12. Glandular trichomes of *Fagonia mollis*. (a) Part of a secretory cell, showing, at its base, numerous mitochondria containing myelin-like structures, modified, elongated mitochondria (MM) and osmiophilic droplets above them. (b) Mitochondria with myelin-like structures between elongated modified mitochondria. (c) Upper part of a secretory cell. Granular lipophilic material occurs between the cytoplasm and the loosely structured cell wall. The secreted substance (SS) is present outside the top of the cell. (d) Osmiophilic droplets are seen between the cell wall and the protoplast, some in invaginations of the plasmalemma. A large droplet opens into the space between the cell wall and the protoplast. (e) Multivesicular bodies occur in invaginations of the plasmalemma. Scale bar = 1 μm. (Reproduced, with permission, from Fahn and Shimony, 1998.)

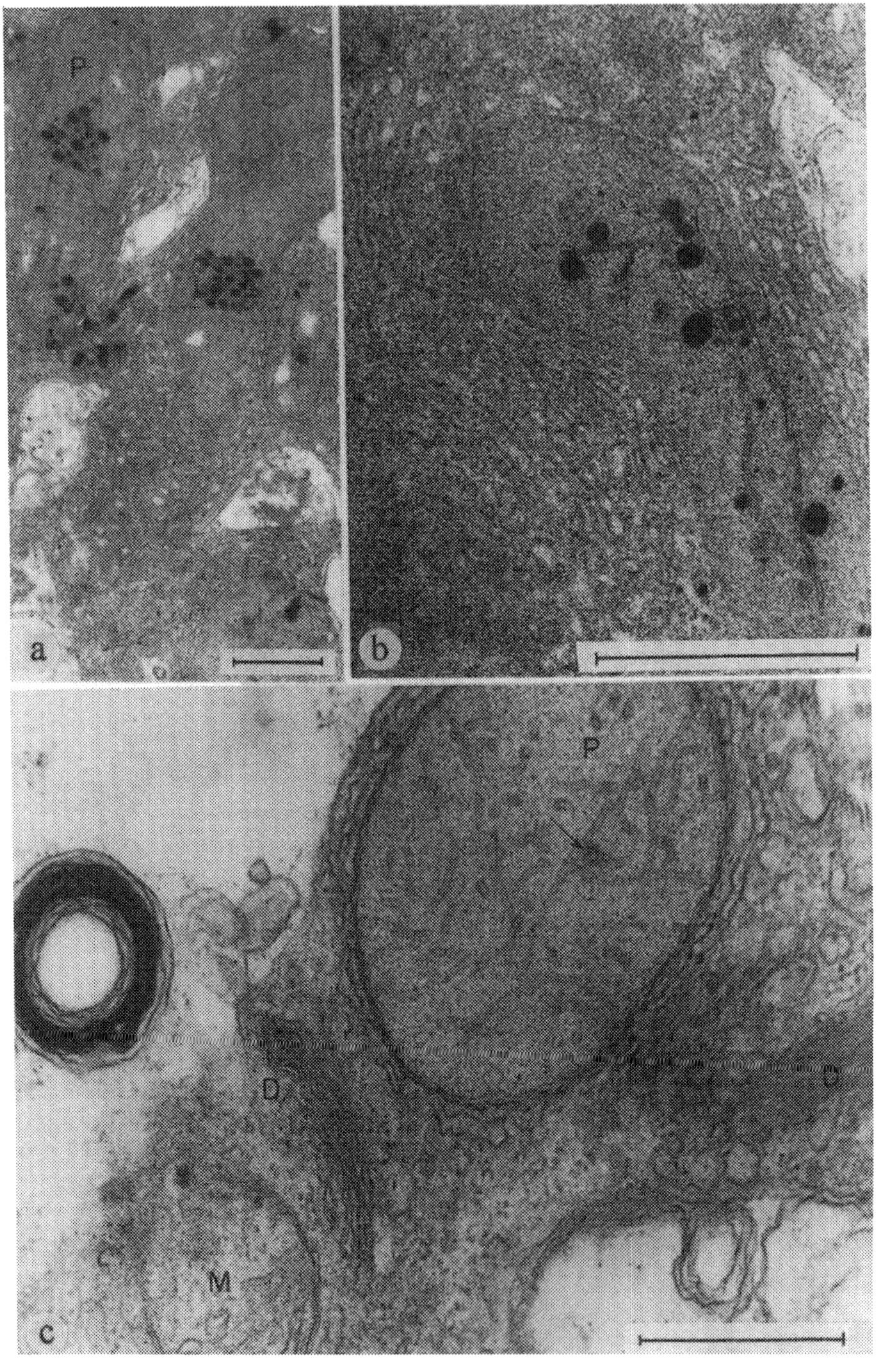
P
a
b
P
D
D
M
c

their contents are released to the outside of the plasmalemma only during the thermogenic activity.

V. SUMMARY AND CONCLUSIONS

The cells involved directly in the process of secretion in plants, the secretory cells, occur singly or in multicellular structures of a variety of forms, and on many plants in the form of trichomes. The study of the processes of secretion deals with a number of problems: (1) movement of precursors of the secreted compound towards the secretory cells; (2) the involvement of the various subcellular compartments in the production of the secreted compounds; (3) the way in which the secreted materials cross the cell membranes; (4) the intracellular transport of secreted compounds; (5) how the secreted compounds cross the outer cell wall and cuticle.

From the results of studies on the secretory tissues discussed in this chapter, the following conclusions can be drawn: in glands eliminating salts and in nectaries, the movement of the salt solution from the xylem, and the pre-nectar from the phloem to the secretory cell, is through the symplast via plasmodesmata.

In all types of secretory cells, numerous mitochondria are evident. In many secretory cells, wall protuberances occur, which result in enlargement of the surface area of the plasmalemma. In salt-secreting cells, numerous small vacuoles or vesicles and multivesicular structures usually occur in the peripheral cytoplasm. In the two-celled salt glands of the Gramineae, long plasmalemma invaginations occur in the basal cell. In the secretory cells of the nectaries, well-developed ER budding-off vesicles occupy much of the cytoplasm. Active dictyosomes are also common. Numerous dictyosomes and Golgi vesicles appear to be present in cells active in the synthesis of polysaccharides. Almost all cell compartments have been suggested to take part in the synthesis of lipophilic compounds. Most commonly associated with secretion are the plastids, which are usually surrounded by ER elements, and the two appear to be interconnected. The appearance of osmiophilic

---

Fig. 13. Glandular trichomes of *Fagonia glutinosa*. (a) A secretory cell with plastids (p) containing plastoglobuli. Scale bar = 1 μm. (b) A plastid surrounded by multilayered periplastidal endoplasmic reticulum (ER) contains branched tubuli bearing plastoglobuli at the ends of the branches. Scale bar = 1 μm. (c) A plastid surrounded by periplastidal ER, lacking plastoglobuli. The inner tubules appear to be orientated to the plastidal envelope and in cross-section they are seen to be built up of globular units (arrow). Electron-dense material appears between some inner tubules, the plastidal envelope and a dictyosome (d). M, mitochondrion; P, plastid. Scale bar = 0.5 μm. (Reproduced, with permission, from Fahn and Shimony, 1998.)

material in different cell compartments differing in electron density, in the same cell, may suggest that different components of the lipophilic material are synthesized by different cell compartments. In two different species of *Fagonia*, the lipophilic material was found to be produced by two different organelles, mitochondria and plastids.

The exudation of the secreted material to the outside of the protoplast in salt glands was suggested by many authors to be eccrine, although in the secretory cells of *Tamarix* and *Avicennia* numerous vesicles and multivesicular bodies were observed close to the plasmalemma. Nectar, polysaccharides and lipophilic compounds were suggested by most researchers to be eliminated from the secretory cells by a granulocrine mechanism. In almost all secretory trichomes, completely cutinized side walls of stalk cells are assumed to block the back-flow of the secreted material to the neighbouring tissue through the apoplast. The intracellular transport of nectar and polysaccharides appears to be carried out by the ER elements and Golgi vesicles. In secretory cells of lipophilic compounds, the process of exudation is, as suggested by most researchers, granulocrine. In some cases, the ultrastructural observations suggested that the ER collects the lipophilic material produced in the various cell compartments and transports them directly to the plasmalemma. In *Fagonia* the lipophilic material appears to be produced in two examined species by two different organelles, mitochondria or plastids.

The wall region through which the secreted material passes to the outside of the cell is usually porous and often rich in pectic substances. The cuticle, as seen in some salt glands and some capitate lipophilic glands, possesses pores. In peltate trichomes secreting lipophilic material, the cuticle usually ruptures by the pressure of the secreted substance.

The views expressed by researchers on the way by which the secreted material passes the plasma membranes, whether by an eccrine or granulocrine mechanism, remain in many cases contradictory. More collaborative research, using new cytological and biochemical methods, may contribute to a better understanding of this problem.

REFERENCES

Amelunxen, F. (1964). Elektronen mikroskopische Untersuchungen an den Drüsenhaaren von *Mentha piperita*. *Planta Medica* **12**, 121–139.

Amelunxen, F. and Arbeiter, H. (1967). Untersuchungen an den Sprizdrüsen von *Dictamnus albus* L. *Zeitschrift für Pflanzenphysiologie* **60**, 156–168.

Ascensão, L. and Pais, M. S. S. (1985). Differenciation et processus sécréteur des trichomes d' *A. campestris ssp maritima* (Compositae). *Annales des Sciences Naturelles, Botanique*, 13e serie **7**, 149–171.

Ascensão, L. and Pais, M. S. S. (1998). The leaf capitate trichomes of *Leonotis leonurus*: histochemistry, ultrastructure, and secretion. *Annals of Botany* **81**, 263–271.

Ascensão, L., Marques, N. and Pais, M. S. (1997). Peltate glandular trichomes of *Leonotis leonurus* leaves. *International Journal of Plant Sciences* **158**, 249–258.
Bahadur, B., Chaturvedi, A. and Swamy, N. R. (1986). Nectar types in Indian plants. *Proceedings of the Indian Academy of Science (Plant Science)* **96**, 41–48.
Baker, D. A., Hall, J. L. and Throp, J. R. (1978). A study of the extrafloral nectaries of *Ricinus communis*. *New Phytologist* **81**, 129–137.
Baker, H. G. and Baker, I. (1973a). Amino-acids in nectar and their evolutionary significance. *Nature* **241**, 543–545.
Baker, H. G. and Baker, I. (1973b). Some autecological aspects of the evolution of nectar-producing flowers, particularly amino acid production in nectar. *In* "Taxonomy and Ecology" (V. H. Heywood, ed.), pp. 243–264. Academic Press, London.
Baker, H. G. and Baker, I. (1975). Studies of nectar-constitution and pollinator–plant coevolution. *In* "Coevolution of Animals and Plants"(L. E. Gilbert and P. H. Raven, eds), pp. 100–140. University of Texas Press, Austin.
Baker, H. G. and Baker, I. (1983a). A brief historical review of the chemistry of floral nectar. *In* "Biology of Nectaries" (B. Bentley and T. Elias, eds), pp. 126–152. Columbia University Press, New York.
Baker, H. G. and Baker, I. (1983b). Floral nectar sugar constituents in relation to pollinator type. *In* "Handbook of Experimental Pollination Biology" (C. E. Jones and R. J. Little, eds), pp. 117–141. Scientific and Academic Edition, Van Nostrand Reinhold, New York.
Balsamo, R. A., Adams, M. E. and Thomson, W. W. (1995). Electrophysiology of the salt glands of *Avicennia germanans*. *International Journal of Plant Sciences* **156**, 658–667.
Behrens, W. J. (1879). Die Nectarien der Blüten. *Flora* **62**, 2–11, 17–27, 49–54, 113–123, 145–153, 233–240, 241–247, 305–314, 369–375, 433–457.
Belin-Depoux, M. and Clair-Maczulajtys, D. (1975). Introduction a l'étude des glandes foliaires de *l'Aleurites moluccana* Willd. II. Aspects histologiques de la glande pétiolaire fonctionnelle. *Revue générale de Botanique* **82**, 119–155.
Belmonte, E., Cardemil, L. and Kalin Arroya, M. T. (1994). Floral nectary structure and nectar composition in *Eccremocarpus scaber* (Bignoniaceae), a hummingbird pollinated plant of Chile. *American Journal of Botany* **81**, 493–503.
Benayoun, J. and Fahn, A. (1979). Intracellular transport and elimination of resin from epithelial duct-cells of *Pinus halepinsis*. *Annals of Botany* **43**, 179–181.
Benner, U. and Schnepf, E. (1975). Die Morphologie der Nectarausscheidung bei Bromeliaceen, Beteiligung des Golgi Apparates. *Protoplasma* **85**, 337–349.
Bentley, B. L. (1976). Plants bearing extrafloral nectaries and the associated ant community, interhabitat differences in the reduction of herbivore damage. *Ecology* **57**, 815–820.
Bieleski, R. L. and Redgwell, R. J. (1980). Sorbitol metabolism in nectaries from flowers of Rosaceae. *Australian Journal of Plant Physiology* **7**, 15–25.
Bonnier, G. (1879). Les nectaires. Thése, Faculty of Science, Paris.
Bosabalidis, A. M. and Thomson, W. W. (1984). Ultrastructural differentiation of an unusual structure lining the anticlinal walls of the inner secretory cells in *Tamarix* salt glands. *Botanical Gazette* **145**, 427–435.
Bosabalidis, A. and Tsekos, I. (1982a). Ultrastructural studies on the secretory cavities of *Citrus deliciosa* Ten. II. Development of the oil-accumulating central space of the gland and process of active secretion. *Protoplasma* **112**, 63–70.
Bosabalidis, A. and Tsekos, I. (1982b). Glandular scale development and essential oil secretion in *Origanum dictamnus* L. *Planta* **156**, 496–504.
Bourett, T. M., Howard, R. J., O'Keefe, D. P. and Hallahan, D. L. (1994). Gland

development on leaf surfaces of *Nepeta racemosa*. *International Journal of Plant Sciences* **155**, 623–632.
Campbell, N. and Thomson, W. W. (1976a). The ultrastructure of *Frankenia* salt glands. *Annals of Botany* **40**, 681–686.
Campbell, N. and Thomson, W. W. (1976b). The ultrastructural basis of chloride tolerance in the leaf of *Frankenia*. *Annals of Botany* **40**, 687–693.
Caspary, R. De (1848). "De Nectariis." Elberfeld.
Cecchi Fiordi, A. and Palandri, M. R. (1982). Anatomic and ultrastructural study of the septal nectary in some *Tillandisia* (Bromeliaceae) species. *Caryologia* **35**, 477–489.
Chapman, V. J. (1944). The morphology of *Avicennia nitida*. *Botanical Journal of the Linnean Society* **52**, 487–533.
Cheniclet, C. and Carde, J. P. (1985). Presence of leucoplasts in secretory cells and of monoterpenes in the essential oil: a correlative study. *Israel Journal of Botany* **34**, 219–238.
Christ, P. and Schnepf, E. (1985). The nectaries of *Cynanchum vincetoxicum* (Asclepiadaceae). *Israel Journal of Botany* **34**, 79–90.
Cruden, R. W., Hermann, S. M. and Peterson, S. (1983). Patterns of nectar production and plant-pollinator coevolution. *In* "Biology of Nectaries" (B. Bentley and T. Elias, eds), pp. 80–125. Columbia University Press, New York.
Dafni, H., Lensky, Y. and Fahn, A. (1988). Flower and nectar characteristics of nine species of Labiatae and their influence on honeybee visits. *Journal of Apicultural Research* **27**, 103–114.
Danilova, M. F. and Kashina, T. K. (1987). Ultrastructure of glandular scales in *Perilla ocymoides* (Lamiaceae) in connection with their possible involvement in the synthesis of steroid hormones and gibberellins. *Botanicheskii Zhurnal* SSSR **72**, 427–435.
Davis, A. R. and Gunning, B. E. S. (1991). The modified stomata of floral nectary of *Vicia faba* L. 2. Stomatal number and distribution as selection criteria for breeding for high nectar sugar production. *Acta Horticulturae* **288**, 329–334.
Davis, A. R. and Gunning, B. E. S. (1992). The modified stomata of the floral nectary of *Vicia faba* L. 1. Development, anatomy and ultrastructure. *Protoplasma* **166**, 134–152.
Davis, A. R., Peterson, R. L. and Shuel, R. W. (1986). Anatomy and vasculature of floral nectaries of *Brassica napus* (Brassicaceae). *Canadian Journal of Botany* **64**, 2508–2516.
Davis, A. R., Peterson, R. L. and Shuel, R. W. (1988). Vasculature and ultrastructure of the floral and stipular nectaries of *Vicia faba* (Leguminosae). *Canadian Journal of Botany* **66**, 1435–1448.
Delpino, F. (1868–1875). Ulteriori osservazioni sulla dicogamia nel regno vegetale. I & II. *Atti della Societas Italiana di Scienze Naturali*, XI, XII.
Dschida, W. J., Platt-Aloia, K. A. and Thomson, W. W. (1992). Epidermal peels of *Avicennia germinans* (L). Stearn: a useful system to study the function of salt glands. *Annals of Botany* **70**, 501–509
Duke, S. O. and Paul, R. N. (1993). Development and fine structure of the glandular trichomes of *Artemisia annua* L. *International Journal of Plant Sciences* **154**, 107–118.
Durkee, L. T. (1982). The floral and extra-floral nectaries of *Passiflora*. II. The extra-floral nectary. *American Journal of Botany* **69**, 1420–1428.
Durkee, L. T., Gaal, D. J. and Reisner, W. H. (1981). The floral and extra-floral nectaries of *Passiflora*. I. The floral nectary. *American Journal of Botany* **68**, 453–462.

Eleftheriou, E. P. and Hall, J. L. (1983). The extra-floral nectaries of cotton. I. Fine structure of the secretory papillae. *Journal of Experimental Botany* **34**, 103–119.

Elias, T. S. and Gelband, H. (1976). Morphology and anatomy of floral and extrafloral nectaries in *Campsis* (Bignoniaceae). *American Journal of Botany* **63**, 1349–1353.

Eriksson, M. (1977). The ultrastructure of the nectary of red clover (*Trifolium pratense*). *Journal of Apicultural Research* **16**, 184–193.

Eymé, J. (1966). Infrastructure des cellules nectarigènes de *Diplotaxis erucoides* D.C., *Helleborus niger* L. et *H. foetidus* L. *Comptes Rendus de l'Academie des Sciences* (Paris) Sér. D **262**, 1629–1632.

Eymé, J. (1967). Nouvelles observations sur l'infrastructure de tissues nectarigènes floraux. *Botaniste* **50**, 169–183.

Fahn, A. (1949). Studies in the ecology of nectar secretion. *Palestine Journal of Botany*, Jerusalem Series **4**, 207–224.

Fahn, A. (1952). On the structure of floral nectaries. *Botanical Gazette* **113**, 464–470.

Fahn, A. (1953). The topography of the nectary in the flower and its phylogenetical trend. *Phytomorphology* **3**, 424–426.

Fahn, A. (1979a). "Secretory Tissues in Plants." Academic Press, London.

Fahn, A. (1979b). Ultrastructure of nectaries in relation to nectar secretion. *American Journal of Botany* **66**, 977–985.

Fahn, A. (1987a). Extrafloral nectaries of *Sambucus niger* L. *Annals of Botany* **60**, 299–308.

Fahn, A. (1988). Secretory tissues in vascular plants. *The New Phytologist* **108**, 229–257.

Fahn, A. and Benayoun, J. (1976). Ultrastructure of resin ducts in *Pinus halepensis*: development, possible sites of resin synthesis, and mode of its elimination from the protoplast. *Annals of Botany* **40**, 857–863.

Fahn, A. and Benouaiche, P. (1979). Ultrastructure, development and secretion in the nectary of banana flowers. *Annals of Botany* **44**, 85–93.

Fahn, A. and Evert, R. F. (1974). Ultrastructure of the secretory ducts of *Rhus glabra* L. *American Journal of Botany* **61**, 1–14.

Fahn, A. and Rachmilevitz, T. (1970). Ultrastructure and nectar secretion in *Lonicera japonica*. *In* "New Research in Plant Anatomy" (N. K. B. Robson, D. F. Cutler and M. Gregory, eds), pp. 51–56. Academic Press, London.

Fahn, A. and Rachmilevitz, T. (1975). An autoradiological study of nectar secretion in *Lonicera japonica* Thunb. *Annals of Botany* **39**, 975–976.

Fahn, A. and Shimony, C. (1996). Glandular trichomes of *Fagonia* L. (Zygophyllaceae) species: structure, development and secreted materials. *Annals of Botany* **77**, 25–34.

Fahn, A. and Shimony, C. (1998). Ultrastructure and secretion of secretory cells of two species of *Fagonia* L. (Zygophyllaceae). *Annals of Botany* **81**, 557–565.

Figier, J. (1968). Localisation infrastructure de la phosphomonoestérase acide dans la stipule de *Vicia faba* L. au niveau du nectaire. Rôles possibles de cet enzyme dans les mecanismes de la sécrétion. *Planta* **83**, 60–79.

Figier, J. (1971). Etude infrastructurale de la stipule de *Vicia faba* L. au niveau du nectaire. *Planta* **98**, 31–49.

Figier, J. (1972). Localisation infrastructure de la phosphatase acide dans les glandes pétiolaires d'Impatiens holstii. Rôles possibles de cette enzyme au cours des sécrétoires. *Planta* **108**, 215–226.

Figueiredo, A. C. and Pais, M. S. (1994). Ultrastructural aspects of glandular cells

from the secretory trichomes and from the cell suspension cultures of *Achillea millefolium* L. ssp. *millefolium*. *Annals of Botany* **74**, 179–190.
Figueiredo, A. C. S. and Pais, M. S. (1992). Ultrastructural aspects of the nectary spur of *Limodorum abortivum* (Orchidaceae). *Annals of Botany* **70**, 325–331.
Findlay, N. and Mercer, F. V. (1971). Nectar production in *Abutilon*. II. Submicroscopic structure of the nectary. *Australian Journal of Biology Sciences* **24**, 657–664.
Frei, E. (1955). Die Innervierung der floralen Nektarien dikotyler Pflanzenfamilien. *Berichte der schweizerischen botanischen Gesellschaft* **65**, 60–114.
Frey-Wyssling, A. (1935). "Die Stoffausscheidung der höheren Pflanzen." Springer, Berlin.
Frey-Wyssling, A. (1955). The phloem supply to the nectaries. *Acta botanica neerlandica* **4**, 358–369.
Frey-Wyssling, A. and Agthe, C. (1950). Nektar ist ausgeschiedener Phloemsaft. *Verhandlungen der schweizerischen naturforschenden Gesellschaft* **130**, 175–176.
Gaffal, K. P., Heimler, W. and El-Gammal, S. (1998). The floral nectary of *Digitalis purpurea* L., structure and nectar secretion. *Annals of Botany* **81**, 251–262.
Goodwin, T. W. and Mercer, E. I. (1983). "Introduction to Plant Biochemistry", 2nd edn. Pergamon Press, Oxford.
Gravano, E., Tani, C., Bennici, A. and Gucci, R. (1998). The ultrastructure of glandular trichomes of *Phillyrea latifolia* L. (Oleaceae) leaves. *Annals of Botany* **81**, 327–335.
Gunning, B. E. S. and Hughes, J. E. (1976). Quantitative assessment of symplastic transport of pre-nectar into the trichomes of *Abutilon* nectaries. *Australian Journal of Plant Physiology* **3**, 619–637.
Haberlandt, G. (1918). "Physiologische Pflanzenanatomie", 5th edn. Engelmann, Leipzig.
Hagemeyer, J. and Waisel, Y. (1988). Excretion of ions (Ca^{2+}, Li^+, Na^+, Cl^-) by *Tamarix aphylla*. *Physiologia Plantarum* **73**, 541–546.
Heinrich, G. (1975a). Über die Lokalisation verschiedener Phosphatase im Nektarien zweier *Aloë*. *Cytobiologie* **11**, 247–263.
Heinrich, G. (1975b). Über den Glucose-Metabolismus in Nektarien zweier *Aloë*-Arten und über den Mechanismus der Pronektar-Sekretion. *Protoplasma* **85**, 351–371.
Heinrich, G. and Schultze, W. (1985). Composition and site of biosynthesis of the essential oil in fruits of *Phellodendron amurense* Rurp. (Rutaceae). *Israel Journal of Botany* **34**, 205–217.
Hill, E. and Hill, B. S. (1976). Elimination processes by glands. Mineral ions. *In* "Encyclopedia of Plant Physiology", new series, Vol. 2. "Transport in Plants", IIB (U. Lüttge and M. G. Pitman, eds), pp. 225–243. Springer, Berlin.
Joel, D. M. and Fahn, A. (1980a). Ultrastructure of resin ducts of *Mangifera indica* (Anacardiaceae). 1. Differentiation and senescence of the shoot ducts. *Annals of Botany* **46**, 225–233.
Joel, D. M. and Fahn, A. (1980b). Ultrastructure of resin ducts of *Mangifera indica* (Anacardiaceae). 2. Resin secretion in the primary stem ducts. *Annals of Botany* **46**, 779–783.
Juniper, B. E., Robins, R. J. and Joel, D. M. (1989). "The Carnivorous Plants." London, Academic Press.
Kalman, F. and Gulyás, S. (1974). Ultrastructure and mechanism of secretion in extrafloral nectaries of *Ricinus communis* L. *Acta Biologica Szeged* **20**, 57–67.
Kartashova, N. N. (1965). "Stroeniei Funktsiya Tsvetka dvudol'nykh rastenii."

Izdatel'stvo, 2nd edn. Tomskogo Universiteta, Tomsk.
Kisser, J. (1958). Der Stoffweschel sekundärer Pflanzenstoffe. *In* "Handbuch der Pflanzenphysiologie", Vol. 10 (W. Ruhland, ed.), pp. 91–131. Springer, Berlin.
Kronestedt, E. C., Robards, A. W., Stark, M. and Olsen, P. (1986) Development of trichomes in the *Abutilon* nectary gland. *Nordic Journal of Botany* **6,** 627–639.
Kronestedt-Robards, E. and Robards, A. W. (1991). Exocytosis in glandular cells. *In* "Endocytosis, Exocytosis and Vesicle Traffic in Plants" (C. R. Hawes, J. O. D. Coleman and D. E. Evans, eds), pp. 199–232. Cambridge, Cambridge University Press.
Lazzaro, M. D. and Thomson, W. W. (1992). Ultrastructural localization of calcium in organic acid secreting trichomes of chickpea (*Cicer arietinum*). *Canadian Journal of Botany* **70**, 2319–2325.
Levering, C. A. and Thomson, W. W. (1971). The ultrastructure of the salt gland of *Spartina foliosa. Planta* **97**, 183–196.
Levin, D. A. (1973). The role of trichomes in plant defense. *Quarterly Review of Biology* **48**, 3–15.
Liphshitz, N. and Waisel, Y. (1974). Existence of salt glands in various genera of the Gramineae. *New Phytologist* **73**, 507–513.
Liphshitz, N., Shomer-Ilan, A., Eshel, A. and Wasel, Y. (1974). Salt glands on leaves of rhodes grass (*Chloris gayana* Kth.). *Annals of Botany* **38**, 459–462.
Lüttge, U. (1961). Über die Zusammensetzung des Nektars und den Mechanismus seiner Sekretion. I. *Planta* **56**, 180–212.
Lüttge, U. (1969). Aktiver Transport (Kurzstreckentransport bei Pflanzen). *Protoplasmatologia* **8**, 1–146.
Lüttge, U. (1971). Structure and function of plant glands. *Annual Review of Plant Physiology* **22**, 23–44.
Lüttge, U. (1977). Nectar composition and membrane transport of sugars and amino acids. A review on the present state of nectar research. *Apidologie* **8**, 305–320.
Lüttge, U. and Schnepf, E. (1976). Elimination processes by glands. Organic substances. *In* "Encyclopedia of Plant Physiology", Transport in Plants, II, N.S. Vol. 2, part B (U. Lüttge and M. G. Pitman, eds), pp. 244–277. Springer, Berlin.
Mahlberg, P. G., Hammond, C. T., Turner, J. C. and Hemphill, J. K. (1984). Structure, development and composition of glandular trichomes of *Cannabis sativa* L. *In* "Biology and Chemistry of Plant Trichomes" (E. Rodriguez, P. L. Healey and I. Mehta, eds), pp. 23–52. Plenum Press, New York, London.
Marginson, R., Sedgley, M. and Knox, R. B. (1985a). Structure and histochemistry of the extrafloral nectary of *Acacia terminalis* (Salisb.) MacBride (Leguminosae, Mimosoideae). *Protoplasma* **127**, 21–30.
Marginson, R., Sedgley, M., Douglas, T. J. and Knox, R. B. (1985b). Structure and secretion of the extrafloral nectaries of Australian acacias. *Israel Journal of Botany* **34**, 91–102.
Maron, R. and Fahn, A. (1979). Ultrastructure and development of oil cells in *Laurus nobilis* L. leaves. *Botanical Journal of the Linnean Society* **78**, 31–40.
Metcalfe, C. R. and Chalk, L. (1983). "Anatomy of the Dicotyledons", 2nd edn, Vol. II, "Wood Structure and Conclusion of the General Introduction". Clarendon Press, Oxford.
Meyberg, M. and Kristen, U. (1981). The nectaries *Aptenia cordifolia*. Ultrastructure, translocation of ^{14}C-labelled sugars, and a possible pathway of secretion. *Zeitschrift für Pflanzenphysiologie* **104**, 139–147.
Mikulska, E. and Zolnierowicz, H. (1976). Ultrastructure of plastids in differentiating

epithelial cells of *Abies homolipis* Sib. and Zucc. leaves. *Biochemie und Physiologie der Pflanzen* **170**, 355–362.
Mohan, J. S. S. and Inamdar, J. A. (1986). Ultrastructure and secretion of extrafloral nectaries of *Plumera rubra* L. *Annals of Botany* **57**, 389–401.
Monteiro, W. R., Fahn, A., Caldeira, W. and Castro, M. M. (1998). Ultrastructural observations on the foliar secretory cavities of *Porophyllum lanceolatum* DC. (Asteraceae). *Flora* **194**, 113–126.
Müller, N. J. C. (1866–1867). Untersuchungen über die Verteilung der Harze, ätherischen Öle, Gummi und Gummiharze, und die Stellung der Sekretionsbehälter in Pflanzenkörper. *Tahrbücher für Wissenschaftliche Botanik* **5**, 387–421.
Nepi, M., Pacini, E. and Willemse, M. T. M. (1996a). Nectary biology of *Cucurbita pepo*: ecophysiological aspects. *Acta Botanica Neerlandica* **45**, 41–54.
Nepi, M., Pacini, E. and Willemse, M. T. M. (1996b). Development and ultrastructure of *Cucurbita pepo* nectaries of male and female flowers. *Annals of Botany* **78**, 95–104.
O'Brien, S. P. O., Loveys, B. R. and Grant, W. J. R. (1996). Ultrastructure and function of floral nectaries of *Chamelaucium uncinatum* (Myrtaceae). *Annals of Botany* **78**, 189–196.
Oross J. W., Leonard, R. T. and Thomson, W. W. (1985). Flux rate and secretion model for salt glands of grasses. *Israel Journal of Botany* **34**, 69–77.
Osmond, C. B., Lüttge, U., West, K. R., Pallaghy, C. K. and Shachar-Hill, B. (1969). Ion absorption in *Atriplex* leaf tissue. II. Secretion of ions to epidermal bladders. *Australian Journal of Biological Sciences* **22**, 797–814.
Paech, K. (1952). Die Differenzierung der Ölzellen und die Bildung des ätherischen les bei *Asarum europaeum*. *Zeitschrift für Botanik* **40**, 53–66.
Pate, J. S. and Gunning, B. E. S. (1972). Transfer cells. *Annual Review of Plant Physiology* **123**, 173–196.
Pedro, L., Campos, P. and Pais, M. S. S. (1991). Ultrastructure of the apical cell of procumbent (Type I) trichomes in *Geranium robertianum* L. (Geraniaceae). *Israel Journal of Botany* **40**, 209–217.
Percival, M. S. (1961). Types of nectar in angiosperms. *New Phytologist* **60**, 235–281.
Pereira da Costa, T. G. C. A. (1985). Bolsas secretoras de *Ruta chalepensis* L. Ph.D. thesis, University of Lisbon.
Pridgeon, A. M. and Stern, W. L. (1983). Ultrastructure of osmopheres in *Restrepia* (Orchidaceae). *American Journal of Botany* **70**, 1233–1243.
Rachmilevitz, T. and Fahn, A. (1973). Ultrastructure of nectaries of *Vinca rosea* L., *Vinca major* L. and *Citrus sinensis* Osbeck cv. *Valencia* and its relation to the mechanism of nectar secretion. *Annals of Botany* **37**, 1–9.
Rachmilevitz, T. and Fahn, A. (1975). The floral nectary of *Tropaeolum majus* L. The nature of the secretory cells and the manner of nectar secretion. *Annals of Botany* **39**, 721–728.
Rachmilevitz, T. and Joel, D. M. (1976). Ultrastructure of the calyx glands of *Plumbago capensis* Thunb. In relation to the process of secretion. *Israel Journal of Botany* **25**, 127–139.
Reed, M. L., Findlay, N. and Mercer, F. V. (1971). Nectar production in *Abutilon*. IV. Water and solute relations. *Australian Journal of Biological Sciences* **24**, 677–688.
Robards, A. W. and Stark, M. (1988). Nectar secretion in *Abutilon*: a new model. *Protoplasma* **142**, 79–91.
Sacchetti, G., Bruni, A., Dall'Olio, G., Nicoletti, M., Di Fabio, A. and Poli, F. (1996). Development and morphology of secretory trichomes of *Calceolaria*

volckmanni (Scrophulariaceae). *Nordic Journal of Botany* **16**, 505–513.
Said, C. (1982). Les nectaires floraux des Crassulacées. Etude morphologique, histologique et anatomique. *Bulletin de la Société de France* **129**, 231–240.
Sawidis, T. (1991). A histochemical study of nectaries of *Hibiscus rosa-sinensis*. *Journal of Experimental Botany* **42**, 1477–1487.
Schnepf, E. (1964). Zur Cytologie und Physiologie pflanzicher Drüsen. 4. Teil. Licht- und elektronenmikroskopische Untersuchungen an Septalektarien. *Protoplasma* **58**, 137–171.
Schnepf, E. (1969a). "Sekretion und Exretion bei Pflanzen." Protoplasmatologia VIII/8. Springer, Vienna, New York.
Schnepf, E. (1969b). Über den Feinbau von Öldresen, IV. Die lgenge von Umbelliferen: *Heracleum sphondylium* und *Dorema ammoniacum*. *Protoplasma* **67**, 375–390.
Schnepf, E. (1972). Tubuläres endoplasmattisches Reticulum in Drüsen mit lipophilen Ausscheidungen von *Ficus*, *Ledum* und *Salvia*. *Biochemie und Physiologie der Pflanzen* **163**, 113–125.
Schnepf, E. and Christ, P. (1980). Unusual transfer cells in the epithelium of the nectaries of *Asclepias curassavica* L. *Protoplasma* **105**, 135–148.
Schnepf, E. and Deichgraber, G. (1984). Electron microscopical studies of nectaries of some *Euphorbia* species. *Akademie der Wissenschaften und der Literatur. Mainz, Mathematisch-Naturwissenschaftlichen Klasse, Tropische und subtropische Pflanzenwelt* **45**, 55–93.
Schrödter, K. (1926). Zur physiologischen Anatomie der Mittelzelle drüsiger Gebilde. *Flora* **120**, 19–86.
Shimony, C. and Fahn, A. (1968). Light- and electron-microscopical studies on the structure of salt glands of *Tamarix aphylla* L. *Botanical Journal of the Linnean Society* **60**, 283–288.
Shimony, C., Fahn, A. and Reinhold, L. (1973). Ultrastructure and ion gradients in the salt glands of *Avicennia marina* (Forsk.) Vierh. *New Phytologist* **72**, 27–36.
Skelding, A. P. and Winterbotham, J. (1939). The structure and development of hydrathodes of *Spartina townsendii* Groves. *New Phytologist* **38**, 69–79.
Skubatz, H., Kunkell, D. D., Patt, J. M., Howald, W. N., Hartman, T. G. and Meeuse, B. J. D. (1995). Pathway of terpene excretion by the appendix of *Sauromatum guttatum*. *Proceedings of the Natural Academy of Sciences USA* **92**, 10 084–10 088.
Sperlich, A. (1939). Exkretionsgewebe. *In* "Handbuch der Pflanzenanatomie", Vol 4, "Das tropische Parenchym" (K. Linsbaur, ed.), pp. 1–185. Borntraeger, Berlin.
Stpiczynska, M. (1997). The structure of nectary of *Platanthera bifolia* L. Orchidaceae. *Acta Societatis Botanicorum Poloniae* **66**, 5–11.
Tacina, F. (1973). La structure optique et électronmicroscopique de la glande nectariefere chez *Cynoglossum officinale* L. *Revue Roumaine de Biologie, Série de Botanique* **18**, 201–209.
Tarkowska, J. A., Zobel, A. M. and Maciek, M. (1981). Development and transfer character of secretory glands in the broad bean (*Vicia faba* L.). *Acta Societatis Botanicorum Poloniae* **50**, 391–398.
Thomson, W. and Platt-Aloia, K. (1979). Ultrastructural transitions associated with the development of the bladder cells of the trichomes of *Atriplex*. *Cytobios* **25**, 105–114.
Thomson, W. W. and Healey, P. L. (1984). Cellular basis of trichome secretion. *In* "Biology and Chemistry of Plant Trichomes" (E. Rodriguez, P. L. Healey and I. Mehta, eds), pp. 95–111. Plenum, New York, London.

Thomson, W. W. and Liu, L. L. (1967). Ultrastructural features of the salt gland of *Tamarix aphylla*. *Planta* **73**, 201–220.

Thomson, W. W., Berry, W. L. and Liu, L. L. (1969). Localization and secretion of salt by the salt glands of *Tamarix aphylla*. *Proceedings of the National Academy of Sciences (USA)* **63**, 310–317.

Thurston, E. L. and Lersten, N. R. (1969). The morphology and toxicology of plant stinging hairs. *Botanical Review* **35**, 393–412.

Tschirch, A. (1906). "Die Harze und die Harzbehälter mit Einschluss der Milchsaft" 2. Aufl. Leipzig.

Van Steveninck, R. F. M., Armstrong, W. D., Peters, P. D. and Hall, T. A. (1976). Ultrastructural localization of ions. III. Distribution of chloride in mesophyll cells of mangrove (*Aegiceras corniculatum* Blanco). *Australian Journal of Plant Physiology* **3**, 367–376.

Vassilyev, A. E. (1969). Submicroscopic morphology of nectary cells and problems of nectar secretion. *Botanicheskii Zhurnal SSSR* **54**, 1015–1031.

Vassilyev, A. E. (1970). On the localization of terpenoids synthesis in plant cells. *Rastitel'nye Resursy* **6**, 29–45.

Vassilyev, A. E. (1971). New data on the ultrastructure of the cells of the flower nectary. *Botanicheskii Zhurnal SSSR* **56**, 1292–1306.

Vassilyev, A. E. (1977). "Functional Morphology of Plant Secretory Cells." Nauka, Leningrad.

Vassilyev, A. E. (1994). Developmental and comparative ultrastructure of glandular hairs in two Urticaceae. I. *Urtica dioica*. *Nordic Journal of Botany* **14**, 531–545.

Vermeer, J. and Peterson, R. L. (1979). Glandular trichomes on the inflorescence of *Chrysanthemum morifolium* cv. Dramatic (Compositae). II. Ultrastructure and histochemistry. *Canadian Journal of Botany* **57**, 714–729.

Vogel, S. (1962). Duftdrüsen im Dienst der Bestaübung. Über Bau und Funktion der Osmophoren. *Akademie der Wissenschaften und der Literatur, Mainz, Abhandlung der Mathematisch-Naturwissenschaftliche* Klasse **10**, 598–763.

Vogel, S. (1966). Scent organs of flowers and their relation to insect pollination. *In* "Proceedings of the 5th World Orchid Conference" (L. R. DeGarmo, ed.), pp. 253–259. Long Beach, CA.

Vogel, S. (1997). Remarkable nectaries: structure, ecology, organophyletic perspectives. I. Substitutive nectaries. *Flora* **192**, 305–333.

Vogel, S. (1998a). Remarkable nectaries: structure, ecology, organophyletic perspectives. II. Nectarioles. *Flora* **193**, 1–29.

Vogel, S. (1998b). Remarkable nectaries: structure, ecology, organophyletic perspectives. III. Nectar ducts. *Flora* **193**, 113–131.

Vogel, S. (1998c). Remarkable nectaries: structure, ecology, organophyletic perspectives. IV. Miscellaneous cases. *Flora* **193**, 225–248.

Waisel, Y. (1972). "Biology of Halophytes." New York, Academic Press.

Wergin, P. W., Elmore, C. D., Hanny, B. W. and Ingber, B. (1975). Ultrastructure of the subglandular cells from the foliar nectaries of cotton in relation to the distribution of plasmodesmata and symplastic transport of nectar. *American Journal of Botany* **62**, 379–411.

Werker, E. and Fahn, A. (1968). Site of resin synthesis in cells of *Pinus halepensis* Mill. *Nature* **218**, 388–389.

Werker, E. and Fahn, A. (1981). Secretory hairs of *Inula viscosa* (L.) Ait: development, ultrastructure, and secretion. *Botanical Gazette* **142**, 461–476.

Werker, E., Ravid, U. and Putievsky, E. (1985a). Structure of glandular hairs and identification of the main components of their secreted material in some species of the Labiatae. *Israel Journal of Botany* **34**, 31–45.

Werker, E., Ravid, U. and Putievsky, E. (1985b). Glandular hairs and their secretions in the vegetative and reproductive organs of *Salvia sclarea* and *S. dominica*. *Israel Journal of Botany* **34**, 239–252.

Whatley, J. M. and Whatley, F. R. (1987). When is a chromoplast? *New Phytologist* **106**, 667–678.

Wollenweber, E. (1984). The systematic implication of flavonoids secreted by plants. *In* "Biology and Chemistry of Plant Trichomes" (E. Rodriguez, P. L. Healey and I. Mehta, eds), pp. 53–69. Plenum, New York, London.

Zandonella, P. and Piolat, C. (1982). Nectaires et secretion du nectar chez *Thunbergia laurifolia* Lindl. (Acanthacées). *Bulletin de la Société botanique de France* **129**, 109–117.

Zer, H. and Fahn, A. (1992). Floral nectaries of *Rosmarinus officinalis* L. Structure, ultrastructure and nectar secretion. *Annals of Botany* **70**, 391–397.

Ziegler, H. (1965). Die Physiologie pflanzlicher Drüsen. *Berichte der deutschen botanischen Gesellschaft* **78**, 466–477.

Ziegler, A. (1960). Zur Anatomie und Protoplasmic der Olidioblaste von *Houttuynia cordata*. *Protoplasma* **51**, 539–562.

Zimmermann, J. G. (1932). Über die extrafloralen Nektarien der Angiospermen. *Beihefte zum botanischen Zentralblatt* **49A**, 99–196.

Monoterpenoid Biosynthesis in Glandular Trichomes of Labiate Plants

D. L. HALLAHAN

Biochemistry and Physiology Department, IACR-Rothamsted, Harpenden, Herts. AL5 2JQ, UK
(Current address: Central Research and Development, The Dupont Company, Biochemical Sciences and Engineering, Wilmington, DE, USA)

I. INTRODUCTION

The plant family Labiatae (=Lamiaceae) contains between 4000 and 5000 species (Mabberley, 1987; Hedge, 1992), classified into over 200 genera. Some

Advances in Botanical Research Vol. 31
ISBN 0-12-005931-2

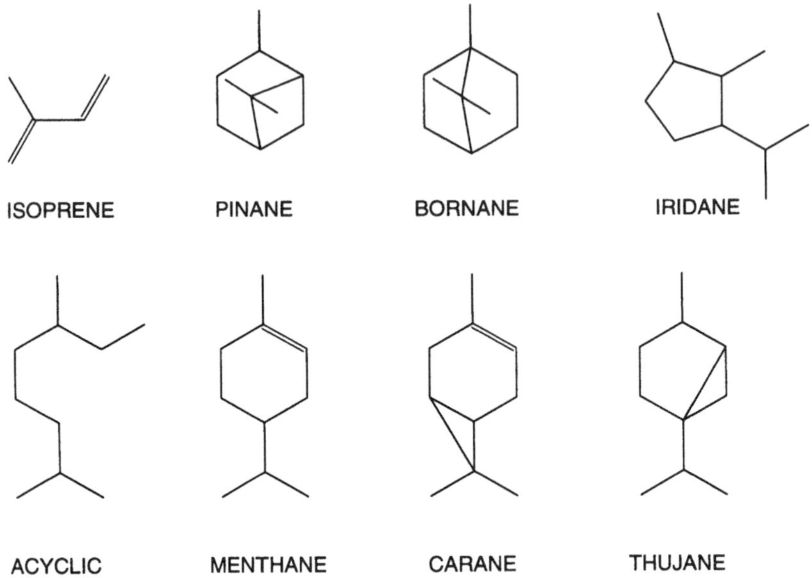

Fig. 1. The structure of isoprene and the carbon skeletons of the principal monoterpenoid classes.

40% of Labiate genera are thought to contain aromatic species (Lawrence, 1992), and among these are many prized for their odours and flavours. These aromatics include familiar culinary herbs such as mint (*Mentha*), sage (*Salvia*), thyme (*Thymus*), marjoram or oregano (*Origanum*), rosemary (*Rosmarinus*) and basil (*Ocimum*), as well as fragrant plants such as lemon balm (*Melissa*), lavender (*Lavandula*) and patchouli (*Pogostemon*). The characteristic flavours and odours of these plants derive from the spectrum of chemical compounds that they accumulate and, in particular, their terpenoid components (Ohloff, 1995; Williams, 1996).

Terpenoids constitute the largest family of natural chemicals known, with over 20 000 individual compounds described to date (Connolly and Hill, 1991). The terpenoid family is very diverse, with structures ranging from simple linear hydrocarbon chains to highly complex derivatised polycyclic molecules (Connolly and Hill, 1991). The unifying feature of this family is that all terpenoids can be hypothetically constructed from multiples of a fundamental five-carbon building block, isoprene (Fig. 1). Apart from the hemiterpenoid isoprene, monoterpenoids (with ten carbon atoms) are the simplest members of this family. Even within this one class of terpenoids, some 38 distinct structural skeletons are known to exist (Mann *et al.*, 1994). The more common of these (Fig. 1) include the acyclic monoterpenoids, some five classes of cyclohexanoid monoterpenoids (menthane, carane, thujane, pinane and bornane skeletons) and the cyclopentanoids (iridane skeleton). From these basic skeletons a huge spectrum of plant monoterpenoids are elaborated, through oxidation, reduction, isomerization and (in the case of secoiridoids) ring-cleavage reactions. Although sesquiterpenoids are often key components (e.g. in patchouli oil), it is generally the monoterpenoids that constitute the principal volatile aromatic constituents of many of the extracted essences (essential oils) prepared from Labiate plants (Ohloff, 1995; Williams, 1996). Figure 2 shows some of the monoterpenoids that have been identified in the essential oils of Labiate plants, to illustrate the structural diversity of this class of compounds.

The constituents of plant essential oils, including these monoterpenoids, are generally classed as "secondary" metabolites. The definition of a secondary metabolite, as a compound not known to be directly involved in well-defined primary metabolic processes, is rather circular but commonly accepted. Examples of primary processes would include (in plants) photosynthesis, respiration and the biosynthesis of fundamental cell constituents (structural proteins and enzymes, nucleic acids, membrane lipids and sterols, hormones, etc.). Where secondary metabolites have proven useful to man, they are often – perhaps more flatteringly – referred to as natural products. Classification of these compounds as "secondary" does not, however, imply that they are not essential. Production of secondary metabolites occurs to some extent in most plant families, often to very high levels. It seems reasonable to assume, therefore, that they must contribute in significant ways to the evolutionary fitness of those plant species that accumulate them (Harrewijn *et al.*, 1995). Although the subject of much debate over the years (Feeny, 1991), the current consensus as to why plants produce this array of secondary metabolites, including monoterpenoids, is that these compounds mediate ecological interactions (generally defensive) with other organisms (recently reviewed in Rhodes, 1994; Harrewijn *et al.*, 1995).

Many plant terpenoids exhibit allelopathic effects towards other plant species, and a variety of monoterpenoids have been shown to be phytotoxic (reviewed by Fischer, 1991). These effects may result from direct exposure to monoterpenoids (either in the atmosphere or, through deposition of leaf litter, in the soil) or to microbial metabolites of the plant-derived compounds present in the soil. It has long been assumed that monoterpenoid secondary metabolites also contribute to defence against herbivory, although relatively little concrete evidence has been produced (Bernays and Chapman, 1994). Harborne (1991) considers the effect of monoterpenoids on mammalian herbivory to be relatively minor, although some species are repelled by monoterpenoid-rich plants. The effects of plant monoterpenoids on insect behaviour, however, are better documented. Laboratory studies have shown that many terpenoids, although not acutely toxic, are generally repellant to insects (Bernays and Chapman, 1994; Harrewijn *et al.*, 1995), and it is accepted that many plant terpenoids play defensive roles. However, the

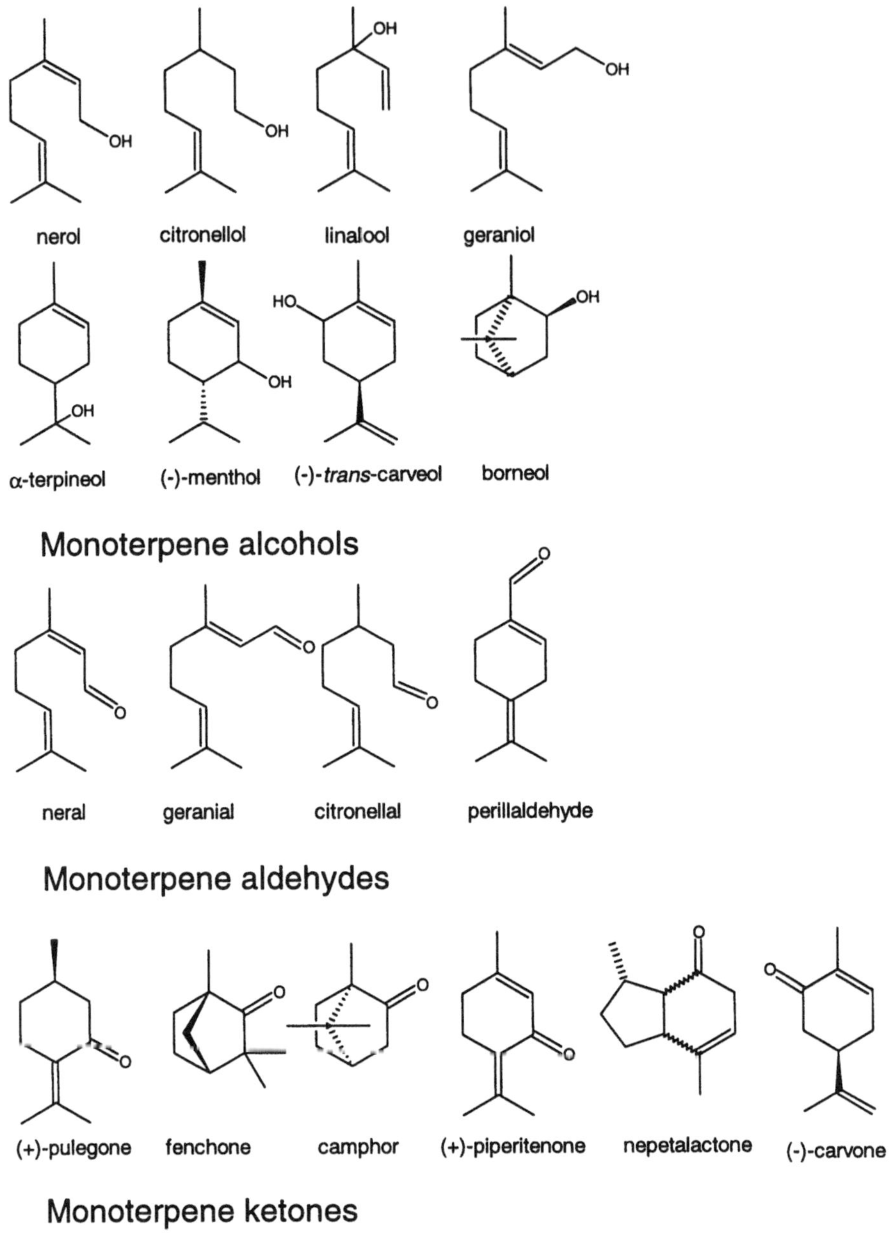

Fig. 2. Some of the monoterpenoid components of Labiate oils.

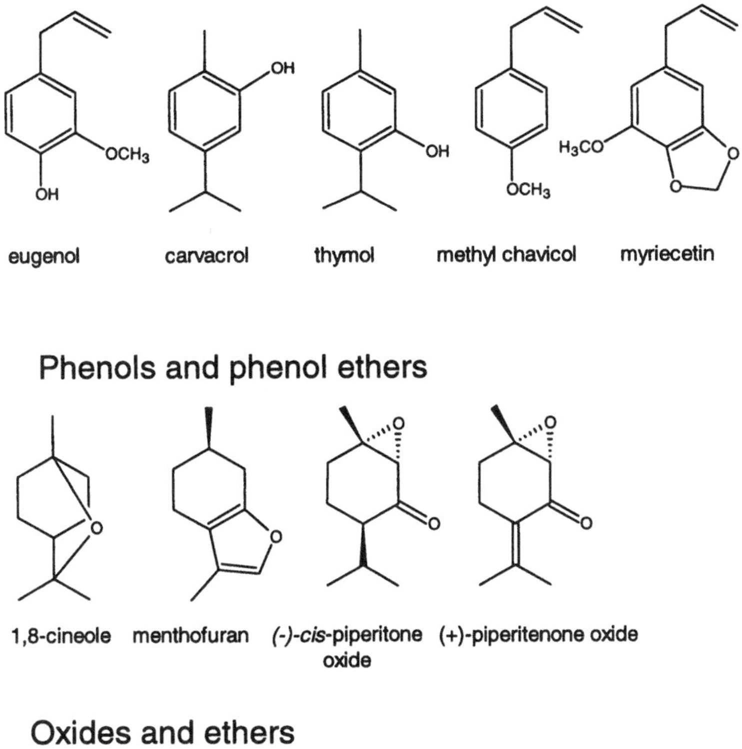

Fig. 2. *continued.*

response to a particular chemical must be dependent on the insect species tested – compounds which attract a pollinator or act as feeding or oviposition stimulants for specialist insect species may well be repellant to generalist species. Mookherjee *et al.* (1992) list many compounds with reported effects on insect behaviour, including monoterpenoids. In their own screening of over 2000 compounds for attractancy or repellancy towards household insect pests (houseflies, mosquitoes), however, they identified relatively few as having any effect on these species. Although citronellol was repellant, the other acyclic monoterpenoids screened (linalool, citral, nerol, geraniol) did not have any significant activity, and carvone, pulegone and α-terpineol were attractants. In any event, repellence of individual insect species by a phytochemical is far from conclusive evidence that the compound is involved in defence (Futuyma and Keese, 1992). The interactions between plants and animals, as mediated by secondary metabolites including monoterpenoids, are often highly complex, and may span multiple trophic levels. Some effects may be the direct result of evolutionary adaptation by plants, but many effects observed under laboratory conditions might equally be artefactual or coincidental.

This complexity, and the difficulty of assigning a role to a particular secondary metabolite accumulated by a plant, is perhaps best illustrated by an example. The catmints (*Nepeta* spp.), members of the Labiatae, synthesize cyclopentanoid (iridoid) monoterpenes (specifically, nepetalactones) as the principal constituents of their essential oil (Regnier *et al.*, 1967; De Pooter *et al.*, 1988; Clark *et al.*, 1997b). Catmint plants are perhaps best known for their excitatory effects on feline behaviour and the active principle of their oil (catnip) has been identified as nepetalactone (McElvain *et al.*, 1942). It seems highly unlikely that the effect of nepetalactones on feline behaviour confers an adaptive advantage on this genus! Cyclopentanoid monoterpenoids and their glycoside conjugates, however, are generally repellant to insects, and are synthesized not only by plants but by several insect species as a defence against predators (Cavill *et al.*, 1956; Eisner, 1964, 1965; Lorenz *et al.*, 1993; Veith *et al.*, 1994; Oldham *et al.*, 1996). Thus the accumulation of these compounds in plants would certainly confer some protection against herbivory. A number of insect species, however, have evolved to forage actively on plants that accumulate iridoid glycosides and sequester these compounds for their own defence (reviewed in Bowers, 1988; Rimpler, 1991). Aphids also possess the capacity to synthesize the cyclopentanoid nepetalactones and nepetalactols, but in this case have evolved to use these compounds as constituents of their sex pheromone (Dawson *et al.*, 1987, 1990; Pickett *et al.*, 1992). These compounds in turn have been shown to act as attractants for aphid parasitoids (Powell *et al.*, 1993; Hardie *et al.*, 1994; Poppy *et al.*, 1997; Boo *et al.*, 1998; Powell *et al.*, 1998). Attempting to tease out simple cause-and-effect relationships from such webs of interactions is clearly a daunting task.

In some cases, secondary metabolites are accumulated to even higher levels (induced) as a result of herbivory (Karban *et al.*, 1997). In wild radish (*Raphanus sativus*; Brassicaceae), herbivory was shown to result in the induction of indole glucosinolates (and increased trichome density) in undamaged leaves, and correlated with increased reproductive fitness (Agrawal, 1998). Elimination of natural enemies from *Arabidopsis thaliana* (Brassicaceae) plants grown in the field had previously been shown to alter the pattern of selection on genetic variation in total glucosinolate concentration and trichome density (Mauricio and Rausher, 1997). Thus a clear correlation between the accumulation of plant secondary metabolites, (insect) herbivory and reproductive fitness has been established. From currently available evidence, it is reasonable to assume that insect–plant coevolution is likely to be an important driving force – perhaps the most important – behind the diversification of secondary metabolite biosynthesis in plants.

As indicated above, many Labiate essential oils are of considerable commercial importance (Weiss, 1997). Around 60 000 tonnes of plant

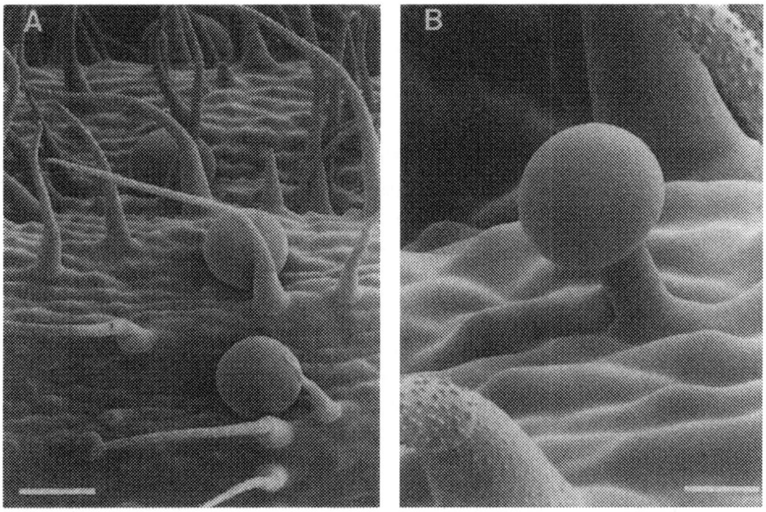

Fig. 3. Trichomes of *Nepeta racemosa*. (A) simple multicellular hairs, small capitate glands and larger peltate glands are present on this species. Scale bar: 50 μm. (B) Capitate gland. Scale bar: 10 μm. Cryogenic scanning electron microscopy was performed by R. Howard and T. Bourett of The DuPont Company.

essential oils are produced annually worldwide (Moyler, 1991). Annual production of Labiate oils has been estimated at some 15 000 tonnes, with *Mentha* oils accounting for some 12 000 tonnes from this total (Lawrence, 1992). The value of these plant products is considerable, and this has justified investment in studies of their biosynthesis and of the feasibility of alternative means of production, such as cell culture. Although cell culture has been of utility for the production of many secondary metabolites (Zenk, 1991), and the production of volatile monoterpenoids in cultured cells has been achieved for some species of *Citrus* (Agrawal *et al.*, 1991), the culture of undifferentiated cells has proved largely unsuccessful as a means of production of (terpenoid) essential oils (Charlwood and Charlwood, 1991; Lawrence, 1992). Monoterpenoids are often phytotoxic at high concentrations and their accumulation would appear to require specialized cell types, most probably glandular trichomes (Brown *et al.*, 1987). The Labiatae are known for the preponderance of hirtous species, and all of the aromatic species of this family possess aerial trichomes (Werker, 1993). The presence among these of secretory or glandular trichomes (Fig. 3) has been recognized since the earliest microscopic observations (reviewed in Uphof, 1962). As will be discussed below, it is now clear that glandular trichomes are the site of synthesis and accumulation of the characteristic monoterpenoid-rich essential oils of the Labiatae. The investment necessitated by the production of such complex structures for presentation of essential oils is perhaps in itself an argument for the importance of these compounds to the plant.

II. MONOTERPENOIDS ACCUMULATED BY LABIATE PLANTS

The distillable essential oils of many of the economically important Labiate species have been intensively studied and are generally highly complex (Lawrence, 1992). Volatile terpenoids, particularly monoterpenoids, constitute the bulk of these distillates. Lavender (*Lavandula angustifolia*, *Lavandula latifolia* and *Lavandula intermedia*) oils, for example, contain over 300 components, with 95% of these being monoterpenoids (Ohloff, 1995). However, 12 compounds (present at 1% or greater by weight) account for 90% of this oil, with a further 26 compounds (present at 0.1–1.0% by weight) accounting for 97% of the oil. Thus, around 250 compounds from this total are present in only trace amounts (Ohloff, 1995). This pattern is common to almost all the aromatic Labiate species whose oils have been studied. Ravid and Putievsky (1983) reported that 97% of the essential oils of *Majorana syriatica*, *Coridothymus capitatus* and *Satureja thymbra* reflected the contribution of 18 compounds. Lawrence (1992) provides a table listing 168 terpenoids present in the oils of 14 oil-rich Labiate species. However, in any one of these oils, on average only nine compounds are present at over 1%.

The analysis of Labiate essential oils is further complicated by the presence of distinct races within a species differing (often markedly) in the composition of their essential oils. Such races are termed chemotypes, and are presumed to have arisen through hybridization within geographically distinct populations of plants. For example, six chemotypes of *Perilla frutescens* have been defined, which exhibit marked variation in their leaf essential oils (Yuba *et al.*, 1995). One of these accumulates only acyclic monoterpenoids, one only cyclohexanoid monoterpenoids, three chemotypes accumulate distinct furanoid monoterpenoids, and one lacks monoterpenoids altogether but yields instead phenylpropanoid compounds.

Labiate essential oils also vary markedly depending on the time of harvest, age of the plant material extracted and post-harvest treatment (Shu and Lawrence, 1997). In *Melissa officinalis*, marked differences were observed in the composition of the essential oil isolated from leaves at the top of the plant from that obtained from leaves towards the base of the plant (Hose *et al.*, 1997). These differences correlated with leaf age rather than position, and most likely reflected metabolism of essential oil constituents during leaf maturation. Voirin and co-workers have examined the effect of age on monoterpenoid composition in leaves of *Mentha* × *piperita*, and have shown variation between the tip and base (Brun *et al.*, 1991). More sophisticated analysis subsequently confirmed that the disappearance of limonene, the accumulation of menthone, and its conversion to menthol and menthyl acetate, as well as the appearance of 1,8-cineole, initiates at the leaf tip and that this metabolism progresses from leaf tip to leaf base as the leaf matures (Voirin and Bayet, 1996). At maturity, all parts of the leaf possess an oil of identical composition. These analyses show that much of the variation in the reported essential oil composition of Labiate plants is most likely due to the harvesting and extraction of bulk material composed of different plant parts at different stages of maturity.

Because of the complexity of these oils, the presence of quite distinct chemotypes within individual species and variation in oil composition during organ development, it is often difficult to tabulate the composition of a "representative" oil for any given species. Many analyses carried out on samples of commercial oils produced from plants grown in different parts of the world show considerable variation, and such samples may often be adulterated. The reader is referred to the publications of Lawrence (1992, 1993, 1995) for detailed treatment of this subject. However, for illustrative purposes the oil composition of some of the Labiate species referred to throughout this review are given in Table I. These include *Nepeta cataria* and *Nepeta racemosa*, whose oils are characterized by cyclopentanoid nepetalactone stereoisomers; *Melissa officinalis*, with a high (although variable – compare data in Lawrence (1993) and Hose *et al.* (1997)) proportion of acyclic monoterpenoid aldehydes, *Mentha* × *piperita*, characterized by *p*-menthane monoterpenoids, and *Salvia officinalis*, which produces *p*-menthane, bornane and thujane monoterpenoids.

III. GLANDULAR TRICHOMES OF THE LABIATAE

A. GLANDULAR TRICHOMES

An association between the presence of glandular trichomes and accumulation of essential oils has been noted since the earliest investigations of plant anatomy (Uphof, 1962). The aerial surfaces of almost all of the aromatic Labiate species examined are covered with trichomes, including non-glandular hairs and glandular or secretory trichomes (Werker *et al.*, 1985; Werker, 1993). Multicellular, non-glandular hairs may be simple or complex (branched) in form. Two forms of glandular trichome are found, termed capitate and peltate trichomes, which are distinguished by their structure and mode of secretion. The catmint *Nepeta racemosa*, for example, possesses both capitate and peltate trichomes in addition to simple hairs, as shown in Fig. 3.

Capitate glandular trichomes are commonly composed of a basal cell, one or more stalk cells, and either one or two cells in the head. In some species, complex capitate trichomes occur that carry a glandular head on one of two branches. Capitate glandular trichomes may be supported by a raised pedestal formed by cells of the epidermis arranged around their basal cell (Ascensão *et al.*, 1995). Secretion of material outside the head cell(s) of capitate trichomes occurs either through a porous cuticle, within a minimal

TABLE I

Principal terpenoid constituents of selected Labiate oils

Compound	*Nepeta cataria*	*Nepeta racemosa*	*Salvia officinalis*	*Melissa officinalis*	*Mentha spicata*	*Mentha × piperita*
Geraniol	—	—	—	0.1	—	—
Nerol	—	—	—	0.1	—	—
Citronellol	—	—	—	0.2	—	—
Geranial	—	—	—	37.2	—	—
Neral	—	—	—	24.1	—	—
Citronellal	—	—	—	0.7	—	—
Linalool	—	0.3	*t*–0.33	0.4	0–0.3	—
Myrcene	*t*–1.5	—	*t*–0.73	—	1.2–5.5	2.48
Geranyl acetate	—	—	—	0.5	—	—
(Z)-β-Ocimene	*t*–0.2	0.6	—	—	0.3–1.0	—
(E)-β-Ocimene	*t*–0.5	1.5	—	—	*t*–0.7	—
α-Pinene	—	—	0.27–3.86	—	0.5–1.2	1.49
β-Pinene	—	—	0.23–2.02	—	0.6–1.5	2.25
Sabinene	—	—	0–0.24	—	0.3–0.9	1.3
p-Cymene	—	—	0.3–0.77	—	0–0.2	1.55
α-Terpinene	—	—	—	—	—	0.86
γ-Terpinene	—	—	—	—	*t*–0.7	1.88
trans-Ocimene	—	—	—	0.2	—	—
Limonene	*t*–0.4	—	*t*–1.42	0.1	3.4–15.8	6.25
Pulegone	—	—	—	—	—	0.69
Menthol	—	—	—	—	0–0.4	21.96
Neomenthol	—	—	—	—	—	2.83
Isomenthol	—	—	—	—	—	5.75
Menthone	—	—	—	—	0–*t*	23.87
Isomenthone	—	—	—	—	0–*t*	4.1
Menthofuran	—	—	—	—	—	3.31
Piperitone	—	—	—	—	*t*	—
Carvone	—	—	—	—	4.9–59.6	—

trans-Carveol	—	—	—	—	0–0.2	—
cis-Carveol	—	—	—	—	0–0.6	—
Dihydrocarveol	—	—	—	—	0.6–7.8	—
cis-Carvyl acetate	—	—	—	—	0.2–5.5	—
Neoisodihydrocarvyl ac.	—	—	—	—	0–20.9	—
Dihydrocarvyl acetate	—	—	—	—	1.2–24.8	—
trans-Dihydrocarvone	—	—	—	—	0–21.2	—
cis-Dihydrocarvone	—	—	—	—	3.1–21.6	—
α-Thujene	—	—	0.04–0.2	—	—	—
α-Thujone	—	—	35.9–45.8	—	—	—
β-Thujone	—	—	4.35–9.6	—	—	—
1,8-Cineole	—	—	4.98–13.4	—	0.5–3.2	6.48
Camphene	—	—	1.14–5.65	—	—	—
Camphor	0.8[a]	—	15.5–21.1	—	—	—
Borneol	—	—	0.74–3.2	—	—	—
Bornyl acetate	—	—	0.82–2.75	—	—	—
(Z)-Salvene	—	—	0.04–0.6	—	—	—
(E)-Salvene	—	—	*t*–0.12	—	—	—
β-Phellandrene	—	3.2	—	—	—	—
Methyl heptenone	—	—	—	0.6	—	—
Nepetalactone	10.6–94.1	40.4	—	—	—	—
Epinepetalactone	0.4–65.2	30.6	—	—	—	—
Neonepetalactone	0.1–0.2	—	—	—	—	—
Dihydronepetalactone	0.2–2.7	—	—	—	—	—
β-Caryophyllene	1.1–6.8	3.7	*t*–3.78	9.5	0.3–2.6	—
Caryophyllene oxide	—	—	0.44–2.17	2.5	—	—
α-Copaene	—	0.2	—	4.0	—	—
β-Elemene	—	—	—	0.2	—	—
β-Bourbonene	—	0.3	—	0.3	*t*–2.1	0.18
Calamenene	—	0.1	—	—	—	—
Germacrene D	—	8.6	—	4.2	0–2.0	5.51

TABLE I *(continued)*

Compound	*Nepeta cataria*	*Nepeta racemosa*	*Salvia officinalis*	*Melissa officinalis*	*Mentha spicata*	*Mentha × piperita*
α-Humulene	—	—	*t*–3.85	0.2	—	—
α-Cadinol	—	—	—	0.3	—	—
δ-Cadinene	—	—	—	1.1	—	—
γ-Cadinene	—	—	—	1.0	—	—
(Z)-3-Hexenol	—	—	—	0.1	—	—
3-Octanol	—	—	—	—	0–0.5	1.71
3-Octanone	—	—	—	0.6	—	—
1-Octen-3-ol	—	0.3	—	1.3	0–0.3	—
Methyl hydrocinnamate	—	0.3	—	—	—	—
Ethyl hydrocinnamate	—	0.1	—	—	—	—

Data for the *Nepeta* spp., *Melissa officinalis* and carvone-rich *Mentha spicata* are taken from Lawrence (1993) and [a]Regnier *et al.* (1967). Those for *Salvia officinalis* are from Piccaglia and Marotti (1993). Data presented for *Mentha × piperita* are from the analysis of peltate trichomes present on 4 cm leaves (Maffei *et al.*, 1989). *t*, trace.

subcuticular space, or with rupture of the cuticle (Werker, 1993). Peltate glandular trichomes are larger, with a single basal cell, a single stalk cell and four or more head cells. In these structures, secreted material accumulates within a large subcuticular space formed by distension of the cuticle and part of the cell wall surmounting the secretory cells (Werker *et al.*, 1985; Werker, 1993). The expanded cuticle of peltate glands gives these structures a diameter of 40–50 μm at maturity.

In addition to epidermal glandular trichomes, the Labiate species patchouli (*Pogostemon cablin*) – uniquely so far as is known – possesses glandular trichomes arising from the palisade layer and developing within the interior of the leaf (Uphof, 1962, and references therein; Henderson *et al.*, 1970). The presence of such internal oil glands may well be the reason why leaves of this species (unlike other Labiate species) require drying or ageing prior to distillation (Weiss, 1997).

The evidence that the material accumulated within the subcuticular cavity of peltate glands of the Labiatae is largely composed of the terpenoid essential oil constituents of these species is compelling, as will be discussed in later sections. The function of the smaller capitate glands, and their possible contribution to distillable essential oil, has not yet been conclusively determined. In one species, *Prasium majus*, which is not aromatic, only capitate glands are present (Werker, 1993), thus their contribution is likely to be minimal. Because of the clear association between peltate glands and accumulation of essential oils in the genus, only this type will be discussed in any detail in this review.

The structure and development of peltate glandular trichomes are discussed in detail elsewhere in this volume (see chapter by Werker). Some examples will serve here to illustrate our knowledge of these structures and their development in the Labiatae. The development of these trichomes has been studied in *Mentha piperita* (Amelunxen, 1965), *Origanum dictamnus* (Bosabalidis and Tsekos, 1982), *Ocimum basilicum* (Werker *et al.*, 1993) and *Nepeta racemosa* (Bourett *et al.*, 1994). From such studies, it would appear that glandular trichomes differentiate from epidermal cells early in the formation of leaf primordia and new glands are produced continuously until cell division ceases at leaf maturity (Werker, 1993). In their studies of the development of the peltate glandular trichomes of *Nepeta racemosa*, Bourett *et al.* (1994) found that peltate glands first appeared on the third pair of leaf primordia from the shoot apex, indicating that these structures differentiate very early during leaf development. At maturity, peltate glands of this species have four cells in the head, surmounted by a smooth distended cuticle of *c.* 50 μm in diameter. Glands at different stages of development (i.e. with two, three or four head cells discernible prior to cuticle expansion) could be observed on these primordia, and the pattern of development was such that glands at the leaf margin were less developed than those nearer the midvein, and those near the tip more developed than those towards the base. The number of glands on successive leaf primordia were counted and shown to increase until reaching around 100 on the adaxial face of the sixth leaf pair from the shoot apex (Bourett *et al.*, 1994). A mature leaf of this species (*c.* 2.5 cm in length) was later shown to bear around 2000 peltate trichomes on the adaxial face, with half this number again present on the abaxial face (Clark *et al.*, 1997b).

A number of changes in head cell ultrastructure have been observed during development of Labiate peltate gland head cells. In *Nepeta racemosa*, large vacuoles present in cells at the earliest stages of secretion gave way to smaller vesicles, and plastids with smooth endoplasmic reticulum (ER) in close proximity became prominent. A later stage was characterized by the presence of smooth membrane cisternae, a marked retraction of the plasma membrane from the cell wall accompanied by blebbing of the membrane, and the appearance of small membrane vesicles between the plasmalemma and cell wall (Bourett *et al.*, 1994). In *Origanum dictamnus*, mitochondria increased in number during the secretory phase, and rough ER cisternae (smooth when in proximity to plastids) were prominent (Bosabalidis and Tsekos, 1982). Peltate trichomes of *Mentha* × *piperita* (Amelunxen, 1965) and *Origanum dictamnus* (Bosabalidis and Tsekos, 1982) exhibited an increase in electron density of stalk cell lateral walls during development, whilst in *Nepeta racemosa*, walls of the head cells adjacent to that of the stalk cell exhibited high electron density (Bourett *et al.*, 1994). The cutinization of these cell walls probably functions in the isolation of the secretory cells from the apoplast (Fahn, 1988).

Most ultrastructural studies of secretory glandular trichome development have focused on the active developmental or secretory phases. The eventual fate of these cells has received only scattered or incidental attention. Uphof (1962, p. 54), summarizing earlier studies, states that, "in older glandular hairs, the nucleus is sometimes already dead, and it may even have disappeared completely, while the plasm is still functioning to some extent, as appears from the fact that secretion goes on. Finally, however, the protoplast also disappears, and then the lumen of the cells merely contains some dead remains or becomes entirely filled with air." At maturity, the head cells of *Nepeta racemosa* peltate glands appeared devoid of large organelles other than nuclei and mitochondria, and the plasma membrane was drastically retracted from the cell wall (Bourett *et al.*, 1994). Although this appearance may well be an aspect of secretion in these cells, these features (retraction and blebbing of the plasmalemma, loss of organelles other than mitochondria) are often found in cells undergoing programmed cell death (McCabe *et al.*, 1997; Pennell and Lamb, 1997). However, programmed cell death is generally accompanied by chromatin condensation and destruction of the nuclear envelope and these were not reported in this study. The disappearance of the nuclear membrane was observed, however, in older head cells of *Origanum dictamnus* peltate glands (Bosabalidis and Tsekos,

1982). An extraplasmic space was formed during development of the secretory cells of *Chrysanthemum morifloium* (Compositae) (Vermeer and Peterson, 1979), and older secretory cells of this species contained only mitochondria, ER cisternae and a degenerate nucleus. The upper cell of these elongated trichomes eventually collapses and, at this stage, only mitochondria were visible. A consistent study of post-secretory phase gland cell development may well reveal evidence of programmed cell death, although it might be difficult to show whether this was a developmentally programmed process as opposed to an effect of accumulation of phytotoxic material. In this regard, it is interesting to note that plant monoterpenoids (namely limonene and perillyl alcohol) have been shown to induce programmed cell death in mammalian carcinoma cells (Mills *et al.*, 1995; Ariazi and Gould, 1996).

B. ACCUMULATION OF MONOTERPENOIDS IN PELTATE GLANDULAR TRICHOMES

Considerable evidence that glandular trichomes are the site of storage of terpenoids in Labiatae and other plant families has accumulated (Kelsey *et al.*, 1984). Early studies of patchouli (*Pogostemon cablin*) indicated a correlation between the number of external glandular trichomes and the sesquiterpenoid content of leaf sections (Henderson *et al.*, 1970). The most detailed studies on the accumulation of monoterpenoids in glandular trichomes of the Labiatae have been carried out on *Mentha* and *Nepeta* spp. Loomis (1967) and Regnier *et al.* (1968) referred to these structures as likely to be the site of storage of the essential oils of these species. Evidence that this was in fact the case was obtained when the contents of these structures were isolated and analysed (Amelunxen *et al.*, 1969).

Maffei *et al.* (1989) reported the analysis of oil collected directly from peltate glandular trichomes of *Mentha × piperita* using fine glass capillaries. Gas chromatography–mass spectrometry (GC–MS) allowed identification of 21 components of this oil (Table I) and analysis of the contents of single trichomes could be achieved. The samples contained a similar spectrum of compounds to that extracted from whole plant material. However, these analyses revealed a high variability in oil composition between glands at different points on the leaf, and between glands present on the abaxial and adaxial leaf surfaces. The authors proposed that this variation was due to sampling of peltate trichomes at different stages of development and suggested that trichomes at different stages of maturation were intermingled on the leaf surface (Maffei *et al.*, 1989). Voirin and Bayet (1996) used a variety of sampling techniques, including the isolation of individual peltate trichomes, in their analysis of developmental changes in oil composition of *Mentha × piperita*. These painstaking experiments showed that the area in which trichomes in close proximity varied in oil composition moved progressively down the leaf from the tip (the most mature part) to the base, in parallel with the appearance and development of peltate trichomes and maturation of the oil. In these studies, individual peltate trichomes were removed from the leaf surface intact and monoterpenoids analysed by GC following desorption-concentration, rather than being directly sampled. Clark *et al.* (1997b) employed microsampling and GC-MS to analyse the composition of material within the subcuticular cavity of peltate glands of *Nepeta racemosa* and *Nepeta cataria*. A microcapillary (tip diameter 15–20 μm) was used to impale glands and the subcuticular cavity contents withdrawn under negative pressure. The density of much longer simple hairs on leaves of these *Nepeta* species precluded sampling from glands of the adaxial epidermis and no effort was made to analyse glands at specific points on the leaves. However, the data indicated that the subcuticular cavity of these glands indeed contained the cyclopentanoid monoterpenoids (nepetalactones), which are the principal components of the essential oil of these species. Quantitative analysis allowed estimation of the amount of nepetalactone accumulated by individual glands of *Nepeta racemosa* to be *c.* 30 ng. This measurement, together with estimations of the number of glands present, also allowed the authors to calculate that the nepetalactone content of leaves of *Nepeta racemosa* could be accounted for by the content of the peltate glandular trichomes alone (Clark *et al.*, 1997b). With *Melissa officinalis*, GC analysis of n-pentane extracts prepared from individual peltate glands (excised using a blade, together with a small area of leaf lamina) again showed the presence of essential oil components (citronellal, methyl citronellate, neral, geranial, geranyl acetate) in these samples (Hose *et al.*, 1997).

It has been assumed that, once the secretory process is complete, the secreted material remains unchanged until its release when the glands are disrupted, e.g. by insect, rain or wind damage (Werker, 1993). Peltate glands of *Leonotis leonurus* and *Salvia aurea* possess a distinct equatorial fracture line, and rupture of the cuticle along this line leads to release of the subcuticular contents (Ascensão *et al.*, 1995; Serrato-Valenti *et al.*, 1997). An analysis of terpenoid turnover in four plant species, together with a review of previously reported data in this regard, has been presented (Gershenzon *et al.*, 1993). In this report, no significant turnover of monoterpenoids, sesquiterpenoids or diterpenoids was detected following exposure of young plants to $^{14}CO_2$ and analysis of labelled compounds. In an experiment with undisturbed, illuminated *Nepeta racemosa* plants (0.5 m^2 total leaf area), GC analysis of volatiles collected by standard headspace sampling techniques over a 6-hour sampling time revealed no trace of nepetalactone (J. Gershenzon and D. L. Hallahan, unpublished data). The detection limit for nepetalactone in this experiment was *c.* 20 ng, indicating that this species did not release nepetalactone at a rate greater than 6.0 ng h^{-1} m^2 leaf area.

In concluding this section, it is clear that sufficient experimental evidence exists to confirm the supposition that peltate glandular trichomes of Labiate plants are the site of accumulation of the principal monoterpenoid components of their essential oils. The role of the capitate secretory trichomes present on these species has yet to be determined with any certainty. Given that peltate glands are the site of accumulation, are these structures also the site of biosynthesis? Before turning to this question, it will be useful briefly to review our current knowledge of the pathways of terpenoid biosynthesis in plants.

IV. CORE PATHWAYS OF MONOTERPENOID BIOSYNTHESIS

Because all terpenoids can, at least hypothetically, be elaborated from isoprene, it was until recently believed that this structural similarity was reflected in a common biosynthetic origin. Thus this array of compounds was assumed to derive from elaboration of a relatively simple isoprenoid precursor synthesized within a single pathway common to all organisms. This assumption was based on the analysis of many terpenoid structures, which led to the formulation of the "biogenetic isoprene rule" (Ruzicka, 1953). In this scheme, acetate is converted to the five-carbon isoprene skeleton, and this is elaborated enzymatically into the 10-, 15- and 20-carbon intermediates geranyl-, farnesyl- and geranylgeranyldiphosphate. These intermediates undergo cyclization and/or chemical modification to yield mono-, sesqui- and diterpenoids, respectively, or self-condensation to yield the 30- and 40-carbon precursors of sterols and carotenoids (Fig. 4).

The synthesis of the active isoprene unit, isopentenyl diphosphate (IPP), was determined to proceed from acetate via the coenzyme A (CoA) ester of 3-hydroxy-3-methylglutarate (HMG-CoA) and mevalonate (MVA) (reviewed in Bach *et al.*, 1990). That this acetate/mevalonate pathway operates in plants has been demonstrated by many studies showing the incorporation of pathway intermediates (particularly MVA) into terpenoids, as well as the presence of acetate/mevalonate pathway enzymes including HMG-CoA reductase, the enzyme responsible for the formation of MVA (reviewed in Gray, 1987; Gershenzon and Croteau, 1990, 1993; McGarvey and Croteau, 1995). However, in many early studies, incorporation of MVA was low (Francis, 1971; Banthorpe *et al.*, 1972). For example, early *in vivo* tracer studies on plants showed that CO_2 was readily incorporated in the light into carotenoids and other chloroplast-located terpenoids, but not into sterols, whereas acetate or MVA was preferentially incorporated into sterols but not carotenoids (Rogers *et al.*, 1968). Additionally, the biosynthesis of certain terpenoids was shown to be insensitive to inhibitors of the mevalonic acid pathway enzymes. Mevinolin, an effective inhibitor of HMG-CoA

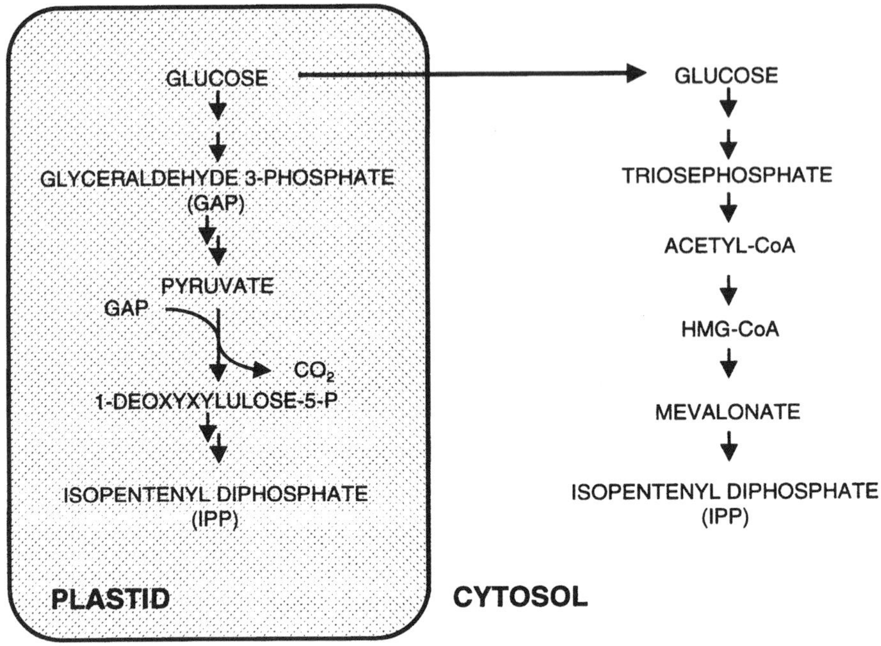

Fig. 4. The two pathways of isopentenyl diphosphate (IPP) biosynthesis in plants. The classical acetate/mevalonate pathway now appears to be responsible for IPP synthesis in the cytosol, whilst a newly discovered pathway operates within plastids.

reductase, blocked synthesis of sterols but not carotenoids (Bach, 1986). Isolated spinach chloroplasts were shown not to incorporate MVA or its phosphorylated derivatives into terpenoids, but did incorporate IPP (Kreuz and Kleinig, 1981). Until recently, for example, it had been assumed that the acetate/mevalonate pathway was the source of nerol/geraniol in cyclopentanoid biosynthesis. Exposure of *Nepeta cataria* seedlings to $^{14}CO_2$ for periods of up to 10 minutes resulted in the incorporation of radioactivity into hexane-soluble compounds, one of which was identified as nepetalactone by co-chromatography on thin layer chromatography (TLC) (Mitchell *et al.*, 1972), but no pathway intermediates were analysed in these experiments. Earlier studies had shown that [2-^{14}C]-2-mevalonate was incorporated into nepetalactone (Regnier *et al.*, 1968). Although incorporation was relatively low (0.001–0.01%), this was comparable with studies of incorporation in other species. It was therefore concluded that nepetalactone biosynthesis in this species involved the acetate/mevalonate pathway. However, in the same year, evidence that [2-^{14}C]-mevalonate was not incorporated into the cyclopentanoid loganin in *Menyanthes trifoliata* was published (Brechbuhler-Bader *et al.*, 1968).

Such discrepancies were interpreted originally as evidence for differential subcellular compartmentation of aspects of terpenoid biosynthesis in plants (Gershenzon and Croteau, 1990), but may now be reinterpreted in the light of the recent discovery of an alternative pathway operating in microbes and plants (Lichtenthaler *et al.*, 1997). This pathway is now known to be responsible for synthesis of plastid-derived isoprenoids, such as β-carotene, lutein, the phytyl chains of chlorophylls, and the prenyl sidechain of plastoquinone in plants and microorganisms, whilst the acetate/mevalonate pathway yields cytosolic steroids (Lichtenthaler *et al.*, 1997). In this newly discovered pathway, pyruvate and glyceraldehyde-3-phosphate act as precursors of isopentenyl diphosphate (Fig. 4). Ultrastructural studies have shown that plastids, within which the pyruvate/glyceraldehyde-3-phosphate pathway occurs, have consistently been implicated in the active secretory phase of glandular trichome development (Fahn, 1988; Kleinig, 1989; Nielsen *et al.*, 1991). Very recent studies on the biosynthesis of monoterpenoids in Labiate plants have conclusively shown that they are derived from IPP via the pyruvate/glyceraldehyde-3-phosphate pathway and not from the classic acetate/mevalonate pathway (Eisenreich *et al.*, 1997; Sagner *et al.*, 1998). *Mentha* × *piperita*, *Thymus vulgaris* or *Mentha pulegium* plantlets were fed with [1-^{13}C]glucose or [U-^{13}C]glucose, and several characteristic monoterpenoids isolated by steam distillation and preparative TLC. ^{13}C enrichment and coupling patterns measured by ^{13}C nuclear magnetic resonance (NMR) showed that the labelling pattern of menthone (from *Mentha* × *piperita*), thymol (from *Thymus vulgaris*) and pulegone (from *Mentha pulegium*) were consistent with an origin from pyruvate/glyceraldehyde-3-phosphate rather than from acetate/mevalonate (Eisenreich *et al.*, 1997). The enzyme that catalyses the first reaction of the pyruvate/glyceraldehyde 3-phosphate pathway, 1-deoxy-D-xylulose-5-phosphate synthase, has been cloned from *Mentha* × *piperita* and expressed in *Escherichia coli* (Lange *et al.*, 1998). The synthase gene contains a putative plastid-targeting sequence, consistent with a role in the biosynthesis of plastid-derived isoprenoids in this species.

Geranyl diphosphate (GPP), the 10-carbon precursor of monoterpenoids, is derived from the condensation of IPP with its isomer dimethylallyl diphosphate (DMAPP) (Fig. 5). DMAPP is produced through the action of IPP isomerase, and its dephosphorylation (catalysed by isoprene synthase, a chloroplast-located enzyme) yields isoprene. The prenyltransferase enzyme, GPP synthase, responsible for the condensation of IPP and DMAPP has been characterized in a number of plant species, including the Labiate *Salvia officinalis* (Croteau and Purkett, 1989). The huge spectrum of monoterpenoids derived from geranyl diphosphate in the Labiatae and other plant genera clearly implies the existence of an equally diverse set of biosynthetic enzymes operating on this common intermediate. This expectation has been amply confirmed by studies of the enzymes of plant terpenoid metabolism. These enzymes, including cyclases, oxidoreductases and hydroxylases, are

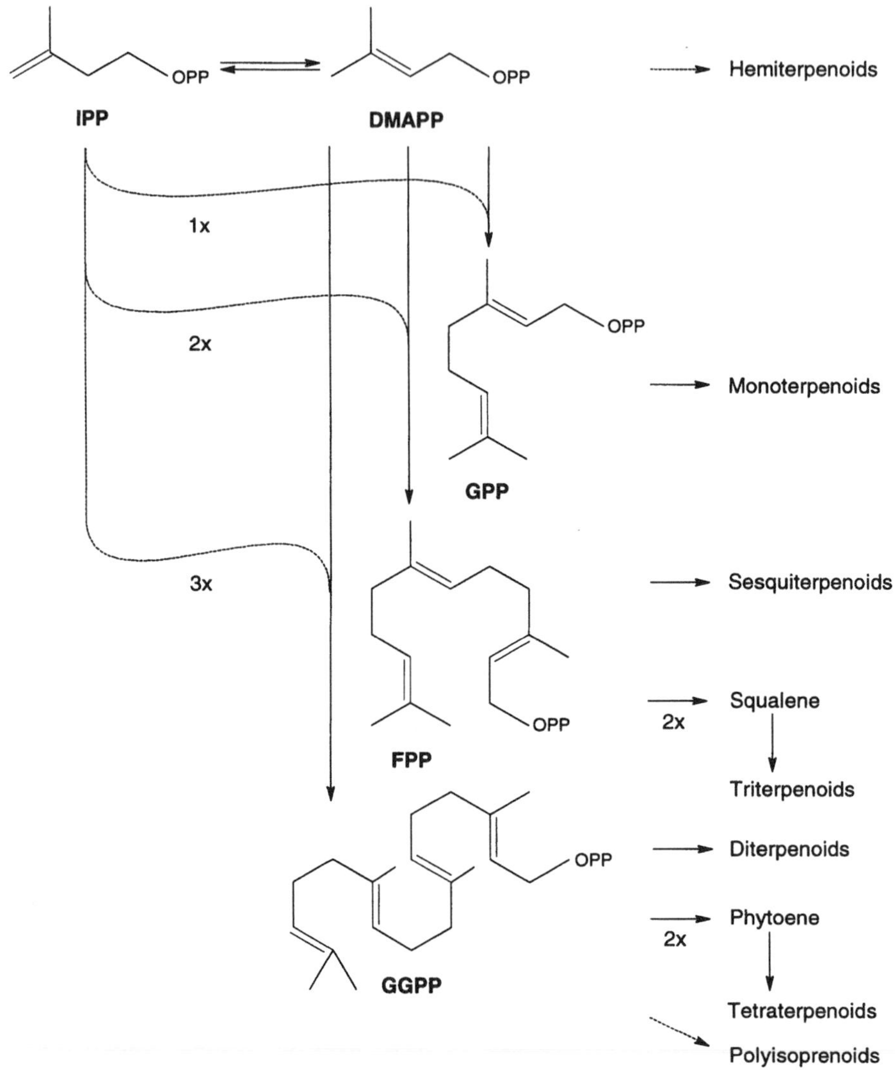

Fig. 5. Biosynthesis of the terpenoid skeletons from isopentenyl diphosphate (IPP). DMAPP, dimethylallyl diphosphate; FPP, farnesyl diphosphate; GGPP, geranylgeranyl diphosphate; GPP, geranyl diphosphate. (Adapted from McGarvey and Croteau, 1995.)

responsible for the elaboration of this common precursor into the diverse monoterpenoids produced by plants, and our knowledge of some of the key enzyme classes is described in the following sections.

V. GLANDULAR TRICHOMES IN STUDIES OF MONOTERPENOID BIOSYNTHESIS

We can now return to the question of whether the monoterpenoids of Labiate essential oils are synthesized within the trichomes known to accumulate them. That we can now answer this question with confidence has come about principally because of the development of techniques for the preparation of biosynthetically active cells derived from glandular trichomes. At the forefront of studies of glandular trichomes of the Labiatae have been Croteau and his co-workers, who developed an efficient technique for isolation of clusters of gland head cells from *Mentha* spp. (Gershenzon *et al.*, 1987, 1991, 1992; McCaskill *et al.*, 1992). Their approach involves the mechanical abrasion of leaves by glass beads, followed by sieving to remove contaminating material. This method has yielded cells capable of the synthesis of monoterpenoids from exogenous precursors and in sufficient quantity to be used for the purification of individual biosynthetic enzymes. It has also been successfully applied to several other Labiate species, including *Salvia officinalis* (Gershenzon *et al.*, 1991). A similar abrasive technique was developed originally for use on *Pelargonium* spp. (Yerger *et al.*, 1992) and has been successfully applied to catmint (Clark *et al.*, 1997a; Hallahan *et al.*, 1998). In this case, finely powdered dry ice is used as the abrasive, and trichomes are again collected after sieving.

Intact glandular trichome secretory cells, isolated by mechanical abrasion of *Mentha* × *piperita* leaves with glass beads, were shown to incorporate high levels of precursors into monoterpenoids, primarily limonene and menthone (Gershenzon *et al.*, 1991, 1992; McCaskill *et al.*, 1992). Incorporation was most efficient with [8-^{3}H]GPP (34.0% from an incubation with 6 μM GPP), although relatively high incorporation of sucrose (0.6% from 2.0 μM; 0.56% from 10 μM), IPP (1.9% from 40 μM), MVA (1.1% from 120 μM) and acetate (0.18% from 16.0 μM) were observed. Although sucrose incorporation appeared to be higher than that of acetate with these cells, since sucrose is equivalent to four acetate units, the incorporation of sucrose was in fact roughly equivalent (McCaskill *et al.*, 1992). Further studies provided evidence to show that monoterpene synthesis in trichome secretory cells of *Mentha* × *piperita* relies on plastid-derived IPP (McCaskill and Croteau, 1995).

Rupture of trichome secretory cells by sonication yields cell-free preparations in which the activity of monoterpenoid biosynthetic enzymes may be measured. Limonene cyclase and hydroxylase activity could be measured in cell-free extracts of *Mentha* trichomes (McCaskill *et al.*, 1992) and monoterpenoid oxidoreductase activities have been measured in extracts of *Nepeta racemosa* trichomes (Hallahan *et al.*, 1998). As discussed in the following sections, one of the principal advantages conferred by an ability to isolate trichomes is that they can be used as starting material for the purification of enzymes (or preparation of trichome-specific populations of cDNAs), and for the localization of enzymatic activities and gene expression. The availability of such preparations has underpinned recent advances in studies of the enzymes of monoterpenoid biosynthesis in the Labiatae and their corresponding genes.

VI. MOLECULAR ENZYMOLOGY OF THE DIVERGENT PATHWAYS OF MONOTERPENOID BIOSYNTHESIS

The essential oils of Labiate plants, as discussed above, contain a variety of monoterpenoid structures, including acyclic and cyclic monoterpenoids. The derivation of the acyclic compounds from GPP has not been studied to the same extent as that of the cyclized derivatives. Dephosphorylation of GPP yields geraniol directly and plants are rich in phosphatase activity, a fact that is problematic in studies of terpenoid biosynthesis using phosphorylated precursors. Whether a specific enzyme has been recruited for this function in those plants that accumulate geraniol and related acyclic monoterpenoids or whether this activity is specifically compartmentalized, has not been established. Isomerization of geraniol to nerol might occur by the action of a redox system or nerol might be formed during dephosphorylation of GPP (Banthorpe *et al.*, 1978). Citronellol is generated by reduction of nerol/geraniol but again no detailed investigations into this activity have been reported. The feeding of radiolabelled geraniol to cell cultures of *Citrus limon*, however, resulted in the rapid formation of both nerol and citronellol (Berger *et al.*, 1990), thus providing evidence that these interconversions do occur in plants.

The formation of the acyclic monoterpenoid myrcene is catalysed by a specific synthase that uses GPP as substrate. This enzyme has been studied in cell-free extracts of grand fir (*Abies grandis*; Pinaceae) and the corresponding cDNA isolated (Bohlmann *et al.*, 1997). This monoterpenoid synthase appears to be synthesized as a preprotein, with an amino-terminal plastid-targeting sequence. Sequence comparison revealed that the monoterpene synthases from *Abies grandis* resemble sesquiterpene (C-15) synthases and diterpene (C-20) synthases from conifers more closely than other monoterpene synthases from angiosperm species, although the significance of this remains unclear.

S-Linalool synthase, an enzyme that stereoselectively converts GPP to *S*-linalool, was purified from *Clarkia breweri* (Onagraceae) (Pichersky *et al.*,

1995). Linalool and linalool oxides are the principal components of the floral scent of this species of plant. The enzyme was operationally soluble and activity was strictly dependent on the presence of a divalent metal co-factor (optimally, Mn^{2+}). Unlike the cyclases involved in cyclohexanoid monoterpenoid biosynthesis (see below), neither *S*- nor *R*-linalyl diphosphates were used as substrates by this enzyme. Antibody production and cloning of the corresponding cDNA allowed the expression of this gene to be examined in flower tissues of *Clarkia breweri*, and it was found that expression was localized to petal epidermal tissue rather than to any specialized scent glands (Dudareva *et al.*, 1996).

The majority of monoterpenoids of the Labiatae, however, are cyclic structures. Cyclohexanoid monoterpenoids (Fig. 6) predominate in the oils of, for example, *Mentha* and *Salvia*. The genus *Nepeta* is characterized by the predominance of cyclopentanoid (iridoid) monoterpenoids, although cyclohexanoid monoterpenoids are also synthesized. The key enzymes responsible for the formation of these cyclized structures from geranyl diphosphate are the monoterpenoid cyclases and subsequent metabolism of the products of these cyclizations gives rise to further diversity.

A. CYCLOHEXANOID MONOTERPENOIDS

The derivation of cyclohexanoid monoterpenoids from direct cyclization of GPP, catalysed by enzymes termed cyclases or synthases, is now well established. These enzymes catalyse the sequential ionization and isomerization of GPP to linaloyl diphosphate (LPP), followed by cyclization (Croteau, 1987). Studies of these enzymes have been thoroughly reviewed in recent years (McGarvey and Croteau, 1995; Bohlmann *et al.*, 1998). Isomerization to LPP prior to cyclization appears to be an essential facet of the activity of these enzymes, as the presence of the *trans*-2,3-double bond of GPP prevents direct cyclization of this compound. The absence of such an impediment in the *cis*-isomer of GPP, neryl diphosphate (NPP) might at first sight make this compound a more likely precursor of monoterpenoids, but studies of multiple cyclases have demonstrated GPP to be their actual substrate. Cyclization of LPP at the terminal double bond yields the α-terpinyl cation, the universal precursor of cyclohexanoid monoterpenoids. This cation is highly reactive and its conversion to the various cyclohexanoid skeletons occurs while bound to the cyclase. Thus the cyclases have in common the ability to generate the α-terpinyl cation from GPP but each differs in the eventual cyclized products that are formed. Interestingly, most of the monoterpenoid cyclases studied thus far produce multiple products, although in each case a single compound predominates. For example, limonene synthase from *Mentha* × *piperita* produces minor proportions of α-pinene, β-pinene and myrcene in addition to limonene (Alonso *et al.*, 1992), even though in *Abies grandis* each of these compounds is synthesized by

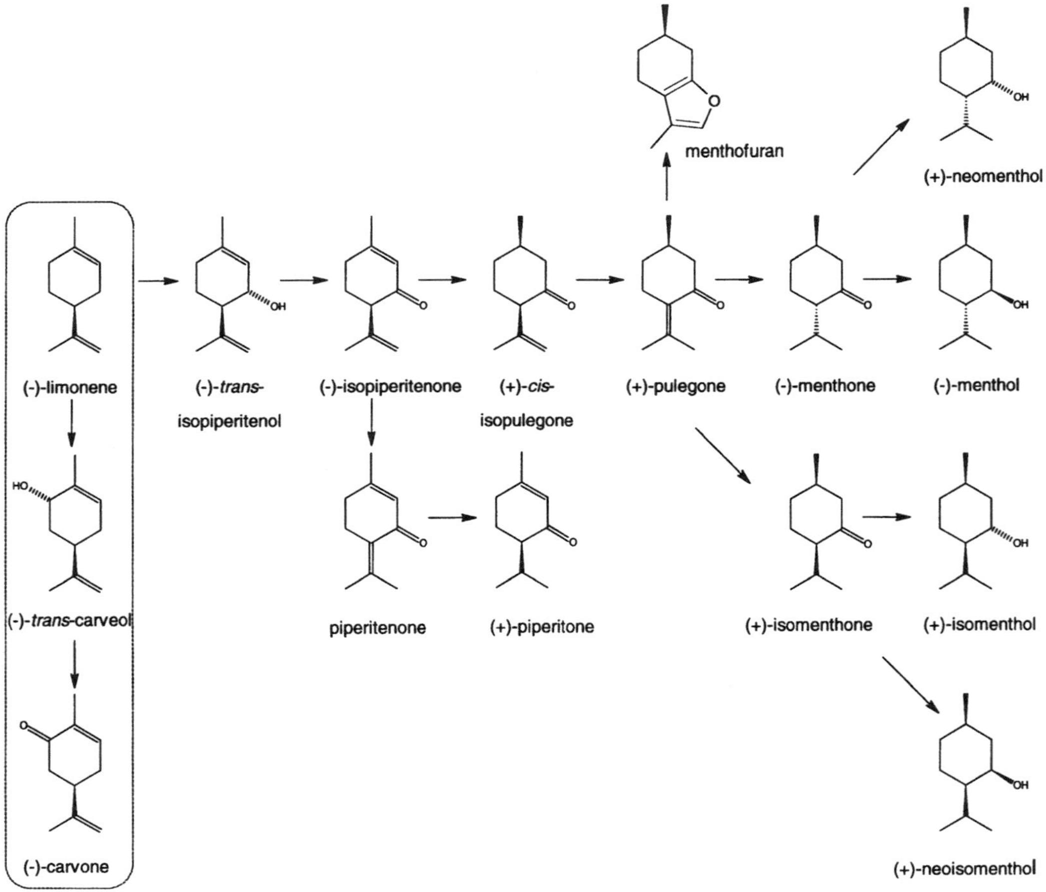

Fig. 6. Biosynthesis of cyclohexanoid *p*-menthane monoterpenoids in *Mentha* spp. The boxed pathway, initiated by hydroxylation of limonene at C-3, occurs in spearmint and results in the accumulation of (−)-carvone. In the peppermint, limonene is hydroxylated at C-6, and subsequent transformations result in an oil containing predominantly menthone, isomenthol, menthol and neomenthol.

individual cyclases (Bohlmann *et al.*, 1997). It is interesting to speculate on whether isoforms within this family of enzymes are evolving into efficient catalysts with high product specificity or whether in fact a degree of redundancy has been advantageous in the face of selection pressure at the whole plant level. This redundancy, however, probably contributes to the large spectrum of cyclohexanoid monoterpenoids commonly found in Labiate essential oils, and the presence of many compounds at trace levels, as remarked on above. For example, analysis of the oil present in peltate glands of young (10 cm in length) leaves of *Mentha piperita* showed that 21 components were present, the majority of which are derived from the activity of limonene synthase (Maffei *et al.*, 1989). These include not just the *p*-menthane monoterpenoids characteristic of this species, but also α-pinene, β-pinene and myrcene. Some 70% of the oil was composed of derivatives of limonene, with the remainder composed of roughly equal proportions of α-pinene, β-pinene and myrcene (9.56% of total oil), 1,8-cineole (4.58%) and various sesquiterpene hydrocarbons (8.23%). Thus despite the apparent complexity of the oil of this species, its cyclic monoterpenoid component conceivably derives from the action of only two enzymes, limonene and cineole synthases.

Several monoterpenoid cyclases have been studied in or purified from cell-free extracts of Labiate leaves or trichomes, including *Mentha* limonene synthases (Alonso *et al.*, 1992; Rajaonarivony *et al.*, 1992), α- and β-pinene synthases (Gambliel and Croteau, 1982) and 1,8-cineole synthase (Croteau and Purkett, 1989; Croteau *et al.*, 1994) from *Salvia officinalis* and γ-terpinene synthase from *Thymus vulgaris* (Alonso and Croteau, 1991). The purification of mono-, sesqui- and diterpenoid cyclases from a number of sources allowed molecular cloning of the corresponding cDNAs, and the sequence information thus provided has accelerated isolation of other members of the terpenoid synthase family in plants (Bohlmann *et al.*, 1998).

Over 30 terpenoid cyclase cDNAs have now been isolated from plants, including limonene synthases from a number of species (*Mentha spicata*, *Mentha candicans*, *Perilla frutescens*, *Abies grandis*), bornyl diphosphate, 1,8-cineole and sabinene synthases from *Salvia officinalis*, and pinene, camphene, terpinolene and myrcene synthases from *Abies grandis* (Bohlmann *et al.*, 1998). Heterologous expression of many of these operationally soluble enzymes in *Escherichia coli* has been achieved, facilitating further enzymatic analyses. In addition to readily providing large quantities of enzyme for analysis, this approach allows analysis of a specific gene product, whereas even ostensibly pure cyclase preparations from plants could well contain multiple closely related enzymes. For example, expression of limonene synthase in *Escherichia coli* has unambiguously shown that this isoform produces multiple products in a ratio identical to that obtained with the native enzyme (Colby *et al.*, 1993). The only detectable product produced from GPP by the heterologously expressed limonene synthase from *Perilla frutescens*, however, was 4(*S*)-limonene (Yuba *et al.*, 1996). *Salvia officinalis* produces a range of cyclized monoterpenes, including members of the *p*-menthane, pinane, thujane, camphane and bornane families. cDNAs encoding three monoterpene synthases from *Salvia officinalis* were isolated, functionally expressed in *Escherichia coli*, and found to encode (+)-bornyl diphosphate, 1,8-cineole and (+)-sabinene synthases (Wise *et al.*, 1998). Each of these heterologously expressed enzymes was found to produce multiple products from GPP. The bornyl diphosphate synthase also produced significant amounts of (+)-α-pinene, (+)-camphene and (+/−)-limonene; the 1,8-cineole synthase produced (+)- and (−)-α-pinene, (+)- and (−)-β-pinene, myrcene and (+)-sabinene, and the (+)-sabinene synthase produced significant quantities of γ-terpinene and terpinolene.

The availability of the cDNAs will allow studies of the regulation of expression of the corresponding genes, as has been carried out on the wound-inducible terpenoid cyclases of *Abies grandis* (Steele *et al.*, 1998) and linalool synthase during flower development in *Clarkia breweri* (Dudareva *et al.*, 1996) and analysis of their roles in terpenoid biosynthesis. For example, it has been possible to show that the limonene synthase gene is absent from chemotypes of *Perilla frutescens*, which do not produce cyclohexanoid monoterpenoids (Yuba *et al.*, 1996). In addition, all the monoterpenoid cyclase cDNAs isolated thus far encode amino-terminal plastid-targeting sequences, again consistent with the plastidial origin of monoterpenes in plants.

Subsequent transformations of the basic skeletons of the cyclohexanoid monoterpenoids are carried out by hydroxylases, oxidoreductases and isomerases. The most detailed work on the secondary transformations of a cyclohexanoid monoterpenoid thus far has been carried out on the derivation of the *p*-menthane monoterpenoids from (−)4*S*-limonene (Fig. 6). Within the genus *Mentha*, the cyclohexanoid monoterpenoids of the essential oils of different species are characterized by the position of oxygenation of the *p*-menthane ring (Lupien *et al.*, 1995). Thus *Mentha* × *piperita*, *Mentha aquatica*, *Mentha arvensis* and *Mentha pulegium* produce *p*-menthane monoterpenoids oxygenated almost exclusively at C3, typified by menthol, whereas in the spearmints (*Mentha spicata*, *Mentha* × *gracilis*), oxygenation occurs almost exclusively at C6, as typified by carvone.

(−)-Limonene 3-hydroxylase activity was characterized in membranes prepared from peppermints, whilst (−)-limonene 6-hydroxylase activity was shown to be present in membranes prepared from spearmint glandular trichomes (Karp *et al.*, 1990; Croteau *et al.*, 1991). These enzymes require NADPH and O_2 for activity, and were characterized as members of the cytochrome P450 superfamily (Mihaliak *et al.*, 1993; Schuler, 1996). Purification of the limonene 6-hydroxylase from *Mentha spicata* glandular trichomes has been reported (Lupien *et al.*, 1995) and the purified enzyme shown to be a cytochrome P450 of molecular mass 57 kDa. After reconstitution of this enzyme with purified *Mentha spicata* cytochrome P450 reductase, conversion of (−)-4*S*-limonene to *trans*-carveol was observed in the presence of NADPH. Sequencing of peptides generated from the purified hydroxylase by proteolysis facilitated the design of oligonucleotide primers, which were used to PCR-amplify a 700 bp cDNA fragment. A cDNA library prepared from peltate glands was then screened using this fragment and a cDNA encoding a full-length cytochrome P450 isolated (Lupien *et al.*, 1995). The sequence of this (and other cytochrome P450 cDNAs isolated as a consequence from other *Mentha* species) has not yet been reported, although at the time of writing a preliminary report of the expression of these cDNAs and their characterization as monoterpenoid hydroxylases has appeared (Haudenschild *et al.*, 1997).

The subsequent enzymes of *p*-menthane monoterpenoid biosynthesis have not been studied to the same extent as the cyclases and hydroxylases, and none have been characterized at the molecular level. Oxidation of isopiperitenol, the product of limonene 6-hydroxylase, to (−)-isopiperitenone, and subsequent isomerization to piperitenone, has been demonstrated using soluble enzyme preparations from *Mentha* × *piperita* leaves (Kjonaas *et al.*, 1985). It has been shown, however, that (+)-*cis*-isopulegone, not piperitenone, is the key intermediate in the conversion of (−)-isopiperitenone to (+)-pulegone in this species (Croteau and Venkatachalam, 1986). (+)-Pulegone reductase activity, yielding (−)-menthone, could be measured in extracts from isolated secretory cells (Gershenzon *et al.*, 1991), and stereospecific dehydrogenase enzymes had previously been shown to catalyse the conversion of (−)-menthone to (−)-menthol and (+)-neomenthol in *Mentha* × *piperita* leaves (Kjonaas *et al.*, 1982). Oxidation of (−)-*trans*-carveol, the product of the limonene 3-hydroxylase, has been demonstrated to occur in cell-free extracts of trichome secretory cells of *Mentha spicata*, and histochemical staining confirmed the presence of this activity within peltate glandular trichomes (Gershenzon *et al.*, 1989).

B. CYCLOPENTANOID MONOTERPENOIDS

In contrast to the cyclohexanoids, the cyclopentanoid (iridoid) monoterpenoids are not derived from direct cyclization of GPP (reviewed in Inouye, 1991; De Luca, 1993). With cyclopentanoids, it has been shown that monoterpenoid alcohols (and not their diphosphorylated derivatives) are the precursors. Geraniol itself, presumably derived from dephosphorylation of GPP, appears to be the first C10 intermediate of this pathway.

In studies of cyclopentanoid biosynthesis in *Catharanthus roseus* (Apocynaceae; formerly *Vinca rosea*), it was shown that hydroxylated derivatives of geraniol (or its isomer nerol) were incorporated into loganin and alkaloids. GPP and NPP, as well as citronellol, were ineffective as precursors for loganin in this species (Battersby *et al.*, 1970; Escher *et al.*, 1970). A pathway involving hydroxylation of geraniol to 8-hydroxygeraniol,* followed by oxidation to 8-oxogeranial and cyclization to iridodial was proposed. Based on the finding that extensive scrambling of label in [9-^{14}C]-8-hydroxynerol occurred during loganin synthesis, it was subsequently proposed that cyclization of 8,9-dioxoneral yielded the cyclopentanoid skeleton. However, it was shown later that 8-oxogeranial was in fact more likely to be the substrate for cyclization in this and other plants (reviewed in Inouye, 1991). Tracer studies using cell-free extracts of *Rauwolfia serpentina* (Apocynaceae) provide the strongest evidence in support of this pathway, demonstrating that 8-oxogeranial is the immediate precursor to iridodial at least in this plant (Uesato *et al.*, 1986a, 1986b). Significant incorporation of iridodial into nepetalactone in *Nepeta cataria* has been demonstrated, indicating the intermediacy of this compound in the biosynthesis of the methylcyclopentanoids produced by this species (Bellesia *et al.*, 1984).

Thus a pathway in which iridodial is derived from geraniol (or nerol) via sequential hydroxylation, oxidation and cyclization can be proposed. There are, however, certain exceptions to this generalized scheme. Although citronellol and its derivatives were shown not to be precursors of cyclopentanoids in *Catharanthus roseus*, tracer studies indicated that 8-hydroxycitronellol, and not 8-hydroxynerol, is the precursor of the cyclopentanoid dolichodial in *Teucrium marum* (Bellesia *et al.*, 1983) and nepetalactone in *Nepeta cataria* (Bellesia *et al.*, 1984). Our understanding of the biosynthetic route to the methylcyclopentane skeleton in plants can therefore be summarized as follows. In some species, the pathway is initiated from the acyclic monoterpene alcohols nerol or geraniol, whereas in others (namely the Labiates *Nepeta cataria* and *Teucrium marum*), the monounsaturated equivalent, citronellol, is preferred. In all cases, biosynthesis is initiated by hydroxylation of these compounds. A single hydroxylation of these compounds at C-8, followed by sequential oxidation, yields the 8-oxo-aldehydes, which may be cyclized to iridodial. This proposed pathway is illustrated in Fig. 7.

The cyclopentanoid biosynthetic pathway elucidated through tracer studies has been supported by the identification and analysis of the enzymes involved. An enzyme catalysing the hydroxylation of nerol and geraniol at C-8 has been characterized in cell-free extracts of *Catharanthus roseus* seedlings (Meehan and Coscia, 1973; Madyastha *et al.*, 1976) and leaves of

*In much of the literature, a different numbering system for the carbon atoms of the acyclic monoterpenoids nerol and geraniol has been used, in which the carbon atom hydroxylated during cyclopentanoid biosynthesis has been designated as C-10. In this review, the numbering shown in Fig. 7 has been used throughout, and this carbon is designated C-8.

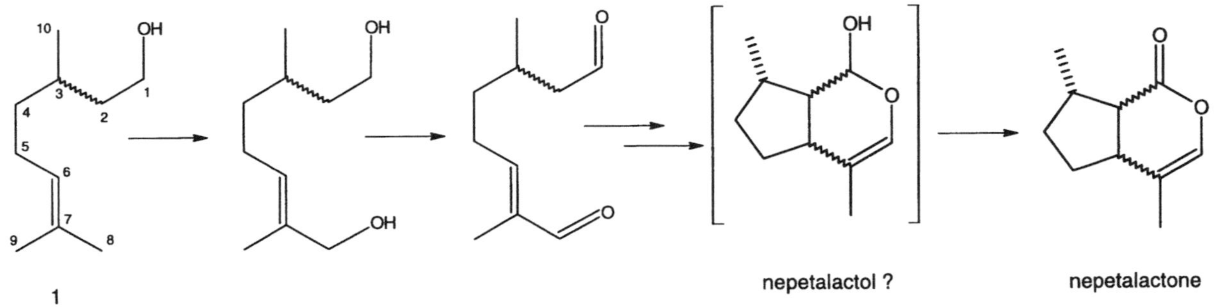

1

Compound 1
~~~~ ; Z double bond = nerol
E double bond = geraniol
saturated = citronellol

Fig. 7. Biosynthesis of cyclopentanoid monoterpenoids in *Nepeta racemosa*.
~~~~

Nepeta racemosa (Hallahan *et al.*, 1992a). In both cases, activity was present in a membrane fraction and exhibited properties indicating that a cytochrome P450 mixed-function oxygenase was responsible. Analysis of monoterpenoid hydroxylation in membrane fractions of *Catharanthus roseus* and *Nepeta racemosa* has shown that hydroxylation of all three potential cyclopentanoid precursors (nerol, geraniol and (3*S*)-citronellol) is catalysed in both species (Hallahan and West, 1995). The purification to homogeneity of this enzyme from cell cultures of *Catharanthus roseus* has been reported and studies of the pure protein have confirmed that it is a member of the cytochrome P450 superfamily of enzymes (Meijer *et al.*, 1993). The purified enzyme was active towards nerol and geraniol, although activity towards citronellol was not reported in this study. A cDNA encoding a cytochrome P450 (designated according to the nomenclature for this superfamily as CYP71A5) has been isolated from *Nepeta racemosa*, and the expression pattern of the corresponding gene is consistent with that which might be expected of an enzyme involved in nepetalactone biosynthesis in this species (Clark *et al.*, 1997a). Transcripts were detected by northern blot hybridization in RNA from young tissues of both *Nepeta racemosa* and *Nepeta cataria*, and specifically located in RNA prepared from isolated trichomes of *Nepeta racemosa* (Clark *et al.*, 1997a). Heterologous expression of the corresponding enzyme is essential for attribution of catalytic activity to the product of a cloned cytochrome P450 gene, and such characterization has not been reported for this isoform. Activity towards these simple monoterpenoids might well be catalysed by multiple members of this large superfamily of enzymes, and indeed such activity has been reported for avocado (*Persea americana*; Lauraceae) CYP71A1, which epoxidizes both geraniol and nerol (Hallahan *et al.*, 1992b, 1994). The hydroxylation of geraniol and nerol at C-8 was found to be catalysed by a cytochrome P450 from *Arabidopsis thaliana* (Brassicaceae) following heterologous expression in baculovirus-infected insect cells (Ohta and Mizutani, 1998). This enzyme (CYP76C1) is expressed in leaves and is strongly induced by light, but its expression is highest in flowers (Mizutani *et al.*, 1998). *Arabidopsis thaliana* is not known to accumulate cyclopentanoid monoterpenoids or derivatives thereof, and thus activity of CYP76C1 towards nerol/geraniol *in vitro* may well be adventitious.

Acyclic monoterpenoid oxidoreductases, catalysing the oxidation of 8-hydroxynerol and 8-hydroxygeraniol to their respective dialdehydes, have also been purified to homogeneity from *Catharanthus roseus* (Ikeda *et al.*, 1991) and *Nepeta racemosa* (Hallahan *et al.*, 1995). The catmint enzyme was active towards nerol, geraniol and (3*S*)-citronellol as well as their 8-hydroxylated derivatives. This enzyme was also present at higher specific activity in cell-free extracts prepared from trichomes of *Nepeta racemosa* as compared to trichome-denuded leaves, implying that biosynthesis of nepetalactones occurs in trichomes of this species (Hallahan *et al.*, 1998).

A cyclase, catalysing the formation of iridodial from 8-oxo-geranial, was studied in extracts prepared from cell cultures of *Rauwolfia serpentina* (Uesato *et al.*, 1986a, 1987). This enzyme was soluble and required NAD(P)H for activity. Activity of the enzyme towards the dialdehyde derivatives of either nerol or citronellol was not described, and further purification of this enzyme for more detailed study has not been reported. Although iridodial and nepetalactol have often been treated as equivalent in the literature, evidence that these structures are not necessarily interconvertible (dependent upon stereochemistry) has been presented (Dawson *et al.*, 1989). However, it would seem plausible that if a cyclase similar to that of *Rauwolfia serpentina* were responsible for the formation of the cyclopentanoid nepetalactones in *Nepeta* spp., nepetalactol might be an intermediate. Analysis of cell-free extracts of *Nepeta racemosa* confirmed that an enzyme was present in this species, which was capable of the NAD^+-dependent oxidation of nepetalactol to nepetalactone, consistent with nepetalactol being an intermediate in the biosynthesis of nepetalactone (Hallahan *et al.*, 1998). It must be pointed out that the intermediacy of nepetalactol was ruled out by Bellesia *et al.* (1984), who recovered label from [1-^{3}H]nerol or [1-^{3}H]citronellol exclusively at C-10 of nepetalactone. Such a hydride shift would not occur during oxidation of nepetalactol or iridodial. However, nepetalactol oxidoreductase activity was detectable only in extracts of leaf trichomes of *Nepeta racemosa* and not in denuded leaf tissue. Given the evidence that peltate glandular trichomes are the sites of synthesis as well as accumulation of essential oil components in the Labiatae, this would support a role for this enzyme in nepetalactone biosynthesis (Hallahan *et al.*, 1998).

Thus the biosynthesis of the free cyclopentanoid nepetalactones accumulated by *Nepeta* spp. would appear to proceed from a monoterpenoid alcohol (nerol or geraniol, but most probably (3*S*)-citronellol) via hydroxylation at C-8, oxidation to a dialdehyde and cyclization of the cyclopentanoid ring through an as-yet poorly understood mechanism. Nepetalactol, if the product of this cyclization, may be oxidized to yield nepetalactone by a trichome-located oxidoreductase. Although the pathway has yet to be clarified, it would appear that both biosynthesis and accumulation of nepetalactones occur within trichomes of these species.

VII. GENETIC MANIPULATION OF MONOTERPENOID BIOSYNTHESIS AND TRICHOME BIOTECHNOLOGY

The isolation of genes encoding monoterpenoid biosynthetic enzymes opens up the possibility of manipulating the monoterpenoid profile of plants through transgenic technology. Such manipulation might be carried out with a number of aims in mind. For example, engineered plants in which accumulation of secondary metabolites has been modified (for overproduction of specific compounds or for their absence) could be used to answer some of the many fundamental questions on the role of secondary metabolites in plant defence.

Modification of the amount (or spectrum) of monoterpenoids, to yield plants of increased value to the flavour and fragrance industries, is the most obvious commercial use of such genes. Such plants might be of equal interest to the gardener because of their novel properties. Manipulations might also be performed without a preconceived idea of their metabolic consequences in the resultant transgenic plants. Such an approach might conceivably yield entirely new and unexpected metabolites, whose properties might be of interest as new flavour or fragrance compounds. However, the industries built up around plant flavour and fragrance compounds are relatively conservative, which is understandable given that their markets are human food and cosmetic products. It is likely, at least initially, that transgenic approaches to creating reproducible yields would be welcomed more than those creating novel products.

As yet, there has been no reported success in selective manipulation of the oil content of gland-bearing plants by molecular genetic means. Trichome-localized oils have been produced efficiently in transformed, differentiated shoot cultures of mints, which carried peltate glands. No attempt was made specifically to manipulate the oil content of these transformed tissues, and they produced an oil identical to that of untransformed material. Transformation and regeneration of lemon-scented geranium (*Pelargonium crispum*; Geraniaceae) plants did affect oil composition but this was an unanticipated side-effect of the transformation process (Pellegrineschi *et al.*, 1994).

The application of transgenic technology depends on an ability to transform and regenerate appropriate plant species in a relatively robust manner. In the main, attempts to manipulate Labiate plants genetically have focused on the mints, presumably because of their economic importance. In an attempt to produce shooty teratomas of *Mentha* × *piperita* cv. citrata, a number of strains of *Agrobacterium tumefaciens* were used to infect plant material (Spencer *et al.*, 1990, 1993). Only one strain out of the ten tested failed to infect this species, although shooty teratomas only developed from those galls induced by infection with the nopaline strains T37 and C58. Similar work was carried out independently with *Mentha* × *piperita* cv. Black Mitchum, *Mentha spicata* and *Mentha* × *gracilis* cv. Baker, using two *Agrobacterium* strains: the nopaline strain C58 and an agropine strain A281. In this report, strain A281 induced gall formation but no differentiation to a shooty morphology occurred, whereas strain C58 failed to induce galls in any of the species examined (Berry *et al.*, 1996). However, these experiments clearly showed that these *Mentha* species were amenable to infection by *Agrobacterium*.

That *Mentha* spp. are also able to regenerate from cultured material has been shown (Vaneck and Kitto, 1990; Sato *et al.*, 1993; Faure *et al.*, 1998).

Caissard *et al.* (1996) reported the optimization of regeneration conditions for leaf explants as well as transient reporter gene expression in *Mentha* × *piperita*. Studies with other Labiate species have also shown that they may be successfully regenerated from cultures (Sanchezgras and Segura, 1988, 1989; Vandenberg *et al.*, 1995).

The production of the first transgenic Labiate (*Mentha* × *piperita*) plants was reported by Niu *et al.* (1998). Whilst biolistic delivery of DNA failed to yield stable transformants, co-cultivation of leaf explants with *Agrobacterium* strain EHA105 yielded kanamycin-resistant calli with a high proportion exhibiting marker gene (β-glucuronidase) activity. Eight plants were recovered from cultured shoots, which polymerase chain reaction and Southern analysis confirmed were independent stable transformants.

This development opens up the possibilities associated with applying transgenic technology to Labiate species in general. However, an additional requirement must probably be met before direct manipulation of Labiate oils becomes feasible – the identification of trichome-specific gene promoters. The use of cell cultures for production of monoterpenoids was dogged by the central problem that terpenoid accumulation in dedifferentiated cells was likely to be phytotoxic, and that only in differentiated, specialized cells could high-level accumulation be achieved. Similarly, to minimize toxicity problems, the application of transgenic technology for manipulation of the oil composition in regenerated plants of the Labiatae will require transgene expression to be specifically targeted to the compartments responsible – peltate glandular trichomes. Although plant epidermal-specific promoters have been reported in the literature, and one such has been shown to control expression in trichomes of transgenic *Nicotiana tabacum* (Solanaceae) (Pyee and Kolattukudy, 1995), none has yet been reported that regulates gene expression in glandular trichomes. However, given the intense interest in genes involved in trichome differentiation and development (see other reviews in this volume) as well as the current rate of success in isolating cDNAs for trichome-localized enzymes involved in the synthesis of terpenoids and other metabolites, it cannot be long before such DNAs become part of the plant metabolic manipulation toolkit.

The ability to manipulate gland biosynthesis through molecular genetic means would in the first instance enable alteration of existing synthetic pathways in plants to commercially desirable ends, as mentioned above. However, glandular trichomes are likely to have evolved specifically for the large-scale synthesis of potentially toxic compounds. Thus the biosynthetic capacity of these structures could be exploited by man for the production of valuable but potentially phytotoxic chemicals in plants. Novel genes might be introduced to divert biosynthetic pathways in gland-bearing plants (e.g. existing agronomically important species, such as potato, hemp, hops, tomato in addition to certain Labiate species) towards a variety of speciality chemicals. For example, the utility of linalool oxides (present in, e.g. lavender oil; Ohloff, 1995) as chiral synthons in the preparation of bioactive tetracyclic ethers and polyether antibiotics has been proposed (Meou *et al.*, 1990). In addition to the introduction of novel genes, silencing of existing genes may well be a prerequisite for production of novel chemicals in these compartments. A better understanding of gland development and the regulation of expression of biosynthetic genes within glands will, however, be necessary before manipulation of their considerable biosynthetic capacity becomes feasible.

VIII. CONCLUSIONS

The aim of this review has been to provide an overview of our understanding of the biosynthetic capacity of Labiate plants, specifically in relation to monoterpenoids, for the interested reader. That this capacity is widespread within the Labiatae, and extensive as regards yield and end-product diversity, should be clear. It is now well established that much of this capacity is achieved solely by secretory glandular trichomes. This understanding only increases our appreciation of the complexity of the processes occurring within these structures, involving transport and sequestration pathways for intermediates and end products in addition to purely synthetic pathways. This complexity at the enzyme level is reflected in the ultrastructural complexity of these organs.

Many of the pathways for synthesis of the cyclohexanoid monoterpenoids produced by Labiate plants have been elucidated, with many enzymes and their corresponding genes now identified. However, little is known of the regulation of these pathways during trichome development, their compartmentation, or their integration with metabolite transport processes. Our understanding of cyclopentanoid monoterpenoid biosynthesis is by comparison incomplete, and the major challenge remaining is the characterization of the enzymes and genes involved in cyclization of monoterpenoid dialdehydes to yield this monoterpenoid skeleton.

Doubtless the isolation of the promoters of cloned biosynthetic enzymes and analysis of their expression (e.g. using promoter/reporter constructs in transgenic plants) will shed light on the regulation of these pathways. In addition, the availability of such promoters will underpin attempts to modify secondary metabolite production in glandular trichomes of transgenic plants for fundamental studies or commercial ends.

ACKNOWLEDGEMENTS

The author would like to acknowledge the excellent assistance provided to him over many years of research at IACR-Rothamsted by Jevon West. IACR receives grant-aided support from the Biotechnology and Biological Sciences Research Council of the United Kingdom. The author is grateful to Professor R. Croteau (Washington State University) and Dr B. J. Hallahan for critical reading of the manuscript.

REFERENCES

Agrawal, A. A. (1998). Induced responses to herbivory and increased plant performance. *Science* **279**, 1201–1202.

Agrawal, R., Patwardhan, M. V. and Gurudutt, K. N. (1991). Formation of flavouring constituents in callus cultures of lime, *Citrus aurantifolia. Biotechnology and Applied Biochemistry* **14**, 265–274.

Alonso, W. R. and Croteau, R. (1991). Purification and characterisation of the monoterpene cyclase gamma-terpinene synthase from *Thymus vulgaris. Archives of Biochemistry and Biophysics* **286**, 511–517.

Alonso, W. R., Rajaonarivony, J. I. M., Gershenzon, J. and Croteau, R. (1992). Purification of 4S-limonene synthase, a monoterpene cyclase from the glandular trichomes of peppermint (*Mentha × piperita*) and spearmint (*Mentha spicata*). *Journal of Biological Chemistry* **267**, 7582–7587.

Amelunxen, F. (1965). Elektronenmikroskopisch untersuchungen an den drüsenschuppen von *Mentha piperita. Planta Medica* **13**, 457–473.

Amelunxen, F., Wahlig, T. and Arbeiter, H. (1969). Über den nachweis des ätherischen öls in isolierten drüsenschuppen von *Mentha piperita* L. *Zeitschrift für Pflanzenphysiologie* **60**, 156–168.

Ariazi, E. A. and Gould, M. N. (1996). Identifying differential gene expression in monoterpene-treated mammary carcinomas using subtractive display. *Journal of Biological Chemistry* **271**, 29 286–29 294.

Ascensão, L., Marques, N. and Pais, M. S. (1995). Glandular trichomes on vegetative and reproductive organs of *Leonotis leonurus* (Lamiaceae). *Annals of Botany* **75**, 619–626.

Bach, T. J. (1986). Hydroxymethylglutaryl-CoA reductase, a key enzyme in phytosterol synthesis? *Lipids* **21**, 82–88.

Bach, T. J., Weber, T. and Motel, A. (1990). Some properties of enzymes involved in the biosynthesis and metabolism of 3-hydroxy-3-methylglutaryl-CoA in plants. *In* "Biochemistry of the Mevalonic Acid Pathway to Terpenoids", Vol. 24. Recent Advances in Phytochemistry (G. H. N. Towers and H. A. Stafford, eds), pp. 1–82. Plenum Press, New York.

Banthorpe, D. V., Charlwood, B. V. and Francis, M. J. O. (1972). The biosynthesis of monoterpenes. *Chemical Reviews* **72**, 115.

Banthorpe, D. V., Modawi, B. M., Poots, I. and Rowan, M. G. (1978). Redox interconversions of geraniol and nerol in higher plants. *Phytochemistry* **17**, 1115–1118.

Battersby, A. R., Brown, S. H. and Payne, T. G. (1970). Biosynthesis of loganin and the indole alkaloids from hydroxygeraniol-hydroxynerol. *Chemical Communications* 827–828.

Bellesia, A., Pagnoni, U. M., Pinetti, A. and Trave, R. (1983). The biosynthesis of dolichdial in *Teucrium marum. Phytochemistry* **22**, 2197–2201.

Bellesia, F., Grandi, R., Pagnoni, U. M., Pinetti, A. and Trave, R. (1984). Biosynthesis of nepetalactone in *Nepeta cataria. Phytochemistry* **23**, 83–87.

Berger, R. G., Akkan, Z. and Drawert, F. (1990). Catabolism of geraniol by cell-suspension cultures of *Citrus limon*. *Biochimica et Biophysica Acta* **1055**, 234–239.

Bernays, E. A. and Chapman, R. F. (1994). "Host-Plant Selection by Phytophagous Insects." Chapman and Hall, New York.

Berry, C., Vaneck, J. M., Kitto, S. L. and Smigocki, A. (1996). Agrobacterium-mediated transformation of commercial mints. *Plant Cell Tissue and Organ Culture* **44**, 177–181.

Bohlmann, J., Steele, C. L. and Croteau, R. (1997). Monoterpene synthases from grand fir (*Abies grandis*) – cDNA isolation, characterization, and functional expression of myrcene synthase, (−)(4S)-limonene synthase, and (−)-(1S,5S)-pinene synthase. *Journal of Biological Chemistry* **272**, 21 784–21 792.

Bohlmann, J., MeyerGauen, G. and Croteau, R. (1998). Plant terpenoid synthases: molecular biology and phylogenetic analysis. *Proceedings of the National Academy of Sciences USA* **95**, 4126–4133.

Boo, K. S., Chung, I. B., Han, K. S., Pickett, J. A. and Wadhams, L. J. (1998). Response of the lacewing *Chrysopa cognata* to pheromones of its aphid prey. *Journal of Chemical Ecology* **24**, 631–643.

Bosabalidis, A. and Tsekos, I. (1982). Glandular scale development and essential oil secretion in *Origanum dictamnus* L. *Planta* **156**, 496–504.

Bourett, T. M., Howard, R. J., O'Keefe, D. P. and Hallahan, D. L. (1994). Gland development on leaf surfaces of *Nepeta racemosa*. *International Journal of Plant Sciences* **155**, 623–632.

Bowers, M. D. (1988). Chemistry and coevolution: iridoid glycosides, plants and herbivorous insects. *In* "Chemical Mediation of Coevolution" (K. C. Spencer, ed.), pp. 133–165. Academic Press, San Diego, CA.

Brechbuhler-Bader, S., Coscia, C. J., Loew, P., von Szczepanski, C. and Arigoni, D. (1968). The chemistry and biosynthesis of loganin. *Chemical Communications* 136.

Brown, J. T., Hegarty, P. K. and Charlwood, B. V. (1987). Monoterpene and sesquiterpene biosynthesis in plant-cells in culture. 4. The toxicity of monoterpenes to plant-cell cultures. *Plant Science* **48**, 195–201.

Brun, N., Colson, M., Perrin, A. and Voirin, B. (1991). Chemical and morphological studies of the effects of ageing on monoterpene composition in *Mentha* × *piperita* leaves. *Canadian Journal of Botany* **69**, 2271–2278.

Caissard, J. C., Faure, O., Jullien, F., Colson, M. and Perrin, A. (1996). Direct regeneration *in vitro* and transient GUS expression in *Mentha* × *piperita*. *Plant Cell Reports* **16**, 67–70.

Cavill, G. W. K., Ford, D. L. and Locksley, H. D. (1956). Iridodial and iridolactone. *Chemistry and Industry* 465.

Charlwood, B. V. and Charlwood, K. A. (1991). Terpenoid production in plant cell cultures. *In* "Ecological Chemistry and Biochemistry of Plant Terpenoids", Vol. 31. Proceedings of the Phytochemical Society of Europe (J. B. Harbourne and F. A. Tomas-Barberan, eds), pp. 95–132. Clarendon Press, Oxford.

Clark, I. M., Forde, B. G. and Hallahan, D. L. (1997a). Spatially distinct expression of two new cytochrome P450s in leaves of *Nepeta racemosa*: identification of a trichome-specific isoform. *Plant Molecular Biology* **33**, 875–885.

Clark, L. J., Hamilton, J. G. C., Chapman, J. V., Rhodes, M. J. C. and Hallahan, D. L. (1997b). Analysis of monoterpenoids in glandular trichomes of the catmint *Nepeta racemosa*. *Plant Journal* **11**, 1387–1393.

Colby, S. M., Alonso, W. R., Katahira, E. J., McGarvey, D. J. and Croteau, R. (1993). 4*S*-Limonene synthase from the oil glands of spearmint (*Mentha spicata*) –

cDNA isolation, characterization, and bacterial expression of the catalytically active monoterpene cyclase. *Journal of Biological Chemistry* **268**, 23 016–23 024.
Connolly, J. D. and Hill, R. A. (1991). "Dictionary of Terpenoids." Chapman and Hall, London.
Croteau, R. (1987). Biosynthesis and catabolism of monoterpenoids. *Chemical Reviews* **87**, 929–954.
Croteau, R., Alonso, W. R., Koepp, A. E. and Johnson, M. A. (1994). Biosynthesis of monoterpenes – partial-purification, characterization, and mechanism of action of 1,8-cineole synthase. *Archives of Biochemistry and Biophysics* **309**, 184–192.
Croteau, R., Karp, F., Wagschal, K. C., Satterwhite, D. M., Hyatt, D. C. and Skotland, C. B. (1991). Biochemical characterization of a spearmint mutant that resembles peppermint in monoterpene content. *Plant Physiology* **96**, 744–752.
Croteau, R. and Purkett, P. T. (1989). Geranyl pyrophosphate synthase – characterization of the enzyme and evidence that this chain-length specific prenyltransferase is associated with monoterpene biosynthesis in sage (*Salvia officinalis*). *Archives of Biochemistry and Biophysics* **271**, 524–535.
Croteau, R. and Venkatachalam, K. V. (1986). Metabolism of monoterpenes – demonstration that (+)-cis-isopulegone, not piperitenone, is the key intermediate in the conversion of (−)-isopiperitenone to (+)-pulegone in peppermint (*Mentha piperita*). *Archives of Biochemistry and Biophysics* **249**, 306–315.
Dawson, G. W., Griffiths, D. C., Janes, N. F., Mudd, A., Pickett, J. A., Wadhams, L. J. and Woodcock, C. M. (1987). Identification of an aphid sex pheromone. *Nature* **325**, 614–616.
Dawson, G. W., Janes, N. F., Mudd, A., Pickett, J. A., Slawin, A. M. Z., Wadhams, L. J. and Williams, D. J. (1989). The aphid sex pheromone. *Pure and Applied Chemistry* **61**, 555–558.
Dawson, G. W., Griffiths, D. C., Merritt, L. A., Mudd, A., Pickett, J. A., Wadhams, L. J. and Woodcock, C. M. (1990). Aphid semiochemicals – a review, and recent advances on the sex pheromone. *Journal of Chemical Ecology* **16**, 3019–3030.
De Luca, V. (1993). Enzymology of indole alkaloid biosynthesis. *Methods in Plant Biochemistry* **9**, 345–368.
De Pooter, H. L., Nicolai, B., De Laet, J., De Buyck, L. F. and Schamp, N. M. (1988). The essential oils of five *Nepeta* species. A preliminary evaluation of their use in chemotaxonomy by cluster analysis. *Flavour and Fragrance Journal* **3**, 155–159.
Dudareva, N., Cseke, L., Blanc, V. M. and Pichersky, E. (1996). Evolution of floral scent in *Clarkia*: novel patterns of *S*-linalool synthase gene expression in the *C. breweri* flower. *Plant Cell* **8**, 1137–1148.
Eisenreich, W., Sagner, S., Zenk, M. H. and Bacher, A. (1997). Monoterpenoid essential oils are not of mevalonoid origin. *Tetrahedron Letters* **38**, 3889–3892.
Eisner, T. (1964). Catnip: its raison d'etre. *Science* **146**, 1318–1320.
Eisner, T. (1965). Defensive spray of a phasmid insect. *Science* **148**, 966–968.
Escher, S., Loew, P. and Arigoni, D. (1970). The role of hydroxygeraniol and hydroxynerol in the biosynthesis of loganin and indole alkaloids. *Chemical Communications*, 823–825.
Fahn, A. (1988). Secretory tissues in vascular plants. *New Phytologist* **108**, 229–257.
Faure, O., Diemer, F., Moja, S. and Jullien, F. (1998). Mannitol and thidiazuron improve in vitro shoot regeneration from spearmint and peppermint leaf disks. *Plant Cell Tissue and Organ Culture* **52**, 209–212.

Feeny, P. (1991). Theories of plant chemical defence – a brief historical survey. *In* "Insects-Plants '89: Proceedings of the 7th International Symposium on Insect–Plant Relationships", Vol. 39. Symposia Biologica Hungarica (A. Szentesi and T. Jermy, eds), pp. 163–175. Akademiai Kiado, Budapest.
Fischer, N. H. (1991). Plant terpenoids as allelopathic agents. *In* "Ecological Chemistry and Biochemistry of Plant Terpenoids", Vol. 31. Proceedings of the Phytochemical Society of Europe (J. B. Harborne and F. A. Tomas-Barberan, eds), pp. 377–398. Clarendon Press, Oxford.
Francis, M. J. O. (1971). Monoterpene biosynthesis. *In* "Aspects of Terpenoid Chemistry and Biochemistry" (T. W. Goodwin, ed.), pp. 29–51. Academic Press, London.
Futuyma, D. J. and Keese, M. C. (1992). Evolution and coevolution of plants and phytophagous arthropods. *In* "Herbivores, their Interactions with Secondary Plant Metabolites" (G. A. Rosenthal and M. R. Berenbaum, eds), pp. 440–475. Academic Press, London.
Gambliel, H. and Croteau, R. (1982). Biosynthesis of (+/−)-alpha-pinene and (−)-beta-pinene from geranyl pyrophosphate by a soluble enzyme system from sage (*Salvia officinalis*). *Journal of Biological Chemistry* **257**, 2335–2342.
Gershenzon, J. and Croteau, R. (1990). Regulation of monoterpene biosynthesis in higher plants. *In* "Biochemistry of the Mevalonic Acid Pathway to Terpenoids", vol. 24. *Recent Advances in Phytochemistry* (G. H. N. Towers and H. A. Stafford, eds), pp. 99–160. Plenum Press, New York.
Gershenzon, J. and Croteau, R. (1993). Terpenoid biosynthesis: the basic pathway and formation of monoterpenes, sesquiterpenes and diterpenes. *In* "Lipid Metabolism in Plants" (T. S. Moore, ed.), pp. 339–388. CRC Press, Boca Raton, FL.
Gershenzon, J., Duffy, M. A., Karp, F. and Croteau, R. (1987). Mechanized techniques for the selective extraction of enzymes from plant epidermal glands. *Analytical Biochemistry* **163**, 159–164.
Gershenzon, J., Maffei, M. and Croteau, R. (1989). Biochemical and histochemical-localization of monoterpene biosynthesis in the glandular trichomes of spearmint (*Mentha spicata*). *Plant Physiology* **89**, 1351–1357.
Gershenzon, J., McCaskill, D., Rajaonarivony, J., Mihaliak, C., Karp, F. and Croteau, R. (1991). Biosynthetic methods for plant natural products: new procedures for the study of glandular trichome constituents. *In* "Modern Phytochemical Methods" (N. H. Fischer, ed.), pp. 347–370. Plenum Press, New York.
Gershenzon, J., McCaskill, D., Rajaonarivony, J. I. M., Mihaliak, C., Karp, F. and Croteau, R. (1992). Isolation of secretory cells from plant glandular trichomes and their use in biosynthetic studies of monoterpenes and other gland products. *Analytical Biochemistry* **200**, 130–138.
Gershenzon, J., Murtagh, G. J. and Croteau, R. (1993). Absence of rapid terpene turnover in several diverse species of terpene-accumulating plants. *Oecologia* **96**, 583–592.
Gray, J. C. (1987). Control of isoprenoid biosynthesis in higher plants. *In* "Advances in Botanical Research", Vol. 14 (J. A. Callow, ed.), pp. 25–92. Academic Press, London.
Hallahan, D. L. and West, J. M. (1995). Cytochrome P450 in plant/insect interactions: geraniol 10-hydroxylase and the biosynthesis of iridoid monoterpenoids. *Drug Metabolism and Drug Interactions* **12**, 369–382.
Hallahan, D. L., Dawson, G. W., West, J. M. and Wallsgrove, R. M. (1992a). Cytochrome P450-catalyzed monoterpene hydroxylation in *Nepeta mussinii*. *Plant Physiology and Biochemistry* **30**, 435–443.
Hallahan, D. L., Nugent, J. H. A., Hallahan, B. J., Dawson, G. W., Smiley, D. W., West,

J. M. and Wallsgrove, R. M. (1992b). Interactions of avocado (*Persea americana*) cytochrome P450 with monoterpenoids. *Plant Physiology* **98**, 1290–1297.

Hallahan, D. L., Lau, S. M. C., Harder, P. A., Smiley, D. W. M., Dawson, G. W., Pickett, J. A., Christoffersen, R. E. and O'Keefe, D. P. (1994). Cytochrome P450-catalyzed monoterpenoid oxidation in catmint (*Nepeta racemosa*) and avocado (*Persea americana*) – evidence for related enzymes with different activities. *Biochimica et Biophysica Acta* **1201**, 94–100.

Hallahan, D. L., West, J. M., Wallsgrove, R. M., Smiley, D. W. M., Dawson, G. W., Pickett, J. A. and Hamilton, J. G. C. (1995). Purification and characterization of an acyclic monoterpene primary alcohol-$NADP^+$ oxidoreductase from catmint (*Nepeta racemosa*). *Archives of Biochemistry and Biophysics* **318**, 105–112.

Hallahan, D. L., West, J. M., Smiley, D. W. M. and Pickett, J. A. (1998). Nepetalactol oxidoreductase in trichomes of the catmint *Nepeta racemosa*. *Phytochemistry* **48**, 421–427.

Harborne, J. B. (1991). Recent advances in the ecological chemistry of plant terpenoids. *In* "Ecological Chemistry and Biochemistry of Plant Terpenoids", Vol. 31. Proceedings of the Phytochemical Society of Europe (J. B. Harborne and F. A. Tomas-Barberan, eds), pp. 399–426. Clarendon Press, Oxford.

Hardie, J., Hick, A. J., Holler, C., Mann, J., Merritt, L., Nottingham, S. F., Powell, W., Wadhams, L. J., Witthinrich, J. and Wright, A. F. (1994). The responses of *Praon* spp parasitoids to aphid sex-pheromone components in the field. *Entomologia Experimentalis et Applicata* **71**, 95–99.

Harrewijn, P., Minks, A. K. and Mollema, C. (1995). Evolution of plant volatile production in insect-plant relationships. *Chemoecology* **5/6**, 55–73.

Haudenschild, C., Lupien, S., Karp, F. and Croteau, R. (1997). Expression in *Escherichia coli* and *Saccharomyces cerevisiae* of the cytochrome P-450 limonene hydroxylase genes from *Mentha* species, either alone or as a fusion protein with the homologous NADPH cytochrome P-450 reductase. *FASEB Journal* **11**, 124–P124.

Hedge, I. C. (1992). A global survey of the biogeography of the Labiatae. *In* "Advances in Labiate Science" (R. M. Harley and T. Reynolds, eds), pp. 7–17. The Royal Botanic Gardens, Kew.

Henderson, W., Hart, J. W., How, P. and Judge, J. (1970). Chemical and morphological studies on sites of sesquiterpene accumulation in *Pogostemon cablin* (Patchouli). *Phytochemistry* **9**, 1219–1228.

Hose, S., Zanglein, A., Van Den Berg, T., Schultze, W., Kubeczka, K.-H. and Czygan, F.-C. (1997). Ontogenetic variation of the essential oil of *Melissa officinalis* L. *Pharmazie* **52**, 247–253.

Ikeda, H., Esaki, N., Nakai, S., Hashimoto, K., Uesato, S., Soda, K. and Fujita, T. (1991). Acyclic monoterpene primary alcohol:$NADP^+$ oxidoreductase of *Rauwolfia serpentina* cells: the key enzyme in biosynthesis of monoterpene alcohols. *Journal of Biochemistry* **109**, 341–347.

Inouye, H. (1991). Iridoids. *Methods in Plant Biochemistry* **7**, 99–143.

Karban, R., Agrawal, A. A. and Mangel, M. (1997). The benefits of induced defenses against herbivores. *Ecology* **78**, 1351–1355.

Karp, F., Mihaliak, C. A., Harris, J. L. and Croteau, R. (1990). Monoterpene biosynthesis – specificity of the hydroxylations of (−)-limonene by enzyme preparations from peppermint (*Mentha piperita*), spearmint (*Mentha spicata*), and perilla (*Perilla frutescens*) leaves. *Archives of Biochemistry and Biophysics* **276**, 219–226.

Kelsey, R. G., Reynolds, G. W. and Rodriguez, E. (1984). The chemistry of biologically active constituents stored in plant glandular trichomes. *In* "Biology and Chemistry of Plant Trichomes" (E. Rodriguez, P. L. Healey and I. Mehta, eds), pp. 187–241. Plenum Press, New York.

Kjonaas, R., Martinkustaylor, C. and Croteau, R. (1982). Metabolism of monoterpenes – conversion of l-menthone to l-menthol and d-neomenthol by stereospecific dehydrogenases from peppermint (*Mentha piperita*) leaves. *Plant Physiology* **69**, 1013–1017.

Kjonaas, R. B., Venkatachalam, K. V. and Croteau, R. (1985). Metabolism of monoterpenes – oxidation of isopiperitenol to isopiperitenone, and subsequent isomerization to piperitenone by soluble enzyme preparations from peppermint (*Mentha piperita*) leaves. *Archives of Biochemistry and Biophysics* **238**, 49–60.

Kleinig, H. (1989). The role of plastids in isoprenoid biosynthesis. *Annual Review of Plant Physiology and Plant Molecular Biology* **40**, 39–59.

Kreuz, K. and Kleinig, H. (1981). On the compartmentation of isopentenyl diphosphate synthesis and utilization in plant cells. *Planta* **153**, 578–581.

Lange, B. M., Wildung, M. R., McCaskill, D. and Croteau, R. (1998). A family of transketolases that directs isoprenoid biosynthesis via a mevalonate-independent pathway. *Proceedings of the National Academy of Sciences USA* **95**, 2100–2104.

Lawrence, B. M. (1992). Chemical components of Labiate oils and their exploitation. *In* "Advances in Labiate Science" (R. M. Harley and T. Reynolds, eds), pp. 399–436. The Royal Botanic Gardens, Kew.

Lawrence, B. M. (1993). Labiate oils – mother nature's chemical factory. *In* "Essential Oils 1988–1991", pp. 188–206. Allured Publishing Corporation, Carol Stream, IL.

Lawrence, B. M. (1995). "Essential Oils 1992–1994." Allured Publishing Corporation, Carol Stream, IL.

Lichtenthaler, H. K., Rohmer, M. and Schwender, J. (1997). Two independent biochemical pathways for isopentenyl diphosphate and isoprenoid biosynthesis in higher plants. *Physiologia Plantarum* **101**, 643–652.

Loomis, W. D. (1967). Biosynthesis and metabolism of monoterpenes. *In* "Terpenoids in Plants" (J. B. Pridham, ed.), pp. 59–82. Academic Press, New York.

Lorenz, M., Boland, W. and Dettner, K. (1993). Biosynthesis of iridodials in the defense glands of beetle larvae (Chrysomelinae). *Angewandte Chemie* (International Edition in English) **32**, 912–914.

Lupien, S., Karp, F., Ponnamperuma, K., Wildung, M. and Croteau, R. (1995). Cytochrome P450 limonene hydroxylases of *Mentha* species. *Drug Metabolism and Drug Interactions* **12**, 245–260.

Mabberley, D. J. (1987). "The Plant Book." Cambridge University Press, Cambridge.

Madyastha, K. M., Meehan, T. D. and Coscia, C. J. (1976). Characterization of a cytochrome P450 dependent monoterpene hydroxylase from the higher plant *Vinca rosea*. *Biochemistry* **15**, 1097–1102.

Maffei, M., Chialva, F. and Sacco, T. (1989). Glandular trichomes and essential oils in developing peppermint leaves. I Variation of peltate trichome number and terpene distribution within leaves. *New Phytologist* **111**, 707–716.

Mann, J., Davidson, R. S., Hobbs, J. B., Banthorpe, D. V. and Harborne, J. B. (1994). "Natural Products: their Chemistry and Biological Significance." Longman Scientific & Technical, London.

Mauricio, R. and Rausher, M. D. (1997). Experimental manipulation of putative selective agents provides evidence for the role of natural enemies in the evolution of plant defense. *Evolution* **51**, 1435–1444.

McCabe, P. F., Levine, A., Meijer, P.-J., Tapon, N. A. and Pennell, R. I. (1997). A programmed cell death pathway activated in carrot cells cultured at low density. *Plant Journal* **12**, 267–280.

McCaskill, D. and Croteau, R. (1995). Monoterpene and sesquiterpene biosynthesis in glandular trichomes of peppermint (*Mentha* × *piperita*) rely exclusively on plastid-derived isopentenyl diphosphate. *Planta* **197**, 49–56.

McCaskill, D., Gershenzon, J. and Croteau, R. (1992). Morphology and monoterpene biosynthetic capabilities of secretory cell clusters isolated from glandular trichomes of peppermint (*Mentha piperita* L). *Planta* **187**, 445–454.

McElvain, S. M., Walters, P. M. and Bright, R. D. (1942). The constituents of the volatile oil of catnip. II. The neutral components. Nepetalic anhydride. *Journal of the American Chemical Society* **64**, 1828–1831.

McGarvey, D. J. and Croteau, R. (1995). Terpenoid metabolism. *Plant Cell* **7**, 1015–1026.

Meehan, T. D. and Coscia, C. J. (1973). Hydroxylation of geraniol and nerol by a monooxygenase from *Vinca rosea*. *Biochemical and Biophysical Research Communications* **53**, 1043–1048.

Meijer, A. H., DeWaal, A. and Verpoorte, R. (1993). Purification of the cytochrome P450 enzyme geraniol 10-hydroxylase from cell cultures of *Catharanthus roseus*. *Journal of Chromatography* **635**, 237–249.

Meou, A., Bouanah, N., Archelas, A., Zhang, X. M., Guglielmentti, R. and Furstoss, R. (1990). Synthesis of all four stereoisomers of enantiomerically pure *cis*- and *trans*-linalyl oxides. *Synthesis*, 752–753.

Mihaliak, C. A., Karp, F. and Croteau, R. (1993). Cytochrome P450 terpene hydroxylases. *Methods in Plant Biochemistry* **9**, 261–279.

Mills, J. J., Chari, R. S., Boyer, I. J., Gould, M. N. and Jirtle, R. L. (1995). Induction of apoptosis in liver tumors by the monoterpene perillyl alcohol. *Cancer Research* **55**, 979–983.

Mitchell, E. D., Downing, M. and Griffith, G. R. (1972). $^{14}CO_2$ incorporation into nepetalactone. *Phytochemistry* **11**, 3193–3194.

Mizutani, M., Ward, E. and Ohta, D. (1998). Cytochrome P450 superfamily in *Arabidopsis thaliana*: isolation of cDNAs, differential expression, and RFLP mapping of multiple cytochromes P450. *Plant Molecular Biology* **37**, 39–52.

Mookherjee, B. D., Wilson, R. A., Schrankel, K. R., Katz, I. and Butler, J. F. (1992). Semio activity of flavor and fragrance molecules on various insect species. *In* "Bioactive Volatile Compounds from Plants", Vol. 525. ACS Symposium Series (R. Teranishi, R. G. Buttery and H. Sugisawa, eds), pp. 35–48. American Chemical Society, Washington, DC.

Moyler, D. (1991). Natural ingredients for flavours and fragrances. *Chemistry and Industry* 11–14.

Nielsen, M. T., Akers, C. P., Jarlfors, U. E., Wagner, G. J. and Berger, S. (1991). Comparative ultrastructural features of secreting and nonsecreting glandular trichomes of 2 genotypes of *Nicotiana tabacum* L. *Botanical Gazette* **152**, 13–22.

Niu, X., Lin, K., Hasegawa, P. M., Bressan, R. A. and Weller, S. C. (1998). Transgenic peppermint (*Mentha* × *piperita* L.) plants obtained by cocultivation with *Agrobacterium tumefaciens*. *Plant Cell Reports* **17**, 165–171.

Ohloff, G. (1995). "Scent and Fragrances: the Fascination of Odors and their Chemical Perspectives." Springer-Verlag, Berlin.

Ohta, D. and Mizutani, M. (1998). Plant geraniol/nerol hydroxylase and DNA coding therefor. US patent 5,753,507.

Oldham, N. J., Veith, M., Boland, W. and Dettner, K. (1996). Iridoid monoterpene biosynthesis in insects – evidence for a de-novo pathway occurring in the

defensive glands of *Phaedon armoraciae* (chrysomelidae) leaf beetle larvae. *Naturwissenschaften* **83**, 470–473.
Pellegrineschi, A., Damon, J.-P., Valtorta, N., Paillard, N. and Tepfer, D. (1994). Improvement of ornamental characters and fragrance production in lemon-scented geranium through genetic transformation by *Agrobacterium rhizogenes*. *Bio/Technology* **12**, 64–68.
Pennell, R. I. and Lamb, C. (1997). Programmed cell death in plants. *Plant Cell* **9**, 1157–1168.
Piccaglia, R. and Marotti, M. (1993). Characterization of several aromatic plants grown in northern Italy. *Flavour and Fragrance Journal* **8**, 115–122.
Pichersky, E., Lewinsohn, E. and Croteau, R. (1995). Purification and characterization of *S*-linalool synthase, an enzyme involved in the production of floral scent in *Clarkia breweri*. *Archives of Biochemistry and Biophysics* **316**, 803–807.
Pickett, J. A., Wadhams, L. J., Woodcock, C. M. and Hardie, J. (1992). The chemical ecology of aphids. *Annual Review of Entomology* **37**, 67–90.
Poppy, G. M., Powell, W. and Pennacchio, F. (1997). Aphid parasitoid responses to semiochemicals – genetic, conditioned or learnt? *Entomophaga* **42**, 193–199.
Powell, W., Hardie, J., Hick, A. J., Holler, C., Mann, J., Merritt, L., Nottingham, S. F., Wadhams, L. J., Witthinrich, J. and Wright, A. F. (1993). Responses of the parasitoid *Praon volucre* (Hymenoptera, Braconidae) to aphid sex-pheromone lures in cereal fields in autumn – implications for parasitoid manipulation. *European Journal of Entomology* **90**, 435–438.
Powell, W., Pennacchio, F., Poppy, G. M. and Tremblay, E. (1998). Strategies involved in the location of hosts by the parasitoid *Aphidius ervi* Haliday (Hymenoptera: Braconidae: Aphidiinae). *Biological Control* **11**, 104–112.
Pyee, J. and Kolattukudy, P. E. (1995). The gene for the major cuticular wax-associated protein and three homologous genes from broccoli (*Brassica oleracea*) and their expression patterns. *Plant Journal* **7**, 49–59.
Rajaonarivony, J. I. M., Gershenzon, J. and Croteau, R. (1992). Characterization and mechanism of (4*S*)-limonene synthase, a monoterpene cyclase from the glandular trichomes of peppermint (*Mentha* × *piperita*). *Archives of Biochemistry and Biophysics* **296**, 49–57.
Ravid, U. and Putievsky, E. (1983). Constituents of essential oils from *Majorana syriaca*, *Coridothymus capitatus* and *Satureja thymbra*. *Planta Medica* **49**, 248–249.
Regnier, F. E., Waller, G. R. and Eisenbraun, E. J. (1967). Studies on the composition of the essential oils of three *Nepeta* species. *Phytochemistry* **6**, 1281–1289.
Regnier, F. E., Waller, G. R., Eisenbraun, E. J. and Auda, H. (1968). The biosynthesis of metylcyclopentane monoterpenoids – II. Nepetalactone. *Phytochemistry* **7**, 221–230.
Rhodes, M. J. C. (1994). Physiological roles for secondary metabolites in plants – some progress, many outstanding problems. *Plant Molecular Biology* **24**, 1–20.
Rimpler, H. (1991). Sequestration of iridoids by insects. *In* "Ecological Chemistry and Biochemistry of Plant Terpenoids", Vol. 31. Proceedings of the Phytochemical Society of Europe (J. B. Harborne and F. A. Tomas-Barberan, eds), pp. 314–330. Clarendon Press, Oxford.
Rogers, L. J., Shah, S. P. J. and Goodwin, T. W. (1968). Compartmentation of biosynthesis of terpenoids in green plants. *Photosynthetica* **2**, 184–207.
Ruzicka, L. (1953). The isoprene rule and the biogenesis of terpenic compounds. *Experientia* **9**, 357–367.
Sagner, S., Latzel, C., Eisenreich, W., Bacher, A. and Zenk, M. H. (1998). Differential incorporation of 1-deoxy-D-xylulose into monoterpenes and carotenoids in

higher plants. *Chemical Communications* 221–222.
Sanchezgras, M. C. and Segura, J. (1988). Regeneration of *Sideritis angustifolia* (Labiatae) plants from single cell cultures. *Journal of Plant Physiology* **133**, 116–118.
Sanchezgras, M. C. and Segura, J. (1989). Factors affecting plant regeneration from leaf explants of *Sideritis angustifolia* Lag (Labiatae) cultured *in vitro*. *Gartenbauwissenschaft* **54**, 90–93.
Sato, H., Enomoto, S., Oka, S., Hosomi, K. and Ito, Y. (1993). Plant-regeneration from protoplasts of peppermint (*Mentha piperita* L). *Plant Cell Reports* **12**, 546–550.
Schuler, M. A. (1996). Plant cytochrome P450 monooxygenases. *Critical Reviews in Plant Sciences* **15**, 235–284.
Serrato-Valenti, G., Bisio, A., Cornara, L. and Ciarallo, G. (1997). Structural and histochemical investigation of the glandular trichomes of *Salvia aurea* L. leaves, and chemical analysis of the essential oil. *Annals of Botany* **79**, 329–336.
Shu, C. K. and Lawrence, B. M. (1997). Reasons for the variation in composition of some commercial essential oils. *ACS Symposium Series* **660**, 138–159.
Spencer, A., Hamill, J. D. and Rhodes, M. J. C. (1990). Production of terpenes by differentiated shoot cultures of *Mentha citrata* transformed with *Agrobacterium tumefaciens* T37. *Plant Cell Reports* **8**, 601–604.
Spencer, A., Hamill, J. D. and Rhodes, M. J. (1993). *In vitro* biosynthesis of monoterpenes by Agrobacterium transformed shoot cultures of two *Mentha* species. *Phytochemistry* **32**, 911–919.
Steele, C. L., Katoh, S., Bohlmann, J. and Croteau, R. (1998). Regulation of oleoresinosis in grand fir (*Abies grandis*) – differential transcriptional control of monoterpene, sesquiterpene, and diterpene synthase genes in response to wounding. *Plant Physiology* **116**, 1497–1504.
Uesato, S., Ogawa, Y., Inouye, H., Saiki, S. and Zenk, M. H. (1986a). Synthesis of iridodial by cell free extracts from *Rauwolfia serpentina* cell suspension cultures. *Tetrahedron Letters* **27**, 2893–2896.
Uesato, S., Ogawa, Y., Inouye, H. and Zenk, M. H. (1986b). Mechanism for iridane skeleton formation in the biosynthesis of secologanin and indole alkaloids in *Lonicera tatarica*, *Catharanthus roseus* and suspension cultures of *Rauwolfia serpentina*. *Phytochemistry* **25**, 839–842.
Uesato, S., Ikeda, H., Fujita, T., Inouye, H. and Zenk, M. H. (1987). Elucidation of iridodial formation mechanism – partial purification and characterization of the novel monoterpene cyclase from *Rauwolfia serpentina* cell suspension cultures. *Tetrahedron Letters* **28**, 4431–4434.
Uphof, J. C. T. (1962). "Plant Hairs", 2nd edn. Gebruder Borntraeger, Berlin.
Vandenberg, T., Aboumandour, A. A. and Czygan, F. C. (1995). Regeneration of plants and production of volatiles from callus-cultures of *Melissa officinalis* L. 2. Root cultures – growth and accumulation of volatiles. *Journal of Applied Botany – Angewandte Botanik* **69**, 140–144.
Vaneck, J. M. and Kitto, S. L. (1990). Callus initiation and regeneration in mentha. *Hortscience* **25**, 804–806.
Veith, M., Lorenz, M., Boland, W., Simon, H. and Dettner, K. (1994). Biosynthesis of iridoid monoterpenes in insects – defensive secretions from larvae of leaf beetles (coleoptera, chrysomelidae). *Tetrahedron* **50**, 6859–6874.
Vermeer, J. and Peterson, R. L. (1979). Glandular trichomes on the inflorescence of *Chrysanthemum morifolium* cv. Dramatic (Compositae). II. Ultrastructure and histochemistry. *Canadian Journal of Botany* **57**, 714–729.
Voirin, B. and Bayet, C. (1996). Developmental changes in the monoterpene

composition of *Mentha × piperita* leaves from individual peltate trichomes. *Phytochemistry* **43**, 573–580.
Weiss, E. A. (1997). "Essential Oil Crops." CAB International, Wallingford.
Werker, E. (1993). Function of essential oil-secreting glandular hairs in aromatic plants of the Lamiaceae – a review. *Flavour and Fragrance Journal* **8**, 249–255.
Werker, E., Putievsky, E., Ravid, U., Dudai, N. and Katzir, I. (1993). Glandular hairs and essential oil in developing leaves of *Ocimum basilicum* L. (lamiaceae). *Annals of Botany* **71**, 43–50.
Werker, E., Ravid, U. and Putievsky, E. (1985). Structure of glandular hairs and identification of the main components of their secreted material in some species of the labiatae. *Israel Journal of Botany* **34**, 31–45.
Williams, D. G. (1996). "The Chemistry of Essential Oils: an Introduction for Aromatherapists, Beauticians, Retailers and Students." Micelle Press, Weymouth.
Wise, M. L., Savage, T. J., Katahira, E. and Croteau, R. (1998). Monoterpene synthases from common sage (*Salvia officinalis*) – cDNA isolation, characterization, and functional expression of (+)-sabinene synthase, 1,8-cineole synthase, and (+)-bornyl diphosphate synthase. *Journal of Biological Chemistry* **273**, 14 891–14 899.
Yerger, E. H., Grazzini, R. A., Hesk, D., Cox-Foster, D. L., Craig, R. and Mumma, R. O. (1992). A rapid method for isolating glandular trichomes. *Plant Physiology* **99**, 1–7.
Yuba, A., Honda, G., Koezuka, Y. and Tabata, M. (1995). Genetic analysis of essential oil variants in *Perilla frutescens*. *Biochemical Genetics* **33**, 341–348.
Yuba, A., Yazaki, K., Tabata, M., Honda, G. and Croteau, R. (1996). cDNA cloning, characterization, and functional expression of 4s-(−)-limonene synthase from *Perilla frutescens*. *Archives of Biochemistry and Biophysics* **332**, 280–287.
Zenk, M. H. (1991). Chasing the enzymes of secondary metabolism – plant cell cultures as a pot of gold. *Phytochemistry* **30**, 3861–3863.

Current and Potential Exploitation of Plant Glandular Trichome Productivity

S. O. DUKE,[1] C. CANEL,[1] A. M. RIMANDO,[1] M. R. TELLEZ,[1]
M. V. DUKE[2] and R. N. PAUL[2]

*[1]USDA, ARS, Natural Products Utilization Research Unit,
P.O. Box 8048, University, Mississippi 38677, USA*
*[2]USDA, ARS, Southern Weed Science Research Unit,
P.O. Box 350, Stoneville, Mississippi 38776, USA*

I. INTRODUCTION

Plants produce a large number of secondary compounds that are utilized as pharmaceuticals, nutriceuticals, natural pesticides, flavourings, fragrances, and for other non-food or fibre purposes. These compounds have diverse biological activities, presumably as a result of the co-evolution of the

Advances in Botanical Research Vol. 31
ISBN 0-12-005931-2

producing plants with pathogens, herbivores, pollinators and other organisms (Duke, 1994). To maximize interaction with the biotic world, the compounds are often produced on or near the plant surface. Many of them are contained in specialized cells of glandular trichomes, where bioactive natural products can be concentrated for maximum effect and sequestered to avoid autotoxicity. Thus, in many cases, the contents of glandular trichomes of plants have significant value to humans.

Great interest exists in improving the production of high-value plant secondary products (e.g. Ferreira and Duke, 1997), many of which are trichome localized. This chapter discusses botanical, genetic, chemical, biochemical and physiological aspects of the glandular trichomes with respect to how they can be used to improve the production of trichome-synthesized compounds. We begin by summarizing what is known about trichome localization of certain secondary products and its implications for biosynthetic production of these compounds. This is followed by a discussion of the utility of some of the contents of glandular trichomes. The present methods of production and the use of new biotechnologies to improve production of trichome products are then reviewed.

II. TRICHOME LOCALIZATION OF SECONDARY PRODUCTS

A. TRICHOMES AS SITES OF STORAGE

Plant species that have glandular trichomes generally produce relatively large amounts of bioactive secondary products. The generally multicellular protuberances appear to be the primary production sites of many of these compounds. However, there are few cases, such as those of certain terpenes of tobacco (Wagner, 1991), cotton (Bell *et al.*, 1987) and *Artemisia annua* (Duke *et al.*, 1994; Tellez *et al.*, 1999), in which the trichome has clearly been demonstrated to be the exclusive site of accumulation. Some other studies have implied that essential oil components of plants with glandular trichomes are localized on trichomes (e.g. Serrato-Valenti *et al.*, 1997); however, our recent work has demonstrated that, within the same plant, some are and others are not (Tellez *et al.*, 1999).

Peltate glandular trichomes of many plant species are characterized by the tumescent, globular appearance of the cuticle that has split from the secretory cell walls as the subcuticular space engorges with secretory product (e.g. Duke and Paul, 1993; Bourett *et al.*, 1994; Ascensão *et al.*, 1995, 1997; Ferreira and Janick, 1995; Serrato-Valenti *et al.*, 1997). The epidermal localization of trichomes and the subcuticular storage of compounds of interest allow for three approaches to selective extraction: careful removal of the material from the glands *in situ* with a micropipette, isolation of the glands from the plant, and selective solvent extraction.

The micropipette method was used to compare the contents of the glands with surrounding cells of cotton, revealing that cells surrounding secretory glands were devoid of volatile terpenes in which the glands were quite enriched (Elzen *et al.*, 1985). Voirin *et al.* (1993) used a glass capillary to examine the oil from different glands of *Mentha* × *piperita*. A special approach was developed to sample directly the subcuticular cavity of the peltate glandular trichomes of *Nepeta racemosa* (Clark *et al.*, 1997). Analysis of the gland samples by gas chromatography/mass spectroscopy (GC-MS) confirmed the accumulation of the monoterpenoid nepetalactones in the subcuticular cavity and indicated that the accumulation probably occurs exclusively within this type of gland. Using trichome microsampling and high-pressure liquid chromatography (HPLC) analysis, sesquiterpene lactones and flavonoids were identified in the capitate glandular trichomes from the leaf surface of various *Scalesia* species (Spring *et al.*, 1997). Micropreparative techniques, including liquid chromatography/nuclear magnetic resonance spectroscopy (LC-NMR) and liquid chromatography/mass spectrometry (LC-MS), were used to elucidate the structure of the compounds.

Isolation of glandular trichomes from other plant tissues has been achieved through abrasion with glass beads for peppermint, spearmint, garden sage, American wormseed, tansy and wild tomato (Gershenzon *et al.*, 1992), through the use of adhesive tape for thyme (Yamaura *et al.*, 1992), by density gradient centrifugation of homogenized tissue for sagebrush (Slone and Kelsey, 1985), by adhesion of sticky glands of tobacco on touching with a cover glass (Keene and Wagner, 1985), and through removal of the eight secretory cells and extracellular subcuticular space from the stalk cell with a microfeather for *Mentha* × *piperita* (Voirin and Bayet, 1996). Duke *et al.* (1994) found that dipping *Artemisia annua* tissues into chloroform for a few seconds selectively removed certain terpene contents of the plant's peltate glands, without causing significant cellular or structural damage other than the collapse of the cuticle covering the stalk cells of the glands. Further extraction of the entire plant tissues yielded only trace amounts of these products, indicating that they were localized exclusively in the glands. Voirin *et al.* (1993) used a similar technique, in which quick ether washes were used to extract the content of manually preperforated glands while causing little damage to non-perforated glands. This study demonstrated that flavonoid aglycones accumulated in glands of *Mentha* × *piperita*.

Discovery of a mutant of *Artemisia annua* lacking glandular trichomes has provided further evidence that many terpenoids accumulate only in the trichome (Duke *et al.*, 1994; Tellez *et al.*, 1999). This mutant or "chemotype" was discovered due to a lack of aroma and almost complete absence of essential oils. It has normal gross morphology and anatomy, but microscopic

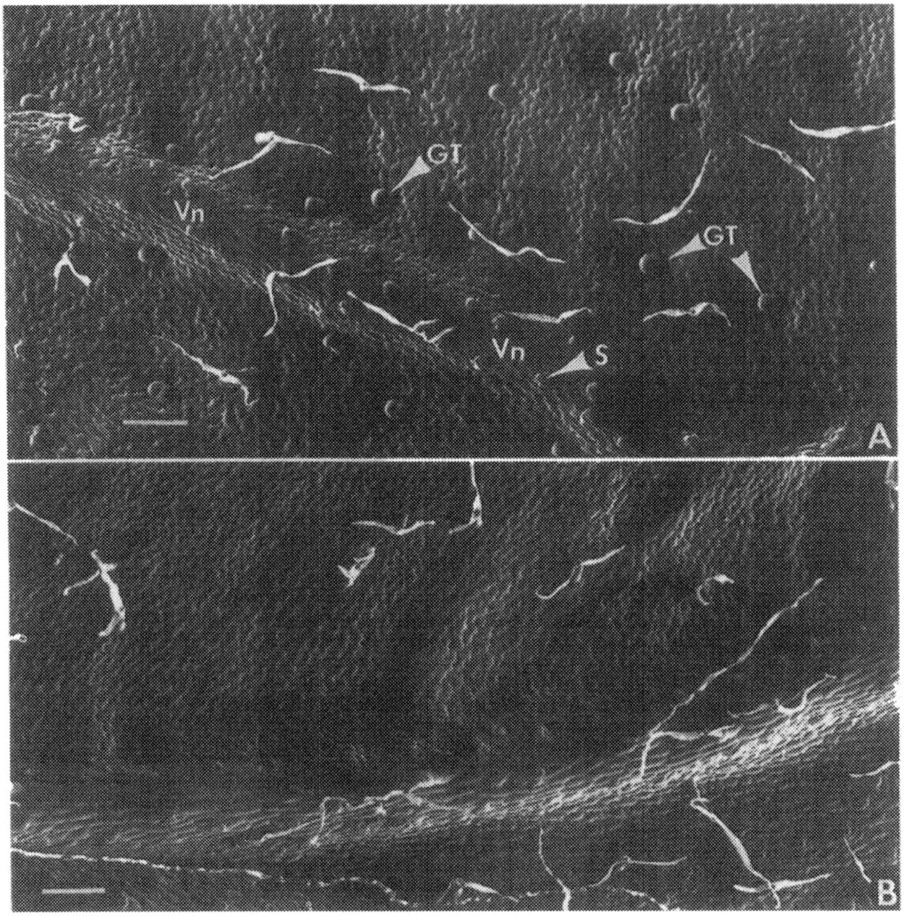

Fig. 1. Scanning electron micrographs of the leaf surfaces of glanded (A) and glandless (B) chemotypes of *Artemisia annua*. Vn, vein; GT, glandular trichome; S, stomata. Scale bars = 100 μm.

examination revealed that it lacks all glandular trichomes (Fig. 1). It completely lacks several sesquiterpenes and most of the monoterpenes found in the glanded wild type (Table I). Several of the compounds (e.g. artemisinin) missing from the glandless mutant are potent phytotoxins. These findings indicate that, in addition to being the exclusive site of storage of most of the monoterpenes and some of the sesquiterpenes of this species, its glandular trichomes may be required to protect the plant from the autotoxic effects of the phytotoxic terpenes.

Cotton possesses glandular trichomes that contain various terpenoids, including gossypol, hemigossypol and a highly visible red pigment, gossy-

TABLE I
The presence of selected terpenoids in glanded and absence in glandless chemotypes of Artemisia annua. *(Data from Duke* et al.*, 1994, and Tellez* et al.*, 1999.)*

Compound	Glanded	Glandless
Artemisinin	+	–
Artemisitene	+	–
Artemisia alcohol	+	–
Artemisia ketone	+	–
Borneol	+	–
α-Campholenal	+	–
1,8-Cineole	+	–
p-Cymene	+	–
Myrtenal	+	–
Myrtenol	+	–
α- and β-pinene	+	–
trans-Pinocarveol	+	–
Pinocarvone	+	–
α-Terpinene	+	–
Terpin-4-ol	+	–

rubilone. Glandless varieties of cotton completely lack gossypol, hemigossypol, and related sesquiterpene and triterpene compounds, as well as gossyrubilone (Bell *et al.*, 1987). Furthermore, most of the volatile terpenes (mostly monoterpenes) of cotton were not found or found in only trace amounts in glandless cotton varieties, compared to genetically similar glanded varieties (Elzen *et al.*, 1985). These are the only published studies of which we are aware containing genetic evidence for the exclusive glandular trichome localization of secondary products.

Older peltate glands often lose their integrity, resulting in loss of secondary product compartmentation. The cuticular covering of the stored secretory product eventually splits, resulting in volatilization of many of the compounds and spreading of others over the plant surface (e.g. Duke and Paul, 1993). Some non-peltate, glandular trichomes exude or secrete secondary products on to the plant surface rather than storing them, e.g. the capitate glands of *Leonatis leonurus* described by Ascensão and Pais (1998).

Vacuoles of cells of glandular trichomes can also be primary sites of storage for compounds of interest. For example, trichomes of certain species of *Hypericum* (Fig. 2) are peltate in superficial appearance but have no subcuticular space filled with secretory products. The anatomy of these structures can vary from a group of slightly protruding cells on the epidermal surface to a stalked, capitate trichome. Electron microscopy shows no subcuticular space, but rather central cells with vacuoles filled with osmiophilic substances (Liszt *et al.*, 1982). We have evidence that the highly biologically active compound hypericin, a polycyclic napthodianthrone, is

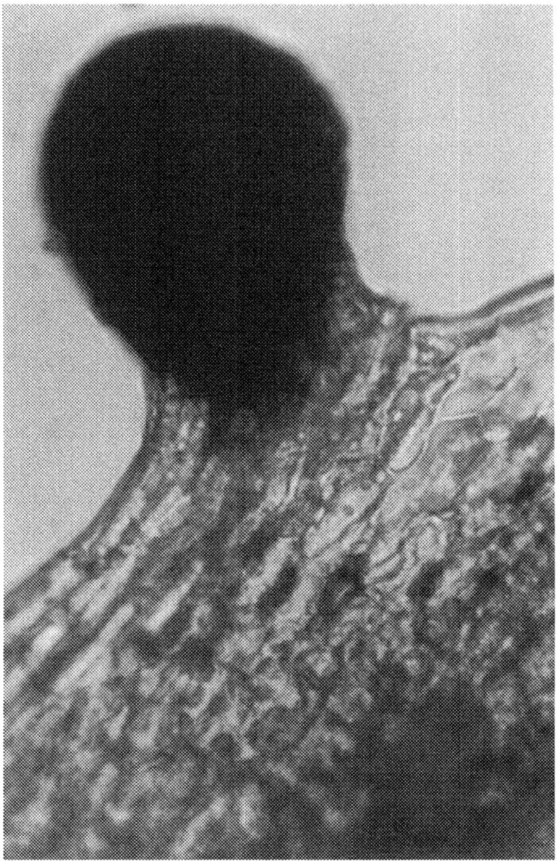

Fig. 2. Light micrograph of a *Hypericum punctatum* trichome.

stored entirely in the vacuoles of the gland cells of *Hypericum punctatum* (M. V. Duke and S. O. Duke, unpublished observations).

These few published studies indicate that relatively little effort has been made to determine the exact cellular location of secondary products in plants. However, the data currently available suggest that many other plant products are found exclusively or almost exclusively in the glandular trichomes of the producing species. This has profound implications for those interested in improving production of these compounds.

B. TRICHOMES AS SITES OF BIOSYNTHESIS

There is little evidence that phenolic or terpenoid secretory products of trichomes are imported by the trichome from other plant tissues. Direct evidence of the biosynthetic competence of glandular trichomes was first found in tobacco. Keene and Wagner (1985) produced compelling evidence indicating that the glandular heads of tobacco trichomes were the only site of biosynthesis of duvatrienediol diterpenes. Cells isolated from these glands were later shown to incorporate radiolabel from carbonate into complex diterpenes (Kandra and Wagner, 1988). Furthermore, biosynthesis of the diterpenes labdenediol and sclareol from *trans*-geranylgeranyl diphosphate was observed in cell-free extracts prepared from leaf midvein epidermal peels of *Nicotiana glutinosa*; these bioactivities were localized exclusively in trichomes (Guo and Wagner, 1995).

The peltate glandular trichomes on the surface of mint leaves are also capable of synthesizing complex isoprenoids. Gershenzon *et al.* (1991, 1992) isolated secretory cells from mint trichomes and showed that these cells were very active in monoterpene biosynthesis and were good sources of key enzymes of monoterpene biosynthesis. The ability of isolated secretory cells of peppermint trichomes to synthesize isoprenoids from the relatively simple glycolytic substrates glucose-6-phosphate and pyruvic acid has been demonstrated (McCaskill and Croteau, 1995). The radiolabel was incorporated into monoterpenes and sesquiterpenes in a ratio closely resembling that observed *in vivo*.

The trichomes of some species appear to be photosynthetically competent, whereas those of others clearly are not, apparently relying completely on underlying tissues as a carbon source. In some species, both cells with chloroplasts and with non-photosynthetic plastids can be found in trichomes. The secretory cells of glandular trichomes of *A. annua* (Duke and Paul, 1993), *Leonotis leonurus* (Ascensão *et al.*, 1997), and tobacco (Neilson *et al.*, 1991) are characterized by specialized, often amoeboid, plastids, including chloroplasts that seldom contain starch. A tobacco genotype without functional plastids in secretory cells produces little, if any, of the terpenoid secretory products of the glands (Neilson *et al.*, 1991). The plastids of gland cells are often enveloped by tubules of endoplasmic reticulum, indicating plastid export of secretory products (e.g. Ascensão *et al.*, 1997). These factors have led to speculation that the secondary terpenoid compounds found in the subcuticular space of peltate glands are synthesized entirely or in part by plastids of trichome cells (Dell and McComb, 1978; Wagner, 1991; Duke *et al.*, 1994). Whether any of the enzymes involved in the biosynthesis of trichome-localized secondary products are plastome encoded is unknown.

The cells of mint and tobacco glands contain numerous plastids, including fully developed chloroplasts in those of the latter. These and other micromorphological features have been implicated in the biosynthesis of secreted products. For example, the synthesis and compartmentation of isoprenoids in pine needles and in glandular trichomes of *Chrysanthemum morifolium* have been linked to the presence of leucoplasts that are in close association with smooth endoplasmic reticulum (Vermeer and Peterson, 1979; Carde *et al.*, 1980).

The biosynthetic capacity of glandular trichomes is not limited to the production of isoprenoids, but also includes that of certain groups of plant phenolics. The detection of the enzymes phenylalanine ammonia lyase and chalcone synthase in the head cells of glands of *Primula kewensis* strongly suggests that these glands possess the capacity to synthesize flavonoids. Chalcone synthase was not found in most other epidermal cells, nor in the glands of a mutant incapable of flavonoid production (Schöpker *et al.*, 1995).

C. IMPLICATIONS

Extraction of natural products sequestered in trichomes has traditionally been accomplished by homogenization of whole tissues (e.g. ElSohly *et al.*, 1990; Vonwiller *et al.*, 1993). Unfortunately, this approach does not consider the anatomy of the plant. Whole plant homogenization to obtain trichome-localized compounds results in a very small ratio of mass of compounds of interest to extracted tissue weight. This approach also increases the probability of undesirable interactions of desired compounds with other plant components, such as enzymatic oxidation (browning) by oxidative enzymes (e.g. polyphenol oxidases and peroxidases). Maceration of tissues breaks cellular compartments, blending enzymes with substrates. Some oxidative enzymes are active in organic solvents. Our finding that all of the artemisinin and one of its precursors, artemisitene, could be extracted by a brief dip in organic solvent to extract only the subcuticular spaces of the glands of *A. annua* (Duke *et al.*, 1994) suggests that extraction of some trichome-localized compounds may be best achieved with a relatively gentle extraction. This rapid and more selective method minimizes the possibility of reaction of trichome-localized compounds with degrading enzymes or other plant constituents. Our studies with fresh leaf tissue of *A. annua* support the view that mixing glandular products with whole plant extractions sometimes reduces yield.

We found with fresh tissue that the dip method consistently yields more artemisinin than a total plant tissue extraction (Duke *et al.*, 1999b). With fresh tissue, homogenization of the tissues yielded only 73.2% of the artemisinin that was extracted by dipping, and only 66.4% of that artemisinin produced by dipping plus homogenization. Dipping alone extracted 90.7% of the amount produced by dipping plus homogenization. This improved extraction efficiency could be found with fresh tissues only, suggesting that intact glands and/or active degradative enzymes may be important in order for effective exploitation of this method. Nevertheless, with a high-value product such as artemisinin, adoption of a method that improves yield by as much as 35% would be cost effective, even if the method is somewhat more involved. We have not compared this method with supercritical carbon dioxide extraction methods, but others have found no advantage between conventional solvent extractions and supercritical carbon dioxide methods (Kohler *et al.*, 1997). Brief extractions with organic solvents without some sort of tissue disruption offered no benefit in our studies with hypericin extraction of the non-peltate, sometimes captitate glandular trichomes of *Hypericum* species (A. M. Rimando and S. O. Duke, unpublished observations).

III. TRICHOME PRODUCTS OF COMMERCIAL INTEREST

A. TERPENES

Kelsey *et al.* (1984) provide a review of compounds found or presumed to be found in glandular trichomes. The secondary products of several biochemical pathways can be found in glandular trichomes of higher plants, although terpenoids seem to predominate and are of primary interest to the chemical industry. Terpenoids and terpenoid derivatives vary from relatively volatile monoterpenes to less volatile sesquiterpene, diterpene and triterpene compounds (Kelsey *et al.*, 1984; Duke *et al.*, 1988; Duke, 1991). Some of these compounds provide the flavours of foliar-derived spices (e.g. oregano) and essential oils used in flavouring (e.g. various mints). There are many culinary, pharmaceutical, pest management and other uses and potential uses of trichome-generated terpenoids. The literature on terpenes from species with glandular trichomes is too extensive to cover in this short review. We will provide a representative example.

From *A. annua* alone, many of the monoterpenes exclusively found in its trichomes (Tellez *et al.*, 1999) have commercial uses, albeit many of them are usually obtained through synthetic routes rather than from natural sources. *A. annua* glandular trichomes contain the following terpenes with commercial uses: *tricyclene* (used in perfumes, plasticizers and intermediates for drugs; Kitano and Tanimoto, 1987); *α-thujene* (used in the synthesis of levocarvone; Jia *et al.*, 1994); *sabinene* (synthesis of perfume components with cosmetic and possible medicinal applications; Retamar *et al.*, 1994); *sabinene hydrate* (as an effective insecticide for cloth; Sato *et al.*, 1979), *1,8-cineole* (as a drug penetration enhancer – Zupan, 1983; Ping *et al.*, 1993; Yamane *et al.*, 1995; in the manufacture of drying solvents – Yanagisawa *et al.*, 1993; in the manufacture of homopolymers – Yamawaki and Kondo, 1988; as light-stabilizing agents to malt beverages – Palamand, 1986; as fragrance raw materials – Opdyke, 1975); *α-terpinene* (binder for the manufacture of thick-layer conductors – Ruplenas *et al.*, 1977; component of universal cleaning agent – Christ, 1997); *E-β-ocimene* (for insect control –

deKramer, 1996; as a food additive – Anon., 1968); *terpinolene* (in the manufacture of coating materials – Saito and Watanabe, 1992; Trumbo *et al.*, 1995; in the formulation of detergents – Lazuchiewicz *et al.*, 1992; Fujeda and Kuroda, 1993; as a binder for conducting paste – Trelinska-Wlazlak *et al.*, 1990); *campholenal* (in the preparation of derivatives for use as detergent fragrances – Bajgrowicz and Frater, 1998); *pinocarveol* (as food additive – Anon., 1973; in the formulation of colognes, soaps, detergents, shampoos, deodorants – Mitchell, 1970; in the synthesis of phytotoxic *β*-dialkylaminoalkyl terpene ethers – Bordenca and Derfer, 1968); *pinocarvone* (in the synthesis of perfume or aroma constituents – Becker, 1980; as a component of cleaners and disinfectants – Rigoni and Travi, 1997; Romano *et al.*, 1998); *borneol* (in the synthesis of pharmaceuticals – Klar *et al.*, 1996; for metal and plastic surface protection – Wang, 1993; as a component of deodorants – Hautmann, 1985; as constituents of soothing aromatic ointments – Nistor *et al.*, 1978); *terpin-4-ol* (as antimicrobial – Raman *et al.*, 1995; as an insecticide for preserving clothes – Ishida and Hayashi, 1994); *p-cymen-8-ol* (in the preparation of intermediates for musk fragrances – Meakins and Motion, 1990; as foaming agents for ore flotation – Noguchi *et al.*, 1974); *α-terpineol* (an aroma chemical of value in the food, flavour and fragrance industry – Singh *et al.*, 1998; in the manufacture of annealable printing pastes for printing on glass surfaces – Kalleder *et al.*, 1998; in the manufacture of insect-repellent detergent compositions for use on solid surfaces – Endlein *et al.*, 1998); *myrtenal* (organoleptic characteristics of flavour materials – Mosciano *et al.*, 1990; as kairomones – Camacho *et al.*, 1998); *verbenone* (as honeybee repellents, for repulsion from pesticide-treated areas – Holtmann, 1991; in the manufacture of insect attractants – Repp *et al.*, 1988; in therapeutical use – Vegezzi, 1982); *carveol* (in the manufacture of liquid crystal materials – Coates *et al.*, 1986); *carvone* (in the synthesis of biologically active natural products – Verstegen-Haaksma *et al.*, 1995; use in agriculture – Hartmans *et al.*, 1995; in perfumes or cosmetics – Ishidoya, 1996); *carvacrol* (in epoxy resin curing compositions – Haberlin and Steinhilber, 1968).

Many other trichome-found terpenes are of significant commercial interest, albeit many of them, especially the monoterpenes, are currently best or solely produced through synthetic methods or from turpentine or citrus oil. Examples of terpenes in this latter category are *α*-pinene, *β*-pinene, 3-carene, the *p*-menthadienes, *p*-cymene, limonene, geraniol and nerol (also from citronella oil), linalool, dehydrolinalool, myrcene, 2-carene, (−)-menthol, camphene, 3,7-dimethyl-1,6-octadiene, alloocimene, dipentene, citral (also from lemongrass oil and *Litsea cubeba* oil), *β*-ionone, citronellol, citronellal (also from citronella oil and *Eucalyptus citriodora*), hydroxy- and methoxy-citronellal, myrcenol, dihydromyrcenol, racemic menthol (Howe-Grant, 1985). Verbenol, verbenone, myrtenol, myrtenal, ocimene and camphor from *α*-pinene; (+)-*trans*-chrysanthemic acid from 3-carene and carvone from limonene can also be synthesized, but are usually obtained from their natural sources (Crotch, 1978). Many sesquiterpene and higher terpenes (with the exception of some carotenoids), however, are usually obtained from natural sources for commercial purposes. An example is the antimalarial sesquiterpene artemisinin found exclusively in the trichomes of *A. annua* (Duke *et al.*, 1994). Other sesquiterpenes found in *A. annua*, although not exclusively in the trichome glands, include: *β-caryophyllene* (used as pheromone in the management of beneficial insects – Pickett *et al.*, 1984; as a component of Vitex oil in the treatment of chronic tracheitis – Fan *et al.*, 1983; in the suppression of gushing in beer – Gardner *et al.*, 1973; and in the synthesis of caryophyllene alcohol – Yunes *et al.*, 1967); *farnesene* (used to inhibit potato sprouting – Kristof and Kristof, 1995; for aphid control – Banfield, 1994; as a hormonal insecticide against aphids and acarids – Gut and VanOosten, 1988; and for augmenting or enhancing the aroma or taste of foodstuffs, chewing-gums, medicinal products and toothpastes – Mookherjee *et al.*, 1984); *germacrene D* (used as a constituent in eye-disease pills – Martens-Lobenhoffer *et al.*, 1998; and as a masking substance of attractants for the cerambycid beetle – Yamasaki *et al.*, 1997); *caryophyllene oxide* (has antimutagenicity activity – Lee *et al.*, 1995; used as fragrance raw material – Opdyke and Letizia, 1983; and as food additives – Anon., 1973); and *α-cadinol* (used as an oral antibacterial and an expectorant – Lu, 1980). These examples are only a small sample of glandular trichome-associated terpenes and their uses.

B. OTHER PRODUCTS

Phenolic compounds from the shikimate pathway are quite common constituents of secretory glands, and many of those found in glands have interesting biological activities. At least some of these compounds have a role in plant defence. The vacuoles of cells in other tissues commonly contain much higher levels of phenolic compounds than are found in peltate glandular trichomes. A likely explanation is that most secretory cells have relatively small vacuoles (e.g. Duke and Paul, 1993) compared to those of other tissues and most phenolic compounds are stored primarily in vacuoles. Compared to terpenes, there has been relatively little interest in commercial exploitation of phenolic compounds localized in glandular trichomes.

In some *Solanum* species, phenolic compounds are released by breakage of specialized glands by insects, resulting in rapid oxidation of the phenolics to quinones by polyphenol oxidase released from gland cell plastids (Kowalski *et al.*, 1990). The phenolic quinones polymerize rapidly enough to "glue" insects to the leaf surface or make their mouth parts unusable. Other secondary compounds of glandular trichomes, such as sugar esters of fatty acids (Neal *et al.*, 1990), play a role in combating plant pests.

As mentioned above, the cells of glandular trichomes of some *Hypericum* species contain hypericin in their vacuoles. Hypericin is a high-value pharmaceutical product. It is active as an antiviral agent, including against the human immunodeficiency virus (HIV) (e.g. Lavie *et al.*, 1995) and has been proposed as a pharmaceutical to prevent macular degeneration (Kimura *et al.*, 1997). It has also been proposed as an anticancer drug (e.g. Koren *et al.*, 1991). Crude preparations of *H. perforatum*, standardized by hypericin content (Upton *et al.*, 1997), are the leading medicinal in Europe for the treatment of mild depression. *H. perforatum* contains other biologically active secondary products, such as hyperforin, but whether these compounds are localized on trichomes is unknown.

C. DISCOVERY OF NEW COMPOUNDS AND NEW USES

Within the past few years, a resurgence of interest in botanical sources of new pharmaceutical, nutriceutical, natural pesticides and pest repellents, and other bioactive compounds, has emerged. Glandular trichomes are a particularly lucrative source of such compounds. The aerial surfaces of 20–30% of vascular plant species possess glandular trichomes (Dell and McComb, 1978; Fahn, 1988; Wagner, 1991); however, the glandular contents of few of these species have been explored. Considering the high probability of biological activity of the contents of glandular trichomes, focusing on species with glandular trichomes and the selective extraction of glandular trichomes could be a component of a strategy for discovery of biologically active compounds from higher plants (Duke *et al.*, 1999a,c).

The selection pressures that caused the evolution of secondary biochemical pathways to produce potent chemical defences also apparently resulted in the evolution of glandular trichomes in many plant species. The general cytotoxicity of many of these compounds necessitated concomitant evolution of mechanisms of secretion and sequestration to avoid autotoxicity, and to deliver the compounds to the proper place for interaction with the biotic environment. One can speculate that glandular trichomes are largely the result of this evolutionary interaction between chemistry and structure, in response to biotic selection pressures. Most natural product discovery research has been driven by their chemistry, with relatively little effort to determine potential useful biological activities. Very few studies have examined plant-derived secondary compounds for a wide range of biological activities. Understanding the chemical ecology of trichomes and their contents can provide valuable clues regarding the potential utility of the compounds sequestered within trichomes, thus indicating what bioassays would be most appropriate.

The functions of glandular trichomes and the compounds they produce are not always clear. There is no evidence to suggest that these structures and their contents help the producing plant by providing a direct advantage in plant–plant competition, nor do they appear to give it a direct reproductive advantage. Unlike other types of trichomes, the relatively sparse coverage of mature plant surfaces by glandular trichomes precludes a direct function as a physical impedance to insects or a means of reducing solar heating or evaporative loss. However, some glandular trichomes produce sticky materials that entrap insects and others can secrete sufficient resinous material to coat the plant surface, particularly in immature tissue, thus indirectly performing these functions (Dell and McComb, 1978; Neal *et al.*, 1990).

Secretion of phytotoxins by glandular trichomes could be of secondary benefit to the producing plant in some cases. For example, one of the first compounds demonstrated to play a role in plant–plant allelopathy was 1,8-cineole (Muller and Muller, 1964). Genetic evidence indicates that 1,8-cineole is found almost entirely, if not entirely, in the glandular trichomes of *Artemisia annua* (Tellez *et al.*, 1999). This and other highly phytotoxic terpenes are probably localized in the glandular trichomes of *Salvia* species (Croteau and Johnson, 1984; Kelsey *et al.*, 1984; Serrato-Valenti *et al.*, 1997) that inhibit their competitors with this and other terpenoid compounds (Muller and Muller, 1964; Muller *et al.*, 1968). Such compounds apparently reach competitors through volatilization from ground litter. This delivery system for phytotoxins does not appear to be as efficient as root exudation of allelochemicals, such as sorgoleone secretion by root hairs of certain *Sorghum* species (Weston *et al.*, 1999).

Glandular trichomes often contain potent phytotoxins that probably play no apparent role in plant–plant allelopathy. For example, artemisinin, a sesquiterpenoid antimalarial drug (Klayman, 1985), is highly phytotoxic to most plants, including the producing species, *Artemisia annua* L. (Duke *et al.*, 1987, 1988; Chen and Leather, 1990; Dayan *et al.*, 1999). Nevertheless, *A. annua* has not been reported to be allelopathic towards other plant species in the field, even though mixing dried leaves in soil suppresses growth of other plants (Lydon *et al.*, 1997). On the contrary, scientists have expended considerable effort in attempts to devise weed management methods for commercial monocultures of this drug plant that are highly susceptible to weed competition (Bryson and Croom, 1991). Artemisinin is found only in the subcuticular space of the glands of *A. annua* (Duke *et al.*, 1994), thus sequestered to avoid autotoxicity. The most likely function of the glandular trichomes of *A. annua* and many other species is as repositories of chemicals that defend against insects and microbial pathogens. Such activity is an indicator that other biological activities might be present, such as the antimalarial activity of artemisinin.

Young developing leaves are particularly vulnerable to attack by insects and pathogens. Glandular trichomes develop exceedingly rapidly on leaf primordia (e.g. Duke and Paul, 1993; Bourett *et al.*, 1994). As they mature and become more sparsely distributed over the expanding leaf surface, they

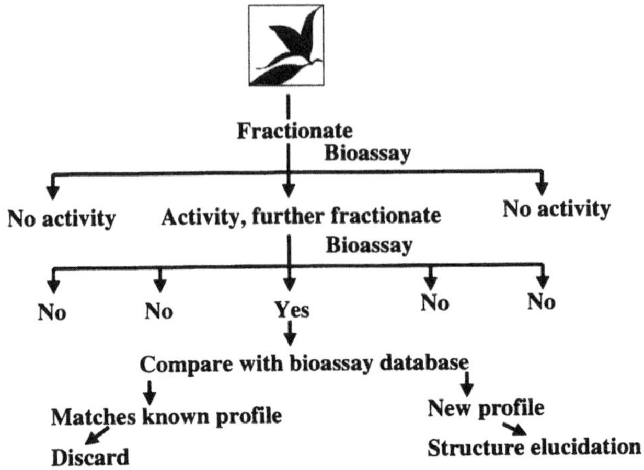

Fig. 3. Bioassay-driven discovery strategy for new phytochemicals.

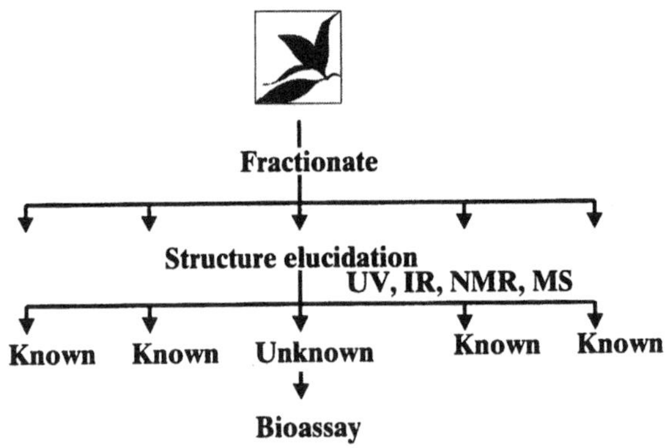

Fig. 4. Instrumentation-driven strategy for new phytochemicals. UV, ultraviolet; IR, infrared; NMR, nuclear magnetic resonance spectroscopy; MS, mass spectroscopy.

may spread their contents on the leaf surface by leakage through the cuticle (e.g. Dell and McComb, 1978; Lin and Wagner, 1994) or by breakage of the cuticle (e.g. Duke and Paul, 1993). Many of the compounds produced by these structures have biological activity that deters damage to the plant by insects and/or plant pathogens (Kelsey *et al.*, 1984). Those compounds protecting against insects have varied types of biological activities, including insecticidal, repellent, behaviour modifying and feeding deterrents. Some of the compounds may play roles in plant–insect interactions other than in protection, such as attraction of pollinators or other beneficial insects.

Many of the compounds produced by glandular trichomes, particularly the terpenoids, are antimicrobial (Kelsey *et al.*, 1984; Duke, 1991). Essential oils containing these compounds have been used for many years as preservatives and disinfectants. It follows that the antifungal and antibacterial properties of these compounds might have some role in plant defences against pathogens. However, constitutive defences of plants against pathogens have not received as much study as inducible defences.

The traditional natural product discovery strategy has been to use bioassay-directed isolation to isolate a compound with the desired activity (Fig. 3). However, with the advent of powerful analytical tandem chromatography/mass spectroscopy or nuclear magnetic resonance instrumentation, more rapid strategies can be used (Fig. 4). With either approach, the appropriate bioassay(s) is crucial.

IV. STRATEGIES FOR IMPROVED PRODUCTION OF TRICHOME PRODUCTS

A. TRADITIONAL METHODS

For many years, the identification of superior varieties and the application of traditional breeding methods have driven the increase in production of trichome-borne phytochemicals and the improvement of the quality of major essential-oil-producing crops such as the mints. Few systematic studies exist of how cultural methods and the environment affect production of trichome-borne compounds. That many compounds of interest are produced largely or exclusively by glandular trichomes has apparently not been considered by those involved in the commercial production of these products.

A recent review of traditional methods to improve production of artemisinin is provided by Laughlin (1994). Artemisinin is an effective antimalarial drug needed to fight *Plasmodium falciparum* strains that have become resistant to the drugs currently used (Klayman, 1985; White, 1994). The plant has been selected and bred for the most productive biotypes and varieties. *A. annua* varieties have been identified that produce about 1% of dry weight of artemisinin (Delabays *et al.*, 1994). Effects of growth regulators, population density, planting date, herbicides and soil characteristics on artemisinin production have been determined. The relative artemisinin yields of different plant parts and the same plant parts at different developmental stages have been studied. All of this has apparently been done without consideration or knowledge of the functioning of the secretory glands in which all of the artemisinin appears to be synthesized.

Another example of this approach is the varieties of *Cannabis sativa* that have been developed, albeit sometimes clandestinely, which produce over 8% of cannabinoids (Lehmann and Brenneisen, 1995). Biotechnological approaches offer new methods to improve on these classical horticultural and breeding methods.

B. CELL AND TISSUE CULTURE

Attempts to produce compounds found in glandular trichomes in undifferentiated plant cell or tissue cultures have been relatively unsuccessful. However, production of secondary products in cell or tissue culture appears to offer potential advantages over agricultural production. These include an absence of pest or weather problems, and the possibility of increasing yield with precisely controlled nutritional, hormonal, elicitor and environmental conditions. Furthermore, inexpensive biosynthetic precursors can sometimes be fed to enhance production of more complicated end products.

In the cases of secretory gland-produced compounds, economical production in such cultures may not be possible unless the mystery of the regulation of the genes controlling synthesis of the compounds is solved. The poor success with artemisinin production in cell and tissue cultures (e.g. Martinez and Staba, 1988; Woerdenbag *et al.*, 1992) is a good example of this problem. Trichome-localized compounds can be produced in shoot cultures that have glands (e.g. Spencer *et al.*, 1990); however, shoot cultures do not offer many of the advantages of cell cultures. Evidence that the enzymes for production of terpenoids localized in peltate glands are found only in gland cells (Croteau and Johnson, 1984; Wagner, 1991; McCaskill *et al.*, 1992; Guo *et al.*, 1994) supports the view that production of trichome-localized compounds in undifferentiated cell or tissue culture is problematic.

An understanding of the regulation of the genes that tie the biosynthesis of these compounds to gland differentiation, including subcellular differentiation of cytoplasm and plastids, may be required to generate these compounds in cell or tissue culture successfully. In tissue cultures, autotoxicity may not be a problem if the phytotoxic compounds can be removed by continuous extraction of media. Continuous removal may also be desirable because cell and tissue cultures often produce enzymes that degrade unsequestered secondary products (e.g. Woerdenbag *et al.*, 1992). Understanding these genes sufficiently to manipulate and transfer them to other organisms could lead to significant advances in the production of compounds of interest by fermentation.

C. TRANSGENICS

The development of efficient vectors and protocols for plant transformation has made it possible to consider the use of recombinant DNA technology to exploit more fully the biosynthetic capacity and plant defence attributes of trichome-bearing plants. The establishment of efficient transformation protocols is an absolute requirement for genetic manipulation. Once limited to a few transformation-friendly Solanaceous species, such as tobacco and tomato, protocols are now available for a large number of plant species. In the past few years, *Agrobacterium*-based protocols have been established and optimized for *Mentha* species and *A. annua* (Vergauwe *et al.*, 1996, 1998; Faure *et al.*, 1998; Niu *et al.*, 1998). The complexity of the manipulation by genetic transgenic methods of genes affecting both trichome formation and secondary product biosynthesis in those trichomes is challenging. The topic of using genetic engineering methods to enhance the quantity and quality of glandular trichome products has been reviewed recently by McCaskill and Croteau (1999).

1. Increased Trichome Number/Density

Whether the increases in productivity obtained by traditional methods are the result of higher trichome densities is not clear. Nevertheless, given the demonstrated metabolic competence of glandular trichomes and the fact that many phytochemicals of interest accumulate in them, it is reasonable to expect that higher production levels will be obtained by increasing the number of these structures.

Most studies of the genetic control of trichome initiation and development have been performed using *Arabidopsis thaliana* as a model, although *Arabidopsis* has only non-glandular trichomes. In *Arabidopsis*, various transcription factors have been shown to direct the implementation of leaf developmental programmes that regulate cell determination and the distribution of specialized cell types, such as stomata and trichomes. Although not yet fully understood, the genetic control of trichome formation in *Arabidopsis* is exerted through the interaction of various transcription factors encoded by the genes *GLABRA1* (*GL1*), *GLABRA2* (*GL2*), *GLABRA3* (*GL3*), *TRANSPARENT TESTA GLABRA* (*TTG*) and *TRIPTYCHON* (*TRY*) (reviewed by Van Lijsebettens and Clarke, 1998, and other authors in this volume).

Mutations in *GL1* and *TTG* block the earliest stages of trichome differentiation, while mutations in *GL2* and *GL3* reduce both trichome number and branching. Overexpression of *GL1*, driven by the constitutive 35S promoter (35S::*GL1*), has shown that only a subset of epidermal cells is competent to respond to the *GL1* signal, indicating that additional factors are required to induce trichome formation (Larkin *et al.*, 1994a). Furthermore, overexpression of *GL1* results in fewer trichomes than in wild-type plants, which suggests that a certain relative ratio of *GL1* to other proteins is required for proper implementation of the developmental plan. Some of these genes affect other aspects of the developmental and metabolic control of the plant. Mutations in *TTG* cause deficiencies in anthocyanin biosynthesis and inhibit the production of seed coat mucilage (Koornneef, 1981). Mutations in *GL1* and *TRY* alter the number of epidermal and mesophyll cells as well as endoreduplication levels in the epidermis, and have therefore been implicated in the control of the cell cycle and cell proliferation (Szymanski *et al.*, 1998).

The interaction among these genes is complex, and shows a level of redundancy suggesting that the plant is capable of exquisitely fine-tuning the distribution of trichomes. In *Arabidopsis*, a mechanism that establishes a minimum distance between trichomes is in place. As a result, the frequency and spacing of initiating trichomes are non-random (Marks, 1997).

The genetic alterations noted above all cause reduced trichome numbers. In contrast, mutations in *TRY* produce "nests" of up to four trichomes in the place of a single trichome (Hülskamp *et al.*, 1994). *Try* mutants lack the lateral inhibition shown by wild-type plants in response to the trichome initiation signal. Such observations indicate that the plant's developmental programme allows a degree of plasticity that can be exploited to increase the production of phytochemicals. Experiments involving the use of the *R* gene, a maize gene encoding a transcriptional regulator, are noteworthy. All aspects of the mutant phenotype of *ttg* plants are complemented by transformation with 35S::*R*. When *R* is overexpressed in *TTG* plants, the number of trichomes increases up to fivefold relative to that found in wild-type plants (Lloyd *et al.*, 1992). In addition, concomitant overexpression of *GL1* and *R* results in the formation of numerous hairs on all shoot epidermal surfaces, including those upon which trichomes do not normally form (Larkin *et al.*, 1994b).

Because of the multifunctionality of some of these gene products, overriding the plant's control over their expression may cause undesirable pleiotropic effects. For example, overexpression of *CPC*, an *Arabidopsis myb* homologue that is involved in epidermal differentiation, causes almost all the epidermal cells of the root to form a root hair, but completely inhibits the formation of trichomes on leaves, stems and sepals (Wada *et al.*, 1997). Furthermore, a line homozygous for 35S::*GL1* and heterozygous for *try* and *cot1*, a mutation in yet another regulator of trichome initiation, produced more cotyledon trichomes than did either parent, but also showed greatly altered leaf shape and epidermal cell morphology (Szymanski *et al.*, 1998). As a result, although perhaps having higher biosynthetic capacity, these plants may suffer from impaired germination and abnormal cell morphology that reduce their viability and fitness.

If regulation of trichome initiation in *Arabidopsis* is in fact an accurate model for plant species of greater economic importance, the prospects for engineering higher productivity by increasing trichome number are promising. Genes with sequence similarity to the *Arabidopsis* regulatory genes mentioned above have been cloned from tomato, snapdragon and other species, but it has not been possible to assign function based on homology alone. A cDNA encoding a homeoprotein homologue with sequence similarity to *GL2* has been cloned from leaves of *A. annua* (C. Canel, unpublished observations). Complementation studies using the *Arabidopsis* mutants may shed light on the function of these genes.

2. *Increased Biosynthetic Capacity of Glandular Trichomes*

Higher trichome productivity may, alternatively or concomitantly, be achieved by increasing the output of specific secondary biosynthetic pathways. Our knowledge of the structure and regulation of the pathways leading to the production of interesting phytochemicals continues to improve. The backbones of major pathways, such as the phenylpropanoid and isoprenoid pathways, are rather well characterized at the enzymatic and genetic levels. Following the lead of flavonoids (reviewed by Forkmann, 1991), the study of which was facilitated by the ability of many of these compounds to absorb visible light, pathways leading to the synthesis of other groups of phenylalanine-derived compounds, such as isoflavonoids (reviewed by Dixon *et al.*, 1995) and lignans (Dinkova-Kostova *et al.*, 1996), have yielded some of their secrets. Of more relevance to trichome-based production, the number of cloned genes encoding biosynthetic enzymes of the isoprenoid pathway continues to rise (Scolnik and Bartley, 1994). Knowledge is sparse, albeit increasing rapidly, about the branches and the specific steps that give rise to isoprenoid metabolites of economic importance (Chappell, 1995; McCaskill and Croteau, 1997, 1999).

The greatest challenge for plant metabolic engineers is to improve the efficiency of photosynthesis and increase the rate of carbon fixation. Until we are able to achieve these goals, all strategies to increase the productivity of secondary metabolic pathways will involve controlling the direction of metabolite fluxes. Overexpression and suppression of individual enzymatic activities or groups thereof both aim at directing the flow of intermediates towards desired end points. An insightful review of the various strategies available for altering metabolite partitioning has been published (Herbers and Sonnewald, 1996).

Because of their influence on flower colour, anthocyanins have long attracted the attention of plant breeders and molecular biologists. Attempts to engineer novel flower colours provided the earliest examples of metabolic engineering in plants, and thus helped devise and test strategies as well as reveal unexpected phenomena, such as sense/antisense suppression (reviewed by Dooner *et al.*, 1991; Courtney-Gutterson, 1994). More recent experiments involving the study of transgenic plants and plant cell cultures have shed light on the influence of key enzymes on the regulation of metabolite flux and the interdependency of secondary metabolic pathways (Yao *et al.*, 1995; Howles *et al.*, 1996; Canel *et al.*, 1998). These studies have shown that coordinated expression of multiple genes is a likely requirement for the successful manipulation of complex pathways. Progress has also been made in this direction. The feasibility of expressing sets of genes from a single promoter (von Bodman *et al.*, 1995) and activating functionally related genes by ectopic expression of transcriptional activators (Grotewold *et al.*, 1998) has been demonstrated.

Trichome-specific promoters are needed to alter the biosynthetic programme of these structures specifically. Trichomes are not only capable of independently synthesizing a variety of phytochemicals from simple precursors, but also show independent regulation of gene expression. In tomato, for example, differential expression of genes encoding polyphenol oxidases has been observed among trichome types (Thipyapong *et al.*, 1997). Furthermore, a trichome-specific form of cytochrome P450, presumably involved in monoterpene metabolism, has been identified in *Nepeta racemosa* (Clark *et al.*, 1997). Despite the identification of differentially expressed genes in trichomes, the regulatory regions that confer them tissue-specific expression have not been cloned.

Specific examples of manipulation of trichome biosynthesis are yet to be reported. Artemisinin production in trichomes of *A. annua* is one potential target. McCaskill and Croteau (1995) have proposed a model for the subcellular compartmentation of isoprenoid synthesis in the glandular trichomes of peppermint, according to which all isopentenyl diphosphate (IPP) is synthesized in the plastids. Like the glandular trichomes of peppermint, those of *A. annua* produce an essential oil that is rich in monoterpenes and contains smaller amounts of sesquiterpenes (Tellez *et al.*, 1999). Based on this model, we hypothesize that the ratio of monoterpenes to sesquiterpenes in the glands of *A. annua* depends on the partitioning of precursor IPP between utilization by plastidic geranyl diphosphate synthase for monoterpene biosynthesis and export to the cytoplasm for sesquiterpene biosynthesis. Were this true, blocking the plastidic pathway in the trichomes could increase the production of artemisinin.

3. *Heterologous Production of the Metabolites of Interest*

Whether to increase the production of specific phytochemicals of economic value or for crop-protection purposes, heterologous expression of biosynthetic pathways is an option that merits consideration. Phytochemicals that confer protection against pathogens, herbivores, ultraviolet radiation and other environmental factors to a producing plant may offer similar protection to a plant where they are not normally found. The backbones of the major secondary biosynthetic pathways are present in most plant species. In some cases, it may be sufficient to direct the flow of metabolic precursors towards the desired target product by introducing a single gene. For example, the introduction of the daffodil gene for phytoene synthase in rice resulted in accumulation of phytoene (an intermediate in vitamin A biosynthesis) in rice endosperm, potentially increasing the nutritional value of this crop (Burkhardt *et al.*, 1997).

The extraction and purification of large quantities of phytochemicals by means of industrial processes are likely to benefit from the use of well-characterized and widely cultivated plant species. In addition to accounting for genetic and metabolic aspects of the biosynthetic process of interest, agronomic and other practical aspects that affect the economics of the production process need to be considered. From the biological point of view, it may make good sense to engineer a biosynthetic pathway in the species that normally produces the target compound. However, plants that have not been domesticated and for which large-scale production systems have not been developed produce many of these phytochemicals. Instead, extensively cultivated and easily transformed species, such as potato, tobacco and tomato, may be induced to do so. Experiments demonstrating the feasibility of heterologous biosynthesis have been reported. For example, transgenic tobacco expressing a coriander gene for acyl-acyl carrier protein desaturase produces petroselinic acid (Cahoon *et al.*, 1992). The synthesis of new hydroxybenzoic acid compounds in cell cultures of tobacco by expression of a bacterial gene encoding chorismate pyruvate lyase has also been demonstrated (Li *et al.*, 1997).

V. CONCLUSIONS AND FUTURE RESEARCH

An ability to manipulate the number and secondary metabolism of glandular trichomes by molecular genetics has many potential economic rewards. Advances in our understanding of the cell biology, genetics, and biochemistry of glandular trichomes, along with advances in plant biotechnology, promise to achieve both more economical production of trichome-derived products and more effective crop protection via trichome-based defences. Improvements in our knowledge of the secondary product metabolism of trichomes are needed. A major advance will come when trichome-specific promoters are available for use in the creation of transgenic plants with enhanced trichome productivity.

Further study of the chemical contents of glandular trichomes and their role in chemical ecology may lead to a better understanding of natural plant protection and to the discovery of additional compounds with useful biological activity. Advances in methods available, such as tandem LC-MS and LC-NMR, are decreasing the time and effort required to identify secondary products from trichomes. The biggest challenge for making full use of trichome-localized compounds is in adequately bioassaying them for a full range of activities. The development of semiautomated microbioassays of all types is improving the ease with which this can be accomplished (Duke *et al.*, 1999c).

REFERENCES

Anon. (1968). Food additives. Synthetic flavoring substances and adjuvants. *Federal Register* **33**(144), 10 569.

Anon. (1973). Food additives. Synthetic flavoring substances and adjuvants. *Federal Register* **38**(95), 12 913–12 914.

Ascensão, L. and Pais, M. S. (1998). The leaf capitate trichomes of *Leonotis leonurus*: histochemistry, ultrastructure and secretion. *Annals of Botany* **81**, 263–271.

Ascensão, L., Marques, N. and Pais, M. S. (1995). Glandular trichomes on vegetative and reproductive organs of *Leonotis leonurus* (Lamiaceae). *Annals of Botany* **75**, 619–626.

Ascensão, L., Marques, N. and Pais, M. S. (1997). Peltate glandular trichomes on *Leonotis leonurus* leaves: ultrastructure and histochemical characterization of secretions. *International Journal of Plant Science* **158**, 249–258.

Bajgrowicz, J. A. and Frater, G. (1998). Preparation of optically pure isomers of campholenic aldehyde derivatives for use as detergent fragrances. (Givaudan-Roure (International) S.A., Switz.) Eur. Pat. Appl., 11 pp., EP 841318 A2 980513, Application: EP 97-118997 971031. Priority: EP 96-117716 961106.

Banfield, M. G. (1994). Indirect aphid control with low concentrations of E-β-farnesene. (Consep Membranes, Inc., USA.) U.S., 5 pp. Cont. of U.S. Ser. No. 669,678, abandoned, US 5308613 A 940503, Application: US 92-945841 920916. Priority: US 91-669678 910314.

Becker, J. J. (1980). Tricyclic lactones and their use as perfume or aroma constituents. (Firmenich S.A., Switz.) Ger. Offen., 13 pp., DE 3016111 801106, Application: DE. Priority: CH 79-3972 790427.

Bell, A. A., Stipanovic, R. D., Elzen, G. W. and Williams, H. J. (1987). Structural and genetic variation of natural pesticides in pigment glands of cotton (*Gossypium*). American Chemical Society Symposium Series **330**, 477–490.

Bordenca, C. and Derfer, J. M. (1968). Phytotoxic β-dialkylaminoalkyl terpene ethers. (SCM Corp.) U.S., 3 pp., US 3397053 680813, Application: US 650811.

Bourett, T. M., Howard, R. J., O'Keefe, D. P. and Hallahan, D. L. (1994). Gland development on leaf surfaces of *Nepeta racemosa*. *International Journal of Plant Science* **155**, 623–632.

Bryson, C. T. and Croom, E. M. (1991). Herbicide inputs for a new agronomic crop, annual wormwood (*Artemisia annua* L.). *Weed Technology* **5**, 117–124.

Burkhardt, P. K., Beyer, P., Wunn, J., Kloti, A., Armstrong, G. A., Schledz, M., Von Lintig, J. and Potrykus, I. (1997). Transgenic rice (*Oriza sativa*) endosperm expressing daffodil (*Narcissus pseudonarcissus*) phytoene synthase accumulates phytoene, a key intermediate of provitamin A biosynthesis. *Plant Journal* **11**, 1071–1078.

Cahoon, E. B., Shanklin, J. and Ohlrogge, J. B. (1992). Expression of a coriander desaturase results in petroselinic acid production in transgenic tobacco. *Proceedings of the National Academy of Sciences USA* **89**, 11 184–11 188.

Camacho, A. D., Pierce, H. D., Jr and Borden, J. H. (1998). Host compounds as kairomones for the western balsam bark beetle Dryocoetes confusus Sw. (Col., Scolytidae). *Journal of Applied Entomology* **122**, 287–293.

Canel, C., Lopes-Cardoso, M. I., Whitmer, S., van der Fits, L., Pasquali, G., van der Heijden, R., Hoge, J. H. C. and Verpoorte, R. (1998). Effects of over-expression of strictosidine synthase and tryptophan decarboxylase on alkaloid production by cell cultures of *Catharanthus roseus*. *Planta* **205**, 414–419.

Carde, J. P., Bernard-Dagan, C. and Gleizes, M. (1980). Membrane systems involved in the synthesis and transport of monoterpene hydrocarbons in pine leaves. *In* "Biogenesis and Function of Plant Lipids" (P. Mazliak, P. Beneviste, C. Costes and R. Douce, eds), pp. 441–444. Elsevier, Amsterdam.

Chappell, J. (1995). Biochemistry and molecular biology of the isoprenoid biosynthetic pathway in plants. *Annual Reviews of Plant Physiology and Plant Molecular Biology* **46**, 521–547.

Chen, P. K. and Leather, G. R. (1990). Plant growth regulatory activities of artemisinin and its related compounds. *Journal of Chemical Ecology* **16**, 1867–1876.

Christ, H. (1997). Universal cleaning agent, its preparation and use. (Christ, Hubert, Germany.) Ger. Offen., 4 pp., DE 19617278 A1 971113, Application: DE 96-19617278 960430.

Clark, I. M., Forde, B. G. and Hallahan, D. L. (1997). Spatially distinct expression of two new cytochrome P450s in leaves of *Nepeta racemosa*: identification of a trichome-specific isoform. *Plant Molecular Biology* **33**, 875–885.

Clark, L. J., Hamilton, J. G. C., Chapman, J. V., Rhodes, M. J. C. and Hallahan, D. L. (1997). Analysis of monoterpenoids in glandular trichomes of the catmint *Nepeta racemosa*. *Plant Journal* **11**, 1387–1393.

Coates, D., Gray, G. W., Lacey, D., Young, D. J. S., Toyne, K. J. and Bone, M. F. (1986). Terpenoid derivatives and liquid crystal materials and devices containing them. (United Kingdom Secretary of State for Defence, London, UK; STC PLC.) PCT Int. Appl., 39 pp., WO 8604328 A1 860731, Application: WO 86-GB46 860124. Priority: GB 85-1999 850126.

Courtney-Gutterson, N. (1994). The biologist's palette: genetic engineering of anthocyanin biosynthesis and flower color. *In* "Genetic Engineering of Plant Secondary Metabolism" (B. E. Ellis, G. W. Kuroki and H. A. Stafford, eds), pp. 93–124. Plenum Press, New York.

Crotch, D. E. (ed.) (1978). "Kirk-Othmer Encyclopedia of Chemical Technology", 3rd edn (M. Grayson, exec. ed.), Vol. 22, pp. 709–759. John Wiley & Sons, New York.

Croteau, R. and Johnson, A. (1984). Biosynthesis of terpenoids in glandular trichomes. *In* "Biology and Chemistry of Plant Trichomes" (E. Rodriguez, P. L. Healey and I. Mehta, eds), pp. 133–185. Plenum, New York.

Dayan, F. E., Hernandez, A., Allen, S. N., Moraes-Cerdeira, R. M., Vroman, J. A., Avery, M. A. and Duke, S. O. (1999). Comparative phytotoxicity of artemisinin and several sesquiterpene analogues. *Phytochemistry* **50**, 607–614.

de Kramer, J., Neumann, U., Klein, U., Meiwald, M. and Krieg, W. (1996). Reducing the effect of pheromones on Lepidoptera, using terpenes. (BASF A.-G., Germany.) Ger. Offen., 5 pp., DE 19520706 A1 961212, Application: DE 95-19520706 950610.

Delabays, N., Jenelten, U., Paris, M., Pivot, D. and Galland, N. (1994). Aspects agronomiques et génétiques de la production d'artémisinine à partir d'*Artemisia annua* L. *Revue suisse de Viticulture, Arboriculture et Horticulture* **26**, 291–296.

Dell, B. and McComb, A. J. (1978). Plant resins – their formation, secretion and possible functions. *Advances in Botanical Research* **6**, 227–316.

Dinkova-Kostova, A. T., Gang, D. R., Davin, L. B., Bedgar, D. L., Chu, A. and Lewis, N. G. (1996). (+)-Pinoresinol/(+)-lariciresinol reductase from *Forsythia intermedia*. Protein purification, cDNA cloning, heterologous expression and comparison to isoflavone reductase. *Journal of Biological Chemistry* **271**, 29 473–29 482.

Dixon, R. A., Harrison, M. J. and Paiva, N. L. (1995). The isoflavonoid phytoalexin pathway: from enzymes to genes to transcription factors. *Physiologia Plantarum* **93**, 385–392.

Dooner, H. K., Robbins, T. P. and Jorgensen, R. A. (1991). Genetic and developmental control of anthocyanin biosynthesis. *Annual Reviews of Genetics* **25**, 173–199.

Duke, M. V., Paul, R. N., Elsohly, H. K., Sturtz, G. and Duke, S. O. (1994). Localization of artemisinin and artimesitene in foliar tissues of glanded and glandless biotypes of *Artemisia annua*. *International Journal of Plant Science* **155**, 365–373.

Duke, S. O. (1991). Plant terpenoids as pesticides. *In* "Handbook of Natural Toxins. Vol. 6. Toxicology of Plant and Fungal Compounds" (R. F. Keeler and A. T. Tu, eds), pp. 269–296. Marcel Dekker, New York.

Duke, S. O. (1994). Commentary: Glandular trichomes – a focal point of chemical and structural interactions. *International Journal of Plant Science* **155**, 617–620.

Duke, S. O. and Paul, R. N. (1993). Development and fine structure of the glandular trichomes of *Artemisia annua* L. *International Journal of Plant Science* **154**, 107–118.

Duke, S. O., Paul, R. N. and Lee, S. M. (1988). Terpenoids from the genus *Artemisia* as potential herbicides. *American Chemical Society Symposium Series* **380**, 318–334.

Duke, S. O., Vaughn, K. C., Croom, E. M. and Elsohly, H. N. (1987). Artemisinin, a constituent of annual wormwood (*Artemisia annua*), is a selective phytotoxin. *Weed Science* **35**, 499–505.

Duke, S.O., Duke, M. V., Paul, R. N., Ferreira, J. F. S., Canel, C., Tellez, M. R., Rimando, A. M. and Smeda, R. J. (1999a). Tissue localization and potential uses of phytochemicals with biological activity. *In* "Recent Progress in Allelopathy, Vol. 1: a Science for the Future" (F. Mascías, J. C. G. Galindo, J. M. G. Molinillo, H. G. Cutler, eds), pp. 211–218. Servicio de Publicaciones-Universidad de Cadiz, Cadiz.

Duke, S. O., Rimando, A. M., Duke, M. V., Paul, R. N., Ferreira, J. F. S. and Smeda, R. J. (1999b). Sequestration of phytotoxins by plants: implications for biosynthetic production. *In* "Natural Products: Agrochemicals and Pharmaceuticals" (H. A. Cutler and S. J. Cutler, eds), pp. 127–136. CRC Press, Boca Raton, FL.

Duke, S. O., Rimando, A. M., Dayan, F. E., Canel, C., Wedge, D. E., Tellez, M. R., Schrader, K. K., Weston, L. A., Smillie, T. J., Paul, R. N. and Duke, M. V. (1999c). Strategies for the discovery of bioactive phytochemicals. *In* "Phytochemicals as Bioactive Agents" (W. R. Bidlack, S. T. Omaye, M. S. Meskin and D. K. W. Jahner, eds). Technomic Publishing Co., Lancaster, PA (in press).

ElSohly, H. N., Croom, E. M., Jr, El-Feraly, F. S. and El-Sherei, M. M. (1990). A large-scale extraction technique of artemisinin from *Artemisia annua*. *Journal of Natural Products* **53**, 1560–1564.

Elzen, G. W., Williams, H. J., Bell, A. A., Stipanovic, R. D. and Vinson, S. B. (1985). Quantification of volatile terpenes of glanded and glandless *Gossypium hirsutum* L. cultivars and lines by gas chromatography. *Journal of Agriculture and Food Chemistry* **33**, 1079–1082.

Endlein, E., Kaiser, R., Saville, D. and Banolas A. M. A. (1998). Insect-repellent detergent compositions for use on solid surfaces. (Johan A. Benckiser G.m.b.H., Germany; Endlein, Edgar; Kaiser, Roger; Saville, Daniele; Banolas Adrogue, Miguel Angel.) PCT Int. Appl., 17 pp., WO 9818891 A1 980507, Application: WO 97-FR1917 971027. Priority: FR 96-13242 961030.

Fahn, A. (1988). Secretory tissues in vascular plants. *New Phytologist* **108**, 229–247.

Fan, J., Qi, D., Liu, M. and Shi, Z. (1983). Sources and utilization of Vitex oil. *Zhongcaoyao* **14**(5), 227–230.

Faure, O., Diemer, F., Moja, S. and Jullien, F. (1998). Mannitol and thidiazuron improve *in vitro* shoot regeneration from spearmint and peppermint leaf disks. *Plant Cell Tissue and Organ Culture* **52**, 209–212.

Ferreira, J. F. S. and Duke, S. O. (1997). Approaches for maximizing biosynthesis of medicinal plant secondary metabolites. *AgBiotech News and Information* **9**, 309N–316N.

Ferreira, J. F. S. and Janick, J. (1995). Floral morphology of *Artemisia annua* with special reference to trichomes. *International Journal of Plant Science* **156**, 807–815.

Forkmann, G. (1991). Flavonoids as flower pigments: the formation of the natural spectrum and its extension by genetic engineering. *Plant Breeding* **106**, 1–26.

Fujeda, T. and Kuroda, M. (1993). Powder detergent compositions containing nonionic surfactants and perfumes. (Kao Corp, Japan.) Jpn. Kokai Tokkyo Koho, 7 pp., JP 05202387 A2 930810 Heisei, Application: JP 92-13975 920129.

Gardner, R. J., Laws, D. R. and McGuinness, J. D. (1973). Suppression of gushing by hop oil. *Journal of the Institute of Brewing* **79**, 209–211.

Gershenzon, J., McCaskill, D. G., Rajaonarivony, J., Mihaliak, C., Karp, F. and Croteau, R. (1991). Biosynthetic methods for natural products: new procedures for the study of glandular trichome constituents. *In* "Recent Advances in Phytochemistry, Vol. 25. Modern Phytochemical Methods" (N. H. Fischer, M. B. Islam and H. A. Stafford, eds), pp. 347–370. Plenum Press, New York.

Gershenzon, J., McCaskill, D., Rajaonarivony, J. I. M., Mihaliak, C., Karp, F. and Croteau, R. (1992). Isolation of secretory cells from plant glandular trichomes and their use in biosynthetic studies of monoterpenes and other gland products. *Analytical Biochemistry* **200**, 130–138.

Grotewold, E., Chamberlin, M., Snook, M., Siame, B., Butler, L., Swenson, J., Maddock, S., St Clair, G. and Bowen, B. (1998). Engineering secondary metabolism in maize cells by ectopic expression of transcription factors. *Plant Cell* **10**, 721–740.

Guo, Z. and Wagner, G. J. (1995). Biosynthesis of labdenediol and sclareol in cell-free extracts from trichomes of *Nicotiana glutinosa*. *Planta* **197**, 627–632.

Guo, Z., Severson, R. F. and Wagner, G. J. (1994). Biosynthesis of the diterpene *cis*-abienol in cell-free extracts of tobacco trichomes. *Archives of Biochemistry and Biophysics* **308**, 103–108.

Gut, J. and Van Oosten, A. M. (1988). Use of farnesene in agriculture or horticulture or in the protection of stocks, or for the preparation of products, and also method for controlling insects and acarids by using farnesene. (Nederlandse Organisatie voor Toegepast-Natuurwetenschappelijk Onderzoek, Neth.) Eur. Pat. Appl., 7 pp., EP 266822 A1 880511, Application: EP 87-201982 871015. Priority: NL 86-2604 861016.

Haberlin, R. J. and Steinhilber, G. W. (1968). Epoxy resin curing compositions. (W. R. Grace and Co.) U.S., 5 pp., US 3366600 680130, Application: US 580730–640602.

Hartmans, K. J., Diepenhorst, P., Bakker, W. and Gorris, L. G. M. (1995). The use of carvone in agriculture: sprout suppression of potatoes and antifungal activity against potato tuber and other plant diseases. *Industrial Crops and Products* **4**, 3–13.

Hautmann, H. (1985). Deodorant. (Globol-Werk G.m.b.H., Fed. Rep. Ger.) PCT Int. Appl., 14 pp., WO 8500980 A1 850314, Application: WO 84-EP268 840904, Priority: DE 83-3331995 830905.

Herbers, K. and Sonnewald, U. (1996). Manipulating metabolic partitioning in transgenic plants. *Trends in Biotechnology* **14**, 198–205.

Holtmann, H. (1991). Honeybee repellents, for repulsion from pesticide-treated areas. (Germany.) Ger. Offen., 5 pp., DE 4012224 A1 911017, Application: DE 90-4012224 900414.

Howe-Grant, M. (1985). "Kirk-Othmer Concise Encyclopedia of Chemical Technology" (J. I. Kroschwitz, exec. ed.), Vol. 22, pp. 1161–1162. John Wiley & Sons, New York.

Hülskamp, M., Miséra, S. and Jürgens, G. (1994). Genetic dissection of trichome cell development in *Arabidopsis*. *Cell* **76**, 555–566.

Ishida, S. and Hayashi, S. (1994). Terpinen-4-ol as an insecticide for preserving clothes. (Nippon Terpen Kagaku Kk, Japan.) Jpn. Kokai Tokkyo Koho, 3 pp., JP 06227906 A2 940816 Heisei, Application: JP 93-20226 930208.

Ishidoya, T. (1996). Use of carvone, anethole, limonene or pulegone to mask unpleasant odor of ethanol in perfumes or cosmetics. (Pola Kasei Kogyo Kk, Japan.) Jpn. Kokai Tokkyo Koho, 5 pp., JP 08188792 A2 960723 Heisei, Application: JP 95-16491 950106.

Jia, W., Song, S. and Pan, G. (1994). Synthesis of thujene nitrosochloride as intermediate for levocarvone. (Dalian Chemical Physics Institute, Chinese Academy of Sciences, Peop. Rep. China.) Faming Zhuanli Shenqing Gongkai Shuomingshu, 5 pp., CN 1092402 A 940921, Application: CN 93-11016 930313.

Kalleder, A., Mennig, M., Schmidt, H. and Suyal, N. (1998). Manufacture of annealable printing pastes for printing on glass surfaces. (Sekurit Saint-Gobain Deutschland G.m.b.H. und Co. K.-G., Germany.) Ger., 4 pp., DE 19650139 C1 980702, Application: DE 96-19650139 961204.

Kandra, L. and Wagner, G. J. (1988). Studies of the site and mode of biosynthesis of tobacco trichome exudate components. *Archives of Biochemistry and Biophysics* **265**, 425–432.

Keene, C. K. and Wagner, G. J. (1985). Direct demonstration of duvatrienediol biosynthesis in glandular heads of tobacco trichomes. *Plant Physiology* **79**, 1026–1032.

Kelsey, R. G., Reynolds, G. W. and Rodriguez, E. (1984). The chemistry of biologically active constituents secreted and stored in plant glandular trichomes. *In* "Biology and Chemistry of Plant Trichomes" (E. Rodriguez, P. L. Healey and I. Mehta, eds), pp. 187–241. Plenum, New York.

Kimura, H., Harris, M. S., Sakamoto, T., Gopalakrishna, R., Gundimeda, U., Cui, J. Z., Spee, C., Hinton, D. R. and Ryan, S. J. (1997). Hypericin inhibits choroidal endothelial cell proliferation and cord formation *in vitro*. *Current Eye Research* **16**, 967–972.

Kitano, H. and Tanimoto, F. (1987). Preparation of tricyclene derivatives for use as perfumes, plasticizers, and intermediates for drugs. (Nippon Petrochemicals Co., Ltd, Japan.) Jpn. Kokai Tokkyo Koho, 5 pp., JP 62226938 A2 871005, Application: JP 86-68977 860326.

Klar, U., Graf, H., Blechert, S. and Sachse, A. (1996). New borneols, process for their preparation and their pharmaceutical use. (Schering A.-G., Germany.) Ger. Offen., 10 pp., DE 19509697 A1 960912, Application: DE 95-19509697 950308.

Klayman, D. L. (1985). Qinghaosu (artemisinin): an antimalarial drug from China. *Science* **228**, 1049–1055.

Kohler, M., Haerdi, W., Christen, P. and Veuthey, J.-L. (1997). Extraction of artemisinin and artemisinic acid from *Artemisia annua* L. using supercritical carbon dioxide. *Journal of Chromatography. A* **785**, 353–360.

Koornneef, M. (1981). The complex syndrome of *ttg* mutants. *Arabidopsis Information Service* **18**, 45–51.

Koren, H., Schenk, G. M., Jindra, R. H., Alth, G., Ebermann, R., Kubin, A., Koderhold, G. and Kreitner, M. (1991). Hypericin in phototherapy. *Journal of Photochemistry and Photobiology* **36B**, 113–119.

Kowalski, S. P., Tingey, W. M. and Steffens, J. C. (1990). Inheritance of polyphenol oxidase in type A glandular trichomes of *Solanum berthaultii*. *Journal of Heredity* **81**, 475–478.

Kristof, E. and Kristof, P. (1995). Method for inhibiting potato sprouting. (Can.) Can. Pat. Appl., 6 pp., CA 2122554 AA 951030, Application: CA 94-2122554 940429.

Larkin, J. C., Oppenheimer, D. G. and Marks, M. D. (1994a). The GL1 gene and the trichome developmental pathway in *Arabidopsis thaliana*. *In* "Plant Promoters and Transcription Factors" (L. Nover, ed.), pp. 261–276. Springer-Verlag, Berlin, Heidelberg.

Larkin, J. C., Oppenheimer, D. G., Lloyd, A., Paparozzi, E. T. and Marks, M. D. (1994b). Roles of the *GLABROUS1* and *TRANSPARENT TESTA GLABRA* genes in *Arabidopsis* trichome development. *Plant Cell* **6**, 1065–1076.

Laughlin, J. C. (1994). Agricultural production of artemisinin – a review. *Transactions of the Royal Society of Tropical Medicine Hygiene* **88**(Suppl. 1), 21–22.

Lavie, G., Mazur, Y., Prince, A. M., Pascual, D., Liebes, L., Levin, B. and Meruelo, D. (1995). Hypericin as an inactivator of infectious viruses in blood components. *Transfusion* **35**, 392–400.

Lazuchiewicz, E., Wilk, J., Zwierzak, K., Lorecka, Z., Kasprzak, P. and Stasiak, L. (1992). Agents for removing oily and greasy stains from fabrics. (Lodzkie Zaklady Chemiczne "Organika", Pol.) Pol., 3 pp., PL 154892 B1 920228, Application: PL 88-270787 880222.

Lee, J. M., Lee, E. J., Bahn, K. N., Kim, J. O. and Ha, Y. L. (1995). Potent antimutagenic activity of caryophyllene oxide for aflatoxin B1 (AEB1) and 2-amino-3-methylimidazo[4,5-f]quinoline (IQ). *Han'guk Nonghwa Hakhoechi* **38**(5), 468–472.

Lehmann, T. and Brenneisen, R. (1995). High performance liquid chromatography profiling of *Cannabis sativa*. *Journal of Liquid Chromatography* **18**, 689–700.

Li, S. M., Wang, Z. X., Wemakor, E. and Heide, L. (1997). Metabolization of the artificial secondary metabolite 4-hydroxybenzoate in *ubiC*-transformed tobacco. *Plant and Cell Physiology* **38**, 844–850.

Lin, Y. and Wagner, G. J. (1994). Surface disposition and stability of pest-interactive, trichome-exuded diterpenes and sucrose esters of tobacco. *Journal of Chemical Ecology* **20**, 1907–1921.

Liszt, K., Kovacs, A. and Fridvalszky, L. (1982). Az orbáncfü (*Hypericum perforatum* L.) levéldifferncialódásának ultrastrukturális vizsgálata. *Herba Hungarica* **21**, 21–36.

Lloyd, A. M., Walbot, V. and Davis, R. W. (1992). *Arabidopsis* and *Nicotiana* anthocyanin production activated by maize regulators *R* and *C1*. *Science* **258**, 1773–1775.

Lu, Y. (1980). Studies on the chemical constituents of the essential oil of *Rhododendron tsinghaiense* Ching. *Hua Hsueh Hsueh Pao* **38**(3), 241–249.

Lydon, J., Teasdale, J. R. and Chen, P. K. (1997). Allelopathic activity of annual wormwood (*Artemesia annua*) and the role of artemisinin. *Weed Science* **45**, 807–811.

Marks, M. D. (1997). Molecular genetic analysis of trichome development in *Arabidopsis*. *Annual Review of Plant Physiology and Plant Molecular Biology* **48**, 137–163.

Martens-Lobenhoffer, J., Behrens-Baumann, W., Loesche, D. and Meyer, F. P. (1998). Identification of a pill for eye-diseases from traditional Chinese medicine. *Pharmazie* **53**(2), 136–137.

Martinez, B. C. and Staba, E. J. (1988). The production of artemisinin in *Artemisia annua* L. tissue cultures. *Advances in Cell Culture* **6**, 69–87.

McCaskill, D. and Croteau, R. (1995). Monoterpene and sesquiterpene biosynthesis in glandular trichomes of peppermint (*Mentha* × *piperita*) rely exclusively on plastid-derived isopentenyl diphosphate. *Planta* **197**, 49–56.

McCaskill, D. and Croteau, R. (1997). Prospects for the bioengineering of isoprenoid biosynthesis. *In* "Advances in Biochemical Engineering/Biotechnology" (Berger, R., ed.), pp. 107–146. Springer Verlag, Berlin.

McCaskill, D. and Croteau, R. (1999). Strategies for bioengineering: the development and metabolism of glandular tissues in plants. *Nature Biotechnology* **17**, 31–36.

McCaskill, D., Gershenzon, J. and Croteau, R. (1992). Morphology and monoterpene biosynthetic capabilities of secreting cell clusters isolated from glandular trichomes of peppermint (*Mentha piperita* L.). *Planta* **187**, 445–454.

Meakins, S. E. and Motion, K. R. (1990). Process for the preparation of alkyl substituted tetralins and indans as intermediates for musk fragrances. (Guest International B.V., Neth.) Eur. Pat. Appl., 5 pp., EP 393742 A1 901024, Application: EP 90-200817 900405. Priority: EP 89-200943 890414.

Mitchell, P. W. D. (1970). Pinocarveol. (International Flavors and Fragrances Inc.) Fr. Demande, 12 pp., FR 2009498 700206, Application: FR. Priority: US 680528–681227.

Mookherjee, B. D., Trenkle, R. W., Chant, B. J., Ouwerkerk, A. V., Kamath, V. and Mussinan, C. J. (1984). Use of isomeric farnesene product-by-process for augmenting or enhancing the aroma or taste of foodstuffs, chewing gums, medicinal products and toothpastes. (International Flavors and Fragrances Inc., USA.) U.S., 34 pp. Cont.-in-part of U.S. 4,376,068, US 4446125 A 840501, Application: US 83-461476 830127. Priority: US 81-292392 810813.

Mosciano, G., Sadural, S., Fasano, M. and Michalski, J. (1990). Organoleptic characteristics of flavor materials. *Perfum Flavorist* **15**(1), 19–20, 22, 24–5.

Muller, W. H. and Muller, C. H. (1964). Volatile growth inhibitors produced by *Salvia* species. *Bulletin of the Torrey Botanical Club* **91**, 327–330.

Muller, W. H., Lorber, P. and Haley, B. (1968). Volatile growth inhibitors produced by *Salvia leucophylla*: effect on seedling growth and respiration. *Bulletin of the Torrey Botanical Club* **95**, 415–422.

Neal, J. J., Tingey, W. M. and Steffens, J. C. (1990). Sucrose esters of fatty acid in glandular trichomes of *Solanum berthaultii* settling and probing by green peach aphid. *Journal of Chemical Ecology* **16**, 487–497.

Neilson, M. T., Akers, C. P., Järlfors, U. E., Wagner, G. J. and Berger, S. (1991). Comparative ultrastructural features of secreting and nonsecreting glandular trichomes of two genotypes of *Nicotiana tabacum* L. *Botanical Gazette* **152**, 13–22.

Nistor, C., Popescu, H., Popescu, V. and Nistor, V. (1978). Soothing aromatic ointment. (Intreprinderea de Produse Cosmetice "Miraj", Rom.) Rom., 2 pp., RO 68717 781130, Application: RO 76-88113 761019.

Niu, X., Lin, K., Hasegawa, P. M., Bressan, R. A. and Weller, S. C. (1998). Transgenic peppermint (*Mentha* × *piperita* L.) plants obtained by cocultivation with *Agrobacterium tumefaciens*. *Plant Cell Reports* **17**, 165–171.

Noguchi, N., Niyama, A. and Fujisaka, M. (1974). Foaming agents for ore flotation. (Mitsubishi Petrochemical Co., Ltd, Japan.) Japan. Kokai, 4 pp., JP 7464502 740622, Application: JP 72-105793 721024.

Opdyke, D. L. J. (1975). Fragrance raw materials. Eucalyptol. *Food and Cosmetics Toxicology* **13**(1), 105–106.

Opdyke, D. L. J. and Letizia, C. (1983). Monographs on fragrance raw materials. Caryophyllene oxide. *Food and Chemical Toxicology* **21**(5), 661–662.

Palamand, S. R. (1986). Controlling light-struck flavor in beverages. (Anheuser-Busch Cos., Inc., USA.) Can., 11 pp., CA 1199517 A1 860121, Application: CA 83-424661 830328.

Pickett, J. A., Dawson, G. W., Free, J. B., Griffiths, D. C., Powell, W., Williams, I. M. and Woodcock, C. M. (1984). Pheromones in the management of beneficial insects. *Proceedings of the British Crop Protection Conference—Pests and Diseases* **1**, 247–254.

Ping, Q. N., Sun, G. Q. and Liu, G. J. (1993). Role of stratum corneum and its lipid in permeability of 5-fluorouracil through human skin. *Yaoxue Xuebao* **28**(12), 924–930.

Raman, A., Weir, U. and Bloomfield, S. F. (1995). Antimicrobial effects of tea-tree oil and its major components on *Staphylococcus aureus*, *Staphylococcus epidermidis* and *Propionibacterium acnes*. *Letters in Applied Microbiology* **21**, 242–245.

Repp, H. D., Stottmeister, U., Dueresch, R., Richter, H. P., Schmidt, J., Doerre, M., Haufe, G., Weber, L. and Pritzkow, T. (1988). Manufacture of verbenol–verbenone mixture with Hansenula polymorpha. (Akademie der Wissenschaften der DDR, Ger. Dem. Rep.) Ger. (East), 3 pp., DD 262875 A1 881214, Application: DD 87-305564 870731.

Retamar, J. A., Elder, H. V., Molli, J. S. and Gasparri, M. G. (1994). α-Thujene, sabinene and thujone as intermediaries in organic synthesis. *Essenze Derivati Agrumari* **64**(1), 61–69.

Rigoni, M. and Trani, M. (1997). Cleaning compositions with effective cleaning and shine performance. (Procter & Gamble Co., USA.) Eur. Pat. Appl., 10 pp., EP 805198 A1 971105, Application: EP 96-870093 960716. Priority: EP 96-870057 960503.

Romano, N., Trani, M. and Wolff, A. (1998). Disinfecting compositions and processes for disinfecting surfaces. (The Procter & Gamble Company, USA.) PCT Int. Appl., 27 pp., WO 9854279 A1 981203, Application: WO 97-US9260 970530.

Ruplenas, V., Petrauskas, G. and Sibailiene, A. (1977). Current-conducting material. (USSR.) U.S.S.R., SU 559285 770525, Application: SU 75-2119218 750402.

Saito, A. and Watanabe, N. (1992). Polymer compositions noncorrosive to metal surface. (Nippon Zeon Co., Ltd, Japan.) Jpn. Kokai Tokkyo Koho, 14 pp., JP 04130160 A2 920501 Heisei, Application: JP 90-413999 901226. Priority: JP 89-339098 891227.

Sato, H., Watanabe, Y., Itoi, H., Higo, M. and Kato, T. (1979). Sublimate of sabinene hydrate. (Lion Dentifrice Co., Ltd, Japan.) Jpn. Kokai Tokkyo Koho, 3 pp., JP 54148183 791120, Application: JP 78-57374 780515.

Schöpker, H., Kneisel, M., Beerhues, L., Robenek, H. and Wiermann, R. (1995). Phenylalanine ammonia-lyase and chalcone synthase in glands of *Primula kewensis* (W. Wats): immunofluorescence and immunogold localization. *Planta* **196**, 712–719.

Scolnik, P. A. and Bartley, G. E. (1994). A table of some cloned plant genes involved in isoprenoid biosynthesis. *Plant Molecular Biology Reporter* **14**, 305–319.

Serrato-Valente, G., Bisio, A., Cornata, L. and Ciarallo, G. (1997). Structural and histochemical investigation of the glandular trichomes of *Salvia aurea* L. leaves, and chemical analysis of the essential oil. *Annals of Botany* **79**, 329–336.

Singh, K. L., Tandon, S., Kahol, A. P., Aggarwal, K. K. and Kumar, S. (1998). Terpineol: an aroma chemical of value in the food, flavor and fragrance industry. *Journal of Medicinal and Aromatic Plant Sciences* **20**, 779–786.

Slone, J. H. and Kelsey, R. G. (1985). Isolation and purification of glandular secretory cells from *Artemisia tridentata* (ssp. *vaseyana*) by Percoll density gradient

centrifugation. *American Journal of Botany* **72**, 1445–1451.
Spencer, A., Hamill, J. D. and Rhodes, M. J. C. (1990). Production of terpenes by differentiated shoot cultures of *Mentha citrata* transformed with *Agrobacterium tumefaciens* T37. *Plant Cell Reports* **8**, 601–604.
Spring, O., Heil, N. and Vogler, B. (1997). Sequiterpene lactones and flavanones in *Scalesia* species. *Phytochemistry* **46**, 1369–1373.
Szymanski, D. B., Klis, D. A., Larkin, J. C. and Marks, M.D. (1998). *cotl*: a regulator of *Arabidopsis* trichome initiation. *Genetics* **149**, 565–577.
Tellez, M. R., Canel, C., Rimando, A. M. and Duke, S. O. (1999). Differential accumulation of isoprenoids in glanded and glandless *Artemisia annua* L. *Phytochemistry* (in press).
Thipyapong, P., Joel, D. M. and Steffens, J. C. (1997). Differential expression and turnover of the tomato polyphenol oxidase gene family during vegetative and reproductive development. *Plant Physiology* **113**, 707–718.
Trelinska-Wlazlak, M., Trelinska-Wlazlak, M. and Dobras, R. (1990). Conducting paste. (Politechnika Wroclawska, Pol.) Pol., 5 pp., PL 151479 B1 900928, Application: PL 87-264126 870213.
Trumbo, D. L., Giddings, C. L. and Wilson, L. R. A. (1995). Terpene-anhydride resins as coating materials. *Journal of Applied Polymer Science* **58**, 69–76.
Upton, R., Graff, A., Williamson, E., Bunting, D., Gatherum, D. M., Walker, E. B., Butterweck, V., Liefländer-Wulf, U., Nahrstedt, A., Wall, A., Winterhoff, H., Cott, J., Marquis, M. C., Shumake, R. L., Stansbury, J. and Petrone, C. (1997). St. John's wort (*Hypericum perforatum*). *In* "American Herbal Pharmacopoeia and Therapeutic Compendium" (R. Upton, ed.), 32 pp. The American Herbal Pharmacopoeia, Santa Cruz, CA.
Van Lijsebettens, M. and Clarke, J. (1998). Leaf development in *Arabidopsis*. *Plant Physiology and Biochemistry* **36**, 47–60.
Vegezzi, D. (1982). Verbenone, myrtenal and pinocarveol and their therapeutical use. (Corvi Mora, Enrico, Switz.) Can., 23 pp. Division of Can. Appl. No. 279,610, CA 1129885 A2 820817, Application: CA 81-369574 810128, Priority: CH 76-7042 760603; CA 77-279610 770601.
Vergauwe, A., Cammaert, R., Vandenberghe, D., Genetello, C., Inze, D., van Montagu, M. and van den Eeckhout, E. (1996). *Agrobacterium tumefaciens*-mediated transformation of *Artemisia annua* L. and regeneration of transgenic plants. *Plant Cell Reports* **15**, 929–933.
Vergauwe, A., van Geldre, E., Inze, D., van Montagu, M. and van den Eeckhout, E. (1998). Factors influencing *Agrobacterium tumefaciens*-mediated transformation of *Artemisia annua* L. *Plant Cell Reports* **18**, 105–110.
Vermeer, J. and Peterson, R. L. (1979). Glandular trichomes on the inflorescence of *Chrysanthemum morifolium* cv. Dramatic (Compositae). II Ultrastructure and histochemistry. *Canadian Journal of Botany* **57**, 714–729.
Verstegen-Haaksma, A. A., Swarts, H. J., Jansen, B. J. M., de Groot, A., Bottema-MacGillavry, N. and Witholt, B. (1995). Application of S-(+)-carvone in the synthesis of biologically active natural products using chemical transformations and bioconversions. *Industrial Crops and Products* **4**, 15–21.
Voirin, B. and Bayet, C. (1996). Developmental changes in the monoterpene composition of *Mentha* × *piperita* leaves from individual peltate trichomes. *Phytochemistry* **43**, 573–580.
Voirin, B., Bayet, C. and Colson, M. (1993). Demonstration that flavone aglycones accumulate in the peltate glands of *Mentha* × *piperita* leaves. *Phytochemistry* **34**, 85–87.
Von Bodman, S. B., Domier, L. L. and Farrand, S. K. (1995). Expression of multiple

eukaryotic genes from a single promoter in *Nicotiana*. *Biotechnology* **13**, 587–591.

Vonwiller, S. C., Haynes, R. K., King, G. and Wang, H.-J. (1993). An improved method for the isolation of quinghao (artemisinic) acid from *Artemisia annua*. *Planta Medica* **59**, 562–563.

Wada, T., Tachibana, T., Shimura, Y. and Okada, K. (1997). Epidermal cell differentiation in *Arapidopsis* determined by a *myb* homolog, *CPC*. *Science* **277**, 1113–1115.

Wagner, G. J. (1991). Secreting glandular trichomes: more than just hairs. *Plant Physiology* **96**, 675–679.

Wang, F. (1993). Borneol compositions for metal and plastic surface protection. (Peop. Rep. China.) Faming Zhuanli Shenqing Gongkai Shuomingshu, 5 pp., CN 1078739 A 931124, Application: CN 93-11185 930618.

Weston, L. A., Nimbal, C. I. and Jeandet, P. (1999). Allelopathic potential of grain sorghum (*Sorghum bicolor* [L.] Moench) and related species. *In* "Principles and Practices in Plant Ecology, Allelopathic Interactions" (Inderjit, K. M. M. Dakshini and C. L. Foy, eds), pp. 467–477. CRC Press, Boca Raton, FL.

White, N. J. (1994). Artemisinin: current status. *Transactions of the Royal Society for Tropical Medicine Hygiene* **88**(Suppl. 1), 3–4.

Woerdenbag, H. J., Pras, N., van Uden, W., De Boer, A, Batterman, S., Visser, J. F. and Malingré, T. M. (1992). High peroxidase activity in cell cultures of *Artemisia annua* with minute artemisinin contents. *Natural Product Letters* **1**, 121–128.

Yamane, M. A., Williams, A. C. and Barry, B. W. (1995). Terpene penetration enhancers in propylene glycol/water co-solvent systems: effectiveness and mechanism of action. *Journal of Pharmacy and Pharmacology* **47**, 978–989.

Yamasaki, T., Sato, M. and Sakoguchi, H. (1997). (−)-Germacrene D: masking substance of attractants for the cerambycid beetle, *Monochamus alternatus* (Hope). *Applied Entomology and Zoology* **32**, 423–429.

Yamaura, T., Tanaka, S. and Tabata, M. (1992). Localization of the biosynthesis and accumulation of monoterpenoids in glandular trichomes of thyme. *Planta Medica* **58**, 153–158

Yamawaki, T. and Kondo, M. (1988). Manufacture of creep-resistant stereoregular 1-butene homopolymer with low residual catalyst content. (Idemitsu Petrochemical Co., Ltd, Japan; Neste Oy.) Jpn. Kokai Tokkyo Koho, 10 pp., JP 63165408 A2 880708 Showa, Application: JP 86-196265 860821.

Yanagisawa, T., Takimoto, H. and Sakata, S. (1993). Drying solvents used in cleaning and manufacture thereof. (Olympus Optical Co., Japan; Kemikaru Tekunorojii Kenkyusho.) Jpn. Kokai Tokkyo Koho, 6 pp., JP 05057105 A2 930309 Heisei, Application: JP 91-246889 910830. Priority: JP 91-179046 910625.

Yao, K., De Luca, V. and Brisson, N. (1995). Creation of a metabolic sink for tryptophan alters the phenylpropanoid pathway and the susceptibility of potato to *Phytophtora infestans*. *Plant Cell* **7**, 1787–1799.

Yunes, R. A., Talenti, E. C. J., Figoli, N. S. and Mazzei, R. J. (1967). Reduction and isomerization of caryophyllene. *Revista de al Facultad de Ingenieria Quimica (Universidad Nacional del Litoral)* **36**, 95–103.

Zupan, J. A. (1983). Use of eucalyptol for enhancing skin permeation of bioaffecting agents. (Merck and Co., Inc., USA.) Eur. Pat. Appl., 37 pp., EP 69385 A2 830112, Application: EP 82-106021 820706, Priority: US 81-280967 810707.

Chemotaxonomy Based on Metabolites from Glandular Trichomes

O. SPRING

Institut für Botanik, Universität Hohenheim, D-70593 Stuttgart, Germany

I. INTRODUCTION

The scientific interest in plant trichomes is based on many grounds. The widespread occurrence and the recognition of the functional importance of trichomes in higher plants were driving forces for their investigation. Studies of the morphology of plant hairs and, later on, of their physiology provoked

Advances in Botanical Research Vol. 31
ISBN 0-12-005931-2

questions on their role for the producing organism itself as well as for interactive processes with the environment.

The morphological diversity of trichomes, leading to an estimated number of more than 300 different forms (Wagner, 1991), has been reviewed widely (e.g. Uphof, 1962; Behnke, 1984). As a consequence of the occurrence of certain types in specific groups of plants, botanists quickly discovered the taxonomic value of trichomes. Comprehensive summaries on the distribution of trichome types in higher plants reach back to the beginning of the twentieth century with the reviews of Solereder (1899) and were regularly updated (e.g. Staudermann, 1924; Metcalf and Chalk, 1957; Hummel and Staesche, 1962).

Despite the intensive use of plant trichomes as micromorphological characters, the chemistry of these trichomes was sparsely consulted to solve taxonomic questions. At first glance, this may appear surprising because the ability of plants to produce taxon-specific phytochemicals in glandular trichomes has been well known for more than a century, and compounds of certain taxa were variously used as fragrance, spices or drugs. Even the general acceptance of the taxonomic usefulness of chemical characters within the last five decades has done little to change the situation of trichome metabolites as a widely ignored source for taxonomic investigations.

Several factors were certainly responsible for preventing chemical trichome analysis from a broad taxonomic use up to now (cf. Spring and Buschmann, 1996). Among these, the lack of techniques for compound extraction, separation and spectroscopic analysis appropriate to sample amounts gained exclusively from trichomes posed the severest limitation. Chemical data, for a long time, were available only from bulk collections of plant material. Obtaining data was a tiresome procedure, which was usually performed by organic chemists and not by taxonomists themselves. This has changed at least partly with the recent progress in the field of chromatography and new on-line techniques, such as liquid chromatography combined with ultraviolet/visual light spectroscopy (LC-UV/VIS), LC combined with mass spectrometry (LC-MS), LC combined with nuclear magnetic resonance spectrometry (LC-NMR), gas chromatography combined with mass spectrometry (GC-MS) and capillary electrophoresis combined with mass spectroscopy (CE-MS). Technological progress in chemistry, however, cannot close the gap in another important prerequisite for the taxonomic use of trichome metabolites: knowledge of the biology of trichome metabolism. This concerns particularly the deficits in our knowledge of the biosynthesis of taxonomically useful phytochemicals (still a general shortcoming of chemotaxonomy), but also includes open questions on the development of glandular trichomes, on their distribution and on the environmental influences on their content. Finally, it cannot be ignored that, with the advent of powerful new tools for the exploration of macromolecules in organisms, the general interest in using phytochemicals in taxonomy has suffered a significant decline. It is the challenging task of this review to show that phytochemical investigations in general, and those on the trichome chemistry in particular, can make a valuable contribution in the future of plant systematics.

II. PHYTOCHEMICALS ASSOCIATED WITH PLANT TRICHOMES

Our experience with plants causing toxic or allergenic reactions on touching them (e.g. *Urtica dioica*) stimulated our awareness of the localization of metabolites at the plant surface long before any scientific interest in trichomes began. Nestler (1904), with the fascinating description of his self-treatment with stems of *Primula obconica*, was one of the early pioneers in the field of trichome chemistry. He obtained the exudate of glandular trichomes selectively by touching them carefully with a microscopic slide and then crystallized the compound by sublimation. The secretion, although identified as the quinone primin only 22 years later, could be described according to its solubility, crystallization and chemical reaction. The biological activity of primin as an irritant of human skin was used by Nestler to trace its occurrence in various plant parts and also to screen 14 other taxa of the genus *Primula* for similar reactions (11 of which were negative). Thus, although primarily interested in the medical aspects of contact dermatitis caused by this plant, Nestler by chance became one of the first scientists to use trichome metabolites for plant taxonomy (Fig. 1).

Glandular trichomes of similar or other types are present in many plant families, and numerous reports exist on the occurrence of certain metabolites secreted by such hairs. However, direct proof for their synthesis or storage in trichomes through micromechanical techniques, such as in the example of *P. obconica*, via gas chromatography analysis for essential oils from isolated trichomes of *Mentha piperita* (Amelunxen *et al.*, 1969), or through histochemical investigation, such as in alkaloids of *Vinca herbacea* (Kartmazova and Lyapunova, 1976), has not been provided in all cases. After a short summary on the role of metabolites from trichomes in plant defence published by Levin (1973), a comprehensive survey on compounds associated with glandular hairs was given by Kelsey *et al.* (1984) with special emphasis on terpenoids. The predominant natural function of biomolecules in glandular hairs as protectants against predators, parasites or competitors together with their localization at the plant surface are conditions that appear to favour lipophilic and volatile compounds, such as terpenoids, as constituents in exudates. However, besides these latter compounds, a broad array of phytochemicals from many different substance classes are meanwhile known to derive from plant trichomes (Table I). Among these are phenolics (Gregory *et al.*, 1986), alkaloids (Kartmazova and Lyapunova, 1976), glycosidated substances (Spring *et al.*, 1994) and many

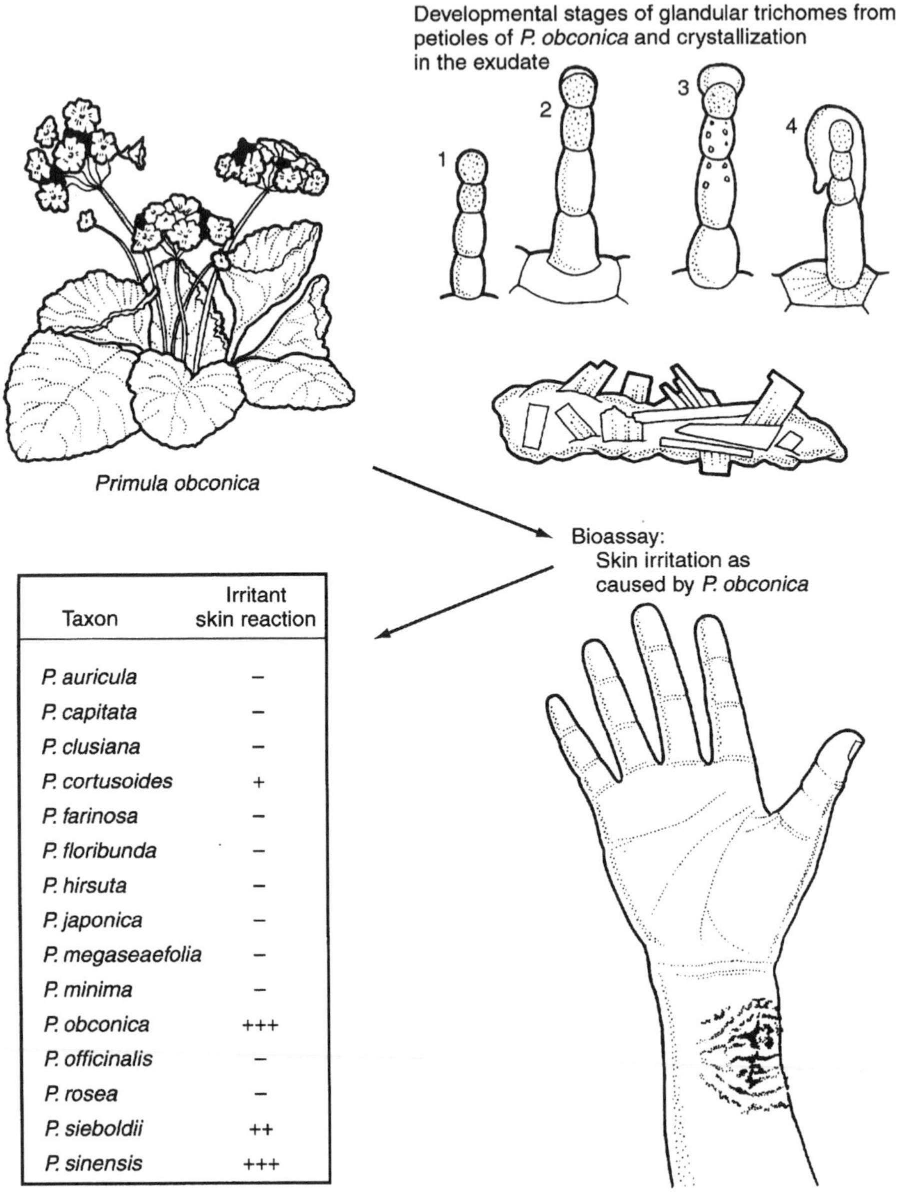

Taxon	Irritant skin reaction
P. auricula	–
P. capitata	–
P. clusiana	–
P. cortusoides	+
P. farinosa	–
P. floribunda	–
P. hirsuta	–
P. japonica	–
P. megaseaefolia	–
P. minima	–
P. obconica	+++
P. officinalis	–
P. rosea	–
P. sieboldii	++
P. sinensis	+++

Fig. 1. Bioassay-guided investigation of irritant compounds (later identified as the quinone primin) in glandular trichomes of *Primula* species, as reported by Nestler (1904). Leaf petioles or leaves of each species were brought into close contact with the skin of the arm of a test person. The irritant reaction was recorded during the following 10–14 days. Four of the 15 taxa showed positive reaction of different intensity (+, ++, +++).

other hydrophilic non-volatiles, which cannot be listed here in detail. In some cases, such as in Solanaceae, even proteins are sequestered in the resin of trichomes, where they function as peroxidase or polyphenol oxidase in the polymerization process of phenolic substances upon rupture of the gland's cuticle by insects (Ryan *et al.*, 1982; Steffens and Walters, 1991). The citation of relatively few plant families in Table I is not necessarily an indication of the lack of certain compounds in other taxa. First, this list is certainly incomplete and, second, the screening for trichome metabolites has so far particularly focused on specific families, such as Lamiaceae, Cannabaceae or Solanaceae, where the occurrence of pharmacologically or ecologically interesting exudates was most obvious.

III. WHICH BIOMOLECULES ARE USEFUL CHARACTERS FOR PLANT TAXONOMY?

Many of the above-mentioned biomolecules were already shown to be useful tools in plant taxonomy. To evaluate their importance in systematics in more detail, however, requires some general remarks on the use of plant metabolites for this scientific field.

Chemotaxonomy in its widest sense covers the use of many types of chemical information provided by organisms through their enormous biosynthetic potential. There has been some controversy about whether this also includes characters gained from proteins and nucleic acids, or whether this aspect necessitates a differentiation into micromolecular and macromolecular taxonomy, respectively. It should, however, be seen that there is a principal difference regarding the quality of information, which is independent of the size of the molecules used. In the case of DNA, RNA or proteins, the taxonomic value lies *a priori* in the sequential order of the monomeric units of the whole molecule, and not in the chemical structure of these units or in the mode of their connection. This is different from the second group of macromolecules that have either structural or metabolic functions in the organism, such as cell wall elements, sugar polymers, lipids, polyphenols or polyterpenes. The structural chemistry of the latter type of macromolecules has essentially contributed to our understanding of the gross relationship between groups of organisms, particularly in the taxonomically higher-ranking taxa (e.g. the presence or absence of murein in prokaryotes; the differentiation of major algae classes by sugar polymers and pigments; chitin as a synapomorphy of the eumycota in comparison to other groups of mycetes; the composition of lignin as used as a phylogenetic character in spermatophyta).

In contrast, chemotaxonomy in a more narrow sense (and in the way it will be used here) predominantly deals with so-called secondary metabolites (or natural products of low molecular weight). Their distribution is usually

TABLE I

Metabolites of different chemical classes associated with trichomes or their resinous exudates in families of flowering plants

Compound	Plant source	Example	References
Monoterpenes	Asteraceae	*Artemisia tridentata*	Kelsey and Shafizadeh (1980)
	Cannabaceae	*Humulus spinosa*	Amelunxen and Arbeiter (1969)
	Lamiaceae	*Mentha piperita*	Sticher and Flück (1968)
	Rutaceae	*Dictamnus albus*	Amelunxen and Arbeiter (1967)
Monoterpene glycosides	Asteraceae	*Tetraneuris acaulis*	Spring *et al.* (1994)
Sesquiterpenes	Asteraceae	*Helianthus annuus*	Spring *et al.* (1992b)
	Cannabaceae	*Humulus lupulus*	Buttery *et al.* (1967)
	Capparidaceae	*Cleome spinosa*	Amelunxen and Arbeiter (1969)
	Lamiaceae	*Leonitis leonurus*	Pedro *et al.* (1991)
	Solanaceae	*Solanum tuberosum*	Gibson and Pickett (1983)
Sesquiterpene lactones	Asteraceae	*Parthenium hysterophorus*	Rodriguez *et al.* (1976)
Sesquiterpene lactone glycosides	Asteraceae	*Hymenoxys robusta*	Spring *et al.* (1994)
Diterpenes	Lamiaceae	*Leonitis leonurus*	Pedro *et al.* (1991)
	Solanaceae	*Nicotiana tabacum*	Keene and Wagner (1985)
Triterpenes	Chloanthaceae	*Newcastelia viscida*	Dell and McComb (1975)
Mixed terpenes	Cannabaceae	*Cannabis sativa*	Willinsky (1973)
Benzufuranes	Asteraceae	*Pappobolus stubelii*	Spring *et al.* (1991)
Myoinositols	Asteraceae	*Tetraneuris acaulis*	Spring *et al.* (1994)
Phenols, phenolic acids	Betulaceae	*Alnus sieboldiana*	Asakawa (1971)
	Solanaceae	*Solanum tuberosum*	Gregory *et al.* (1986)
	(species from additional 20 families gave positive histochemical nitroso reaction for unspecified phenols)		Beckman *et al.* (1972)

Stilbenes	Betulaceae	*Alnus sieboldiana*	Asakawa (1971)
Flavonoids	Asteraceae	*Helianthus maximiliani*	Gershenzon (1984)
	Betulaceae	*Alnus sieboldiana*	Asakawa (1971)
	Zygophyllaceae (and many other taxa)	*Larrea tridentata*	Thompson *et al.* (1979) (for references, see Wollenweber, 1984)
Tannins	Urticaceae	*Urtica membranacea*	Corsi and Maffei (1992)
	Zygophyllaceae (and many other taxa)	*Larrea tridentata*	Thompson *et al.* (1979) (for references, see Kelsey *et al.*, 1984)
Quinones	Hydrophyllaceae	*Phacelia minor*	Reynolds and Rodriguez (1981)
	Lamiaceae	*Coleus aquaticus*	Ruedi and Eugster (1971)
	Primulaceae	*Primula obconica*	Nestler (1904)
Lignans	Zygophyllaceae	*Larrea tridentata*	Thompson *et al.* (1979)
Amines	Urticaceae	*Urtica membranaceae*	Corsi and Maffei (1992)
Alkaloids	Apocynaceae	*Vinca herbacea*	Kartmazova and Lyapunova (1976)
	Solanaceae	*Nicotiana* species	Thurston *et al.* (1966)
Sugars (polysaccharides)	Solanaceae	*Solanum tuberosum*	Steffens and Walters (1991)
	Urticaceae	*Urtica membranacea*	Corsi and Maffei (1992)
	Zygophyllaceae	*Fagonia glutinosa*	Fahn and Shimony (1996)
Proteins (peroxidase, phenoloxidase)	Solanaceae	*Solanum tuberosum*	Ryan *et al.* (1982); Steffens and Walters (1991)

restricted to certain groups of organisms. Therefore, they mostly serve for the solution of taxonomic problems in lower-ranking taxa, usually at or below the family level. The major taxonomic value of such plant metabolites is clearly shown in the huge structural diversity of which we still have only relatively limited knowledge. "We live in a chemical world ... a world dominated by color, by scent and by taste" was Swain's opening sentence of a previous review on so-called secondary compounds of plants in which he mentioned details of more than 10 000 low-molecular-weight biomolecules from higher plants and fungi (Swain, 1977). Just 20 years later, this number has been enlarged many times and may now have reached 10 times this amount.

Many natural products of low molecular weight, listed above as being associated with glandular trichomes, are unquestionably important taxonomic characters. Flavonoids, sesquiterpene lactones and diterpenes especially have been employed intensively to solve taxonomic and phylogenetic questions. Their occurrence in plants has been reviewed in detail (e.g. Fischer *et al.*, 1979; Harborne and Mabry, 1982; Seaman, 1982; Bohm, 1988; Seaman *et al.*, 1990), at least as long as the authors were able to keep up with the number of newly identified structures. With respect to the topic of this review, it should be stated, however, that in nearly all cases, the occurrence of certain compounds in a taxon was deduced from the extraction of bulk samples and not from specific cells such as plant trichomes.

IV. STRATEGIES AND OBSTACLES IN CHEMOTAXONOMY

The option of gaining concentrated metabolites from the plant surface by using more selective extraction methods has so far been widely ignored in the field of chemotaxonomy. The reasons for this may be seen in the relatively complicated way in which chemical data for plant systematics were collected in the past and in the previous lack of analytical techniques to deal with the extremely small quantities of samples obtainable from plant trichomes. "A really significant use of chemical characters requires a relatively sound knowledge of the chemistry, biochemistry and distribution of natural products on the part of the taxonomist, and of biological practices and problems on the part of the chemist" was stated by Hegnauer (1986) with respect to the necessity of a close collaboration between the two scientific disciplines in chemotaxonomy. However, even given that, it appears that some strategic errors were responsible for some stagnation in the development of chemotaxonomy. To explain this, it is necessary to explain the common strategy briefly and to trace its deficits.

The structural diversity of low-molecular-weight plant metabolites together with their specific distribution often permits their use for the classification down to the level of species, subspecies or even chemical races. This is, at the same time, one of the most critical points in the practice of chemotaxonomy. While for morphological characters usually hundreds of plants from different populations are compared, chemical data are often based on the investigation of a single collection. If sampling of data is done so sparsely that variation due to modificatory influences (i.e. phenotypic variation due to environmental conditions and not due to genetic reasons) or developmental processes remains undiscovered, the results can be misleading, although the chemical information itself may be correct. Such a situation can be seen as a consequence of the complicated and often expensive way of gaining chemical data when compared with the elaboration of morphological, anatomical or cytological characters. However, it could also be argued that taxonomists have just missed an opportunity to develop or to improve their own methodology for compound identification that could make them independent from the traditional way of structure elucidation used in organic chemistry. A simplified technique to gain chemical information is a requirement to enable comparison of data on a much broader basis as it is necessary for the investigation of taxonomic relationships (cf. Spring and Buschmann, 1996).

Whether a compound is useful to solve taxonomic problems or not depends on several specific attributes, such as: (1) a broad distribution in nature; (2) sufficient concentration in the plant for uncomplicated detection; (3) simple extractability and separation with chromatographic techniques; and (4) a high level of chemical stability in order to prohibit compound degradation during the storage of plant material or during the preparation process. The chemical nature itself is only of secondary importance for taxonomic application. This may appear strange, since almost all chemotaxonomic reports contain a wealth of chemical information and there is no question that the value of the data reaches a superior level with the knowledge of the structural identity of the compounds in a plant. However, when we compare it, for instance, with molecular systematics, nobody would raise doubts about the taxonomic value of DNA restriction fragments or randomly amplified sequences, although we may have no idea of the genetic information they inherit. The first question is whether we are able to ascertain the presence or absence of a certain character in an organism or not. After that, it is a superior task to elucidate the precise sequence in nucleic acids of a fragment and perhaps the genetic information it is coding for.

Chemotaxonomy, in the past, has often followed an inverse strategy by making the second step before the first. Thus, structure elucidation, which is the most complicated and time-consuming part in phytochemical work, was given higher priority than the elaboration of techniques that could prove the presence or absence of a character in different plant samples. To a considerable degree this can be explained by the fact that the data were not elaborated by taxonomists themselves. Often the characters were "side products" of chemical investigations that focused primarily on other goals than plant taxonomy. Because large amounts of plant material were necessary to provide sufficient yields of purified metabolites for spectroscopic structure elucidation, broad sampling in different populations of a taxon as required in plant systematics was often impossible. Sample selection was frequently a matter of pharmacological, physiological or ecological interest, or depended on random accessibility rather than taxonomic importance.

There are some fields in which attempts to overcome these problems were made successfully. They are all based on the principle of indirect compound identification. Perhaps one of the oldest attempts reaches back to the nineteenth century. Nylander (1866, 1867) recognized that specific staining reactions of lichens with alkalic solutions were a reliable means to differentiate between taxa that showed very low variation in morphological characters of their thalli. Meanwhile, we know of several hundreds of different metabolites from lichens among which are aliphatic acids, quinones, depsides, polyols or triterpenes that can be used in chemotaxonomy. Of course, the methods to prove their occurrence have changed considerably as may be seen from a recent study on depsides in Umbilicariaceae by means of high-performance liquid chromatography (HPLC) and NMR spectrometry (Narui *et al.*, 1998).

As a further example, phenolics in general and particularly flavonoids have been employed very intensively for plant classification. Standardized and simple fingerprinting techniques for their detection by means of two-dimensional thin layer chromatography as used by Smith and Levin (1963) for *Asplenium* species or by Alston (1967) in Fabaceae were applied to compare compound patterns of plants. Schilling and colleagues (1988; Schilling, 1989) used column chromatography on Sephadex LH 20 for flavonoid purification in *Viguiera* species and identified compounds by comparison of UV/VIS data as suggested by Markham (1982).

Besides flavonoids, sesquiterpene lactones, polyacetylenes and chromenes of the Asteraceae were investigated intensively in order to solve taxonomic problems. Seaman (1982) not only reviewed the occurrence of more than 1400 different sesquiterpene lactones reported from *ca.* 850 species but also listed useful guidelines for their application in taxonomy. Payne and coworkers were among the first to investigate infraspecific variation of these constituents and the influence of hybridization in *Ambrosia* on the base of an increased number of samples in various plant populations (Payne, 1976). Again, this was achieved by screening plant extracts with thin layer chromatography (TLC) and using reference compounds for the identification of metabolites instead of purifying compounds for spectroscopic structure elucidation (Payne *et al.*, 1973). Similar investigations were reported by Kelsey *et al.* (1973) for sesquiterpene lactones in *Artemisia* (cf. Kelsey and Shafizadeh, 1979), by Mabry *et al.* (1975) for flavonoids and sesquiterpene lactones in *Vernonia*, and by Proksch and Rodriguez (1982) for chromenes in Asteraceae, to give a few examples. In addition, various attempts were reported to work out standard procedures for compound identification on TLC by means of staining reactions (Geissmann and Griffin, 1971; Picman *et al.*, 1980).

The number of known structures in phytochemistry has increased rapidly within the last two decades. Thus, at the end of the 1980s, Harborne (1989) noted not less than 4000 different flavonoid structures, and similar increases in the identification of new structures were reported for almost all classes of compounds. This urged the employment of more sophisticated separation techniques for compound fingerprinting. Gas chromatography (GC) and high-pressure liquid chromatography (HPLC) were shown to be useful means for the separation of plant extracts and were employed for the investigation of sesquiterpene lactones (Marchand *et al.*, 1983), flavonoids (Harborne and Boardley, 1984) and many other substance classes. As the most recent development in separation techniques, capillary electrophoresis (CE) was shown to be a powerful tool for the analysis of plant secondary metabolites (Tomas-Barberan, 1995).

V. REQUIREMENTS FOR AND ADVANTAGES OF TRICHOME EXUDATE ANALYSIS FOR STUDIES IN PLANT SYSTEMATICS

With HPLC, CE and GC, taxonomists nowadays have tools at hand that help to overcome a major obstacle for gaining chemical information from glandular trichomes, namely the low yield of substances from trichome extracts. These techniques allow compound detection in the nanogram range. For sesquiterpene lactones it was shown that micromechanical sampling of 10–20 glands from *Helianthus annuus* afforded an extract sufficiently concentrated to trace all compounds in HPLC diagrams that were previously identified by standard phytochemical procedures in a bulk sample of the same plants (Spring, 1989). The method was based on the standardization of the extraction and separation of compounds so that plant samples could be compared by fingerprinting (Spring, 1991). This was achieved by using specific columns, solvents and detection wavelengths. The application of certain combinations of these parameters allowed differentiation of compounds with very similar physicochemical behaviours. Most peak assignments in these studies were made by using reference compounds. Spectroscopic measurements for structure elucidation were limited to a few cases of specific interest (Fig. 2). This strategy helped to save time and to avoid structure elucidation of compounds that were already known. In this way, more than 2000 samples from over 300 taxa of the Heliantheae have been screened in the past 10 years in order to establish a chemotaxonomic survey, mainly in the Helianthinae (cf. Spring and Buschmann, 1996). The method was also used to test the reliability of sesquiterpene lactones as taxonomic characters. Thus seasonal variation in compound patterns, the distribution of metabolites in different plant organs, the influence of environmental effects and the stability of compounds after sample harvesting were studied (for references, see Spring, 1991).

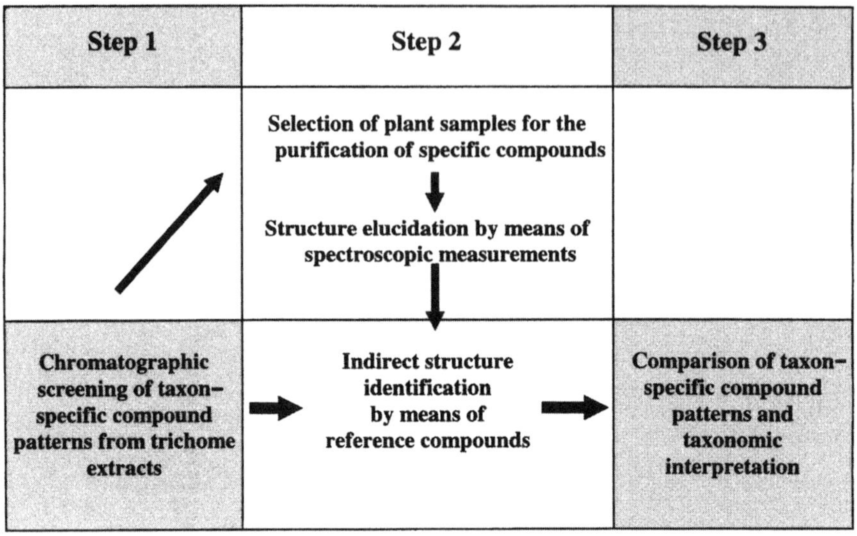

Fig. 2. Strategy applied for the chemotaxonomic investigation of species of the Asteraceae based on sesquiterpene lactones from capitate glandular trichomes (Spring *et al.*, 1992a).

Trichome microsampling in Asteraceae requires no preparation steps other than dissolving the contents of the collected hairs in the solvent used for HPLC analysis. This is not just interesting from the economical point of view, but also prohibits oxidative degradation and spontaneous derivatization of sensitive metabolites, which is important to avoid misleading data due to preparation artifacts.

The potential of chemical trichome analysis lies with the concentrated localization of specific biomolecules in one compartment. A similar situation may only be seen in the storage of metabolites in intracellular structures, such as vacuoles or lipid bodies. However, in none of these alternatives are the metabolites presented at the plant surface, and ready for collection without destroying the remaining tissue and spoiling the extract with the many kinds of phytochemicals present in a plant. In contrast to metabolites located in intracellular compartments, compounds from glandular trichomes are often sequestered and deposited extracellularly. This allows an analysis of the final products because, once they were synthesized, they are prevented from undergoing subsequent enzymatic alterations. On the other hand, it should be assured in both cases that oxidative degradation of plant products caused by non-uniform treatment of the harvested organisms prior to the extraction does not result in misleading data.

Selecting plant trichomes instead of whole plants for chemotaxonomic studies can provide the opportunity to check plants microscopically for the possible lack of the characters under investigation. This means that, if microscopic analysis fails to observe glandular trichomes, it is mostly invaluable to perform extensive phytochemical investigation of bulk samples with respect to the selected compounds. Thus, in many reports that published the lack of sesquiterpene lactones in Asteraceae, the negative result would have been predicted considering the existence (or better the absence) of glands in the plant material used. Vice versa, the experience from micromorphological and developmental studies may help to select specific plant organs for chemotaxonomic studies that promise good and stable results. Approximately 30–40% of the taxa in Helianthinae, for instance, lack glandular trichomes on leaves or stems. More than 90%, however, present them on the outer tips of the anthers (so-called anther appendages) (Fig. 3). Upon opening of the flowers and after harvesting the plants, this source of information happens to disappear by simply falling off. Since it is known that developed glands are already present in immature organs and in flowers with a closed corolla (Weber and Deufel, 1951; Spring and Bienert, 1987), the chemical characters can be found by dissecting anther appendages from young flowers (O. Spring, unpublished observations).

These were examples taken from a closely related group of plants and it is self-evident that the distribution of glands has to be checked independently for each taxon. In some cases they may be found unexpectedly on parts such as bracts, receptacles or the outer surface of the corolla as recently reported for *Artemisia annua* (Ferreira and Janick, 1995). They may occur on seeds and have even been found lined up at the bottom of the capitulum as in *Inula helenium* (Fig. 3). The abundance of glandular trichomes on floral organs is also quite common in other families. In *Leonotis leonurus* (Lamiaceae), for instance, different types of glandular trichomes (capitae, peltate) were shown to have a specific distribution between the calyx, corolla, stamen and ovary (Ascensão *et al.*, 1995).

Careful microscopic investigation prior to trichome microsampling is also useful in order to distinguish between different types of glandular trichomes possibly occurring in the same plant. In some Solanaceae, two types of very similar-appearing trichomes produce totally different metabolites (Ryan *et al.*, 1982; Steffens and Walters, 1991). Glands with a longer stalk produce glycolipids and (E)-β-farnesene, while the shorter ones store ortho-diphenolics and polyphenoloxidase. Leaves of sunflower, in addition to sesquiterpene lactone-leading capitate glands, bear uniserate non-capitate glandular trichomes, which produce sesquiterpenes (Spring *et al.*, 1992b).

The small amount of plant material necessary for the screening process with trichome microsampling is perhaps the most important advantage of this technique. Collection of data can be performed with herbarium specimens without destroying the vouchers. This is an invaluable key to make uncounted numbers of plants accessible for chemotaxonomic investigations.

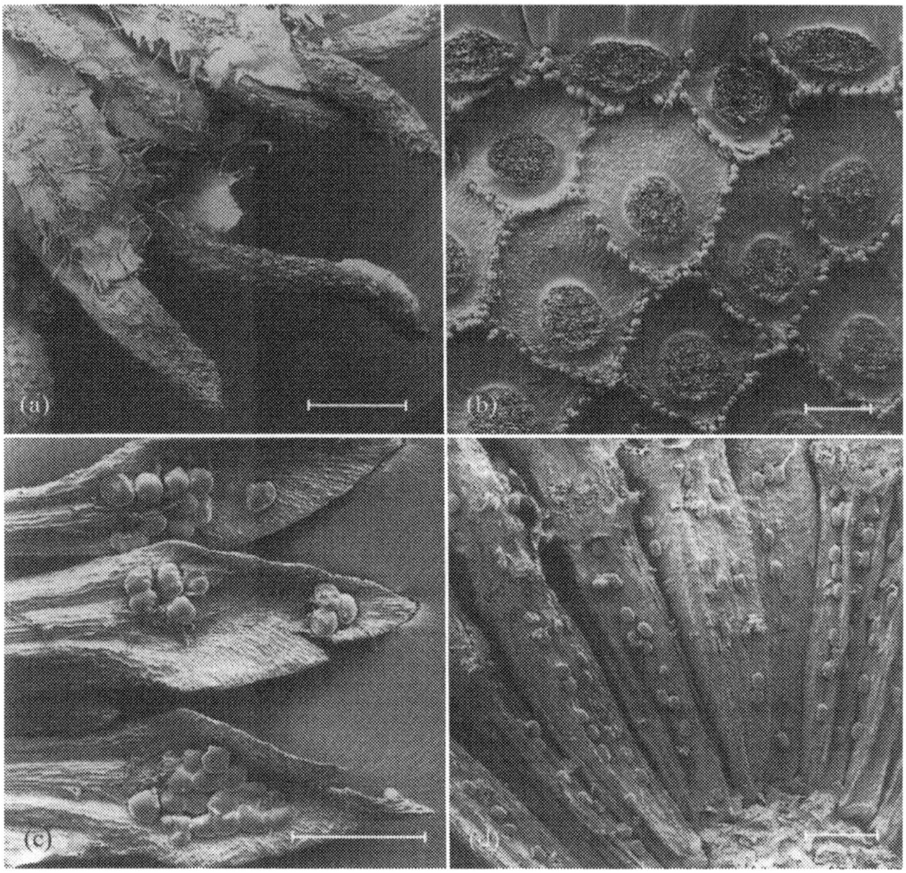

Fig. 3. Scanning electron micrographs showing distribution of capitate glandular trichomes in floral parts of various Asteraceae: (a) on the abaxial surface of bracts and ray flowers of *Chrysanthemum parthenium* (scale bar = 500 μm); (b) on the bottom of the capitulum of *Inula helenium* (achenes removed; scale bar = 200 μm); (c) on the anther appendages of *Viguiera robusta* (scale bar = 200 μm); (d) on the achenes of *Tanacetum vulgare* (scale bar = 200 μm).

In many cases, collection of fresh plant material of complete genera for the comparison of their chemical characters is impossible. This may be due to the widespread distribution of taxa, their scarcity or because some may already be extinct (cf. Spring, 1991). Of course, the stability of the compounds under investigation determines how long herbarized plants can be used for chemotaxonomic studies and this has to be established in each case. At least for some flavonoids and sesquiterpene lactones this time frame was shown to be amazingly wide, even reaching back to the beginning of the twentieth century. In a recent investigation of the genus *Scalesia* from Galapagos Islands, almost no difference was observed between two plants, collected in 1905 and 1967, respectively (Spring *et al.*, 1998).

VI. PERSPECTIVES

A. PROBLEMS AND CHANCES WITH STRUCTURE IDENTIFICATION IN ANALYTICAL STUDIES

In comparison to standard phytochemical investigations, indirect methods for compound identification always carry a higher risk of misassignments. Even very careful employment of HPLC or GC will not prevent mistakes, especially in the substance classes where thousands of compounds have more or less identical skeletal structures with only minute differences in specific substituents. This concern may be allayed somewhat by the experience that the diversity of structures found within a certain taxon is usually proportional to its taxonomic ranking. Nevertheless, it is necessary from time to time to check indirect structure assignments with the classical means of structure elucidation using spectroscopic measurements. The result of the screening process from the taxon under investigation is often a helpful guide for the selection of the plant that provides the largest number of compounds upon extraction.

Recent developments in spectroscopic on-line techniques have facilitated the process of structure identification. Thus, for instance, coupling of diode-array ultraviolet detectors with HPLC for the measurement of on-line UV/VIS spectra was shown to be very effective for compound identification of phenolics in *Equisetum* species (Veit *et al.*, 1995). GC-MS has meanwhile become a routine method for the identification of terpenoids, fatty acids and many other volatile compounds in complex plant extracts. On the other hand, the necessity to derivatize those metabolites that are otherwise unsuitable for gas chromatographic separation is a disadvantage in some cases. LC-MS coupling and capillary electrophoresis linked to MS (CE-MS) have recently become important alternative means for the identification of natural products (Gu *et al.*, 1997; Heinze *et al.*, 1998; Stöckigt, 1998). Isomeric structures in many classes of natural compounds, however, are indistinguishable by mass spectrometric analysis and require additional measurements by NMR spectroscopy. In the past, milligram quantities of a compound were necessary for a proton NMR spectrum of satisfactory quality. This has been reduced significantly with the new generations of NMR facilities. Coupling of HPLC separation with ^{1}H-NMR (LC-NMR) has recently been successfully employed for structure elucidation of sesquiterpene lactones in crude plant extracts of Asteraceae at concentrations below 100 μg (Spring *et al.*, 1995; Vogler *et al.*, 1998). Technical progress in spectrometry has now reached a level where direct structure identification and even elucidation is possible from extracts gained by trichome microsampling. Given that the average content of a compound is about 50 ng per gland, as shown for sesquiterpene lactones in sunflower (Spring *et al.*, 1987), approximately 500 glands are sufficient to obtain a reasonable proton NMR spectrum for structure elucidation.

B. BIOSYNTHETIC PATHWAYS OF NATURAL PRODUCTS: FUTURE KEY CHARACTERS IN CHEMOSYSTEMATICS

Since the early 1980s, a range of studies has been published focusing on pathways of plant metabolites and their employment in plant systematics. In contrast to the analysis of the occurrence of certain compounds in plants with the aim of classification, the steps leading to their production can be seen as a key to reconstruct the phylogenetic origin of a taxon. Chemical evolution has become an important field in plant systematics (for references, see Gottlieb, 1989, 1990). Unfortunately, biosynthetic pathways of many compounds have not yet been clarified and therefore theoretical models were suggested to be used for chemotaxonomic analysis (e.g. calculation of oxidation levels and skeletal specialization in flavonoids and sesquiterpene lactones of Asteraceae; Emerenciano *et al.*, 1987).

It was shown that plant trichomes can play an important role in the investigation of biosynthetic pathways of natural compounds. Microscopic observation of the development of young glandular trichomes, as initiated by Polites (1946) on sesquiterpene lactone-producing *Cnicus benedictus* and various *Artemisia*, or by Stahl (1956) on *Achillea millefolium* during the formation of cham-azulene, pointed at the role of basal cells of trichomes for the localization of enzymes responsible for the formation of the products that were sequestered at the tip of the trichome. Croteau and his coworkers, in a series of publications, reported on the enzymes involved in the biosynthesis of monoterpenes in mints, and on their cellular and subcellular localization (Gershenzon and Croteau, 1989, 1993; Gershenzon *et al.*, 1989; Alonso *et al.*, 1992; Ponnamperuma and Croteau, 1996). A mechanized technique for the selective extraction of these proteins from gland cells was used to study the terpenoid metabolism in *Mentha* species (Gershenzon *et al.*, 1987; McCaskill *et al.*, 1992). A similar attempt by Yerger *et al.* (1992) tried to adapt trichome isolation to the requirements of a broad selection of taxa from different plant families. Meanwhile, glandular trichomes also served for the study of monoterpenoid synthesis in thyme (*Thymus vulgaris*) by means of ^{14}C-labelled precursors (Yamaura *et al.*, 1991) and for the conversion of sesquiterpenes in *Rosa rugosa* (Hashidoko and Urashima, 1995).

Although currently just at its beginning, the use of trichomes for the investigation of biosynthesis of plant products is most likely to play an important role in future phytochemistry. It is the most convenient way to obtain enzymes in high purity, and knowledge in the sequence of enzymatic steps leading to the final plant product sequestered by a trichome will become a key issue for our understanding of the phylogeny of biomolecules. This will push the interpretation of chemical characters in plant systematics to a new level.

VII. SUMMARY

Plant metabolites associated with trichomes were up to now consulted little for the investigation of taxonomic relationships. This is unexpected given the fact that plant trichomes provide a unique means of making biomolecules of specific pathways accessible at the plant surface in a concentrated form. In cases where compounds from glandular trichomes have been investigated, they have proved to be reliable taxonomic characters. In combination with microscopic studies, the presence or absence of metabolites in a plant or in specific organs is often predictable, thus avoiding unnecessary chemical analysis. The high resolution of modern chromatography allows the development of fingerprinting techniques for compound assignment of extracts gained directly from trichomes via microsampling. The previous inability of spectroscopic structure identification with analytical sample amounts is nowadays partly compensated. On-line techniques such as GC-MS, LC-MS, CE-MS and LC-NMR require sample amounts adjusted to the concentrations found in trichomes. Trichomes were also shown to be useful tools for recent trends in chemosystematics considering information from biosynthetic pathways in plants. Enzymes extracted from living glandular cells were successfully employed to study the biosynthesis of terpenoids. Considering these circumstances, chemotaxonomy based on metabolites from glandular trichomes could become an interesting section of plant systematics – it just has to be attempted.

ACKNOWLEDGEMENTS

The author is grateful to Dr H. Buschmann, Department of Biology, University of Bath, UK for helpful comments and discussion, as well as to B. Rozynek and B. Curth, Institut für Botanik, Universität Hohenheim, for technical assistance.

REFERENCES

Alonso, W. R., Rajaonarivony, J. I. M., Gershenzon, J. and Croteau, R. (1992). Purification of 4S-Limonene synthases, a monoterpene cyclase from the glandular trichomes of peppermint (*Mentha* × *piperita*) and spearmint (*Mentha spicata*). *Journal of Biological Chemistry* **267**, 7582–7587.

Alston, R. E. (1967). Biochemical systematics. *In* "Evolutionary Biology", Vol. 1 (T. Dobzhansky, M. K. Hecht and W. C. Steer, eds), pp. 197–305. Appleton-Century-Crofts, New York.
Amelunxen, F. and Arbeiter, H. (1967). Untersuchungen an den Spritzdrüsen von *Dictamnus albus* L. *Zeitschrift für Pflanzenphysiologie* **58**, 49–69.
Amelunxen, F. and Arbeiter, H. (1969). Untersuchungen an den Drüsenhaaren von *Cleome spinosa* L. *Zeitschrift für Pflanzenphysiologie* **61**, 73.
Amelunxen, T., Wahlig, T. and Arbeiter, H. (1969). Über den Nachweis des ätherischen Öls in isolierten Drüsenhaaren und Drüsenschuppen von *Mentha piperita* L. *Zeitschrift für Pflanzenphysiologie* **61**, 68–72.
Asakawa, Y. (1971). Chemical constituents of *Alnus sieboldiana* (Betulaceae). II. The isolation and structure of flavonoids and stilbenes. *Bulletin of the Chemical Society of Japan* **44**, 1794.
Ascensão, L., Marques, N. and Pais, M. S. (1995). Glandular trichomes on vegetative and reproductive organs of *Leonotis leonurus* (Lamiaceae). *Annals of Botany* **75**, 619–626.
Beckman, C. H., Mueller, W. C. and McHardy, W. E. (1972). The localization of stored phenols in plant hairs. *Physiological Plant Pathology* **2**, 69–74.
Behnke, H.-D. (1984). Plant trichomes – structure and ultrastructure: general terminology, taxonomic applications, and aspects of trichome-bacteria interaction in leaf tips of *Dioscorea*. *In* "Biology and Chemistry of Plant Trichomes" (E. Rodriguez, P. L. Healey and I. Mehta, eds), pp. 1–22. Plenum Press, New York, London.
Bohm, B. A. (1988). The minor flavonoids. *In* "The Flavonoids: Advances in Research since 1980" (J. B. Harborne, ed.), pp. 329–388. Chapman and Hall, London.
Buttery, R. G., Lundin, R. E. and Ling, L. (1967). Characterization of some C15 constituents of hop. oil. *Journal of Agricultural and Food Chemistry* **15**, 58–63.
Corsi, G. and Maffei, F. (1992). *Urtica membranacea* pearl glands. I. Morpho-ontogenetic and histochemical aspects. *Phyton* **32**, 235–245.
Dell, B. and McComb, A. J. (1975). Glandular hairs, resin production, and habitat of *Newcastelia viscida* E. Pritzel (Dicrastylidaceae). *Australian Journal of Botany* **23**, 373–379.
Emerenciano, V. P., Ferreira, Z. S., Kaplan, M. A. C. and Gottlieb, O. R. (1987). A chemosystematic analysis of tribes of Asteraceae involving sesquiterpene lactones and flavonoids. *Phytochemistry* **26**, 3103–3115.
Fahn, A. and Shimony, C. (1996). Glandular trichomes of *Fagonia* L. (Zygophyllaceae) species: structure, development and secreted materials. *Annals of Botany* **77**, 25–34.
Ferreira, J. F. S. and Janick, J. (1995). Floral morphology of *Artemisia annua* with special reference to trichomes. *International Journal of Plant Science* **156**, 807–815.
Fischer, N. H., Olivier, E. J. and Fischer, H. D. (1979). "The Biogenesis and Chemistry of Sesquiterpene Lactones." Springer-Verlag, Vienna.
Geissmann, T. A. and Griffin, T. J. (1971). Sesquiterpene lactones: acid-catalyzed color reactions as an aid in structure determination. *Phytochemistry* **10**, 2475–2485.
Gershenzon, J. (1984). The terpenoid chemistry of Helianthus, series Corona-solis and its ecological and systematic application. Dissertation, University of Texas, Austin.
Gershenzon, J. and Croteau, R. (1989). Regulation of monoterpene biosynthesis in higher plants. *In* "Biochemistry of the Mevalonic Acid Pathway to Terpenoids"

(G. H. N. Towers and H. A. Stafford, eds), pp. 99–160. Recent Advances in Phytochemistry, Vol. 24.
Gershenzon, J. and Croteau, R. (1993). Terpenoid biosynthesis: the basic pathway and formation of monoterpenes, sesquiterpenes, and diterpenes. *In* "Lipid Metabolism in Plants" (T. S. Moore, ed.), pp. 340–388. CRC Press, Boca Raton.
Gershenzon, J., Duffy, M. A., Karp, F. and Croteau, R. (1987). Mechanized techniques for the selective extraction of enzymes from plant epidermal glands. *Analytical Biochemistry* **163**, 159–164.
Gershenzon, J., Maffei, M. and Croteau, R. (1989). Biochemical and histochemical localization of monoterpene biosynthesis in the glandular trichomes of spearmint (*Mentha spicata*). *Plant Physiology* **89**, 1351–1357.
Gibson, W. S. and Pickett, J. A. (1983). Wild potato repels aphids by release of aphid alarm pheromone. *Nature* **302**, 608.
Gottlieb, O. R. (1989). The role of oxygen in phytochemical evolution towards diversity. *Phytochemistry* **28**, 2545–2558.
Gottlieb, O. R. (1990). Phytochemicals: differentiation and function. *Phytochemistry* **29**, 1715–1724.
Gregory, P., Ave, D. A., Bouthyette, P. Y. and Tingey, W. M. (1986). Insect-defensive chemistry of potato glandular trichomes. *In* "Insects and the Plant Surface" (B. Juniper and S. Southwood, eds). Edward Arnold, London.
Gu, Z.-M., Zhou, D., Wu, J., Shi, G., Zeng, L. and McLaughlin, J. L. (1997). Screening for annonaceous acetogenins in bioactive plant extracts by liquid chromatography/mass spectrometry. *Journal of Natural Products* **60**, 242–248; and references therein.
Harborne, J. B. (1989). "Introduction to Ecological Biochemistry." Academic Press, London.
Harborne, J. B. and Mabry, T. J. (eds) (1982). "Advances in Flavonoid Research, 1975–1980." Chapman and Hall, London.
Harborne, J. B. and Boardley, M. (1984). Use of high-performance liquid chromatography in the separation of flavonolglycosides and flavonol sulphates. *Journal of Chromatography* **299**, 377–385.
Hashidoko, Y. and Urashima, M. (1995). Efficient preparation of browning-free glandular trichome tissues from the surface of leaves of *Rosa rugosa* Thunb. *Plant Cell Physiology* **36**, 127–132.
Hegnauer, R. (1986). Phytochemistry and plant taxonomy – an essay on the chemotaxonomy of higher plants. *Phytochemistry* **25**, 1519–1535.
Heinze, S., Hülsmann, H., Schlegel, R. and Gräfe, U. (1998). Advances of screening for natural compounds using mass spectrometric tools. *In* "Natural Product Analysis" (P. Schreier, M. Herderich, H.-U. Humpf and W. Schwab, eds), pp. 177–186. Vieeg Verlag, Braunschweig.
Hummel, K. and Staesche, K. (1962). Die Verbreitung der Haartypen in den natürlichen Verwandschaftsgruppen. *In* "Plant Hairs" (J. C. Uphof, ed.), pp. 209–250. Gebrüder Bornträger, Berlin.
Kartmazova, L. S. and Lyapunova, P. W. (1976). The anatomic structure and localization of alkaloids in the vegetative organs of the herbaceous periwinkle (*Vinca herbacea* Waldst. et Kit.). *Farmatsiya* (Mosc.) **25**, 22; *Biological Abstracts* **63**, 46–422.
Keene, C. K. and Wagner, G. J. (1985). Direct demonstration of duvatrienediol biosynthesis in glandular heads of tobacco trichomes. *Plant Physiology* **79**, 1026–1032.
Kelsey, R. G. and Shafizadeh, F. (1979). Sesquiterpene lactones and systematics of the genus *Artemisia*. *Phytochemistry* **18**, 1591–1611.

Kelsey, R. G. and Shafizadeh, F. (1980). Glandular trichomes and sesquiterpene lactones of *Artemisia nova* (Asteraceae). *Biochemical Systematics and Ecology* **8**, 371–375.
Kelsey, R. G., Morris, M. S., Bhadane, N. R. and Shafizadeh, F. (1973). Sesquiterpene lactones of *Artemisia*: TLC analysis and taxonomic significance. *Phytochemistry* **12**, 1345–1350.
Kelsey, R. G., Reynolds, G. W. and Rodriguez, E. (1984). The chemistry of biologically active constituents secreted and stored in plant glandular trichomes. *In* "Biology and Chemistry of Plant Trichomes" (E. Rodriguez, P. L. Healey and I. Mehta, eds), pp. 187–241. Plenum Press, New York.
Levin, D. A. (1973). The role of trichomes in plant defense. *Quarterly Review of Biology* **48**, 1–15.
Mabry, T. J., Abdel-Baset, Z., Padolina, W. G. and Jones, S. B. (1975). Systematic implications of flavonoids and sesquiterpene lactones in species of *Vernonia*. *Biochemical Systematics and Ecology* **2**, 185–192.
Marchand, B., Behl, H. M. and Rodriguez, E. (1983). Application of high-performance liquid chromatography for analysis and isolation of sesquiterpene lactones. *Journal of Chromatography* **265**, 97–104.
Markham, K. R. (1982). "Techniques of Flavonoid Identification." Academic Press, London.
McCaskill, D., Gershenzon, J. and Croteau, R. (1992). Morphology and monoterpene biosynthetic capabilities of secretory cell clusters isolated from glandular trichomes of peppermint (*Mentha piperita* L.). *Planta* **87**, 445–454.
Metcalf, C. R. and Chalk, L. (1957). "Anatomy of the Dicotyledons", Vols I and II. Clarendon Press, Oxford.
Narui, T., Sawada, K., Takatsuki, S., Okuyama, T., Culberson, C. F., Culberson, W. L. and Shibata, S. (1998). NMR assignments of depsides and tridepsides of the lichen family Umbilicariaceae. *Phytochemistry* **48**, 815–822.
Nestler, A. (1904). "Hautreizende Primeln. Untersuchungen über die Entstehung, Eigenschaften und Wirkungen des Primelhautgiftes." Verlag Gebr. Bornträger, Berlin.
Nylander, W. (1866). Circa novum in studio lichenum criterium chemicum. *Flora (Germany)* **49**, 198.
Nylander, W. (1867). Hypochlorite of lime and hydrate of potash, two new criteria in the study of lichens. *Journal of the Linnean Society of London, Botany* **9**, 358.
Payne, W. W. (1976). Biochemistry and species problems in *Ambrosia* (Asteraceae – Ambrosieae). *Plant System Evolution* **125**, 169.
Payne, W. W., Geissman, T. A., Lucas, A. J. and Saitoh, T. (1973). Chemosystematics and taxonomy of *Ambrosia chamissonis*. *Biochemical Systems* **1**, 21–33.
Pedro, L. G., Barroso, J. G., Marques, N. T., Ascensão, L., Pais, M. S. and Scheffer, J. J. (1991). Composition of the essential oil from sepals of *Leonitis leonurus* R. Br. *Journal of Essential Oil Research* **3**, 451–453.
Picman, A. K., Ranieri, R. L., Towers, G. H. N. and Lam, J. (1980). Visualization reagents for sesquiterpene lactones and polyacetylenes on thin-layer chromatograms. *Journal of Chromatography* **189**, 187–198.
Politis, J. (1946). Sur la formation des glucosides amers dans les poils glanduleux de certain plantes. *Comptes Rendues* **222**, 910–911.
Ponnamperuma, K. and Croteau, R. (1996). Purification and characterization of an NADPH-Cytochrome P450 (Cytochrome c) reductase from spearmint (*Mentha spicata*) glandular trichomes. *Archives of Biochemistry and Biophysics* **329**, 9–16.
Proksch, P. and Rodriguez, E. (1982). High performance liquid chromatography of

chromenes and benzofurans from the genus *Encelia* (Asteraceae). *Journal of Chromatography* **240**, 543–546.

Reynolds, G. and Rodriguez, E. (1981). Prenylated hydroquinons: contact allergens from trichomes of *Phacelia minor* and *P. parryi*. *Phytochemistry* **20**, 1365–1366.

Rodriguez, E., Dillon, M. O., Mabry, T. J., Mitchell, J. C. and Towers, G. H. N. (1976). Dermatologically active sesquiterpene lactones in trichomes of *Parthenium hysterophorus* L. (Compositae). *Experientia* **32**, 236–238.

Ruedi, P. and Eugster, C. H. (1971). Struktur von Coleone C, einem neuen Blattfarbstoff aus *Coleus aquaticus* Curcke. *Helvetica Chimica Acta* **54**, 1606.

Ryan, J. D., Gregory, P. and Tingey, W. M. (1982). Phenolic oxidase activities in glandular trichomes of *Solanum berthaultii*. *Phytochemistry* **21**, 1885.

Schilling, E. E. (1989). External flavonoid aglycons of *Viguiera* series *Viguiera* (Asteraceae: Helianthinae). *Biochemical Systematics and Ecology* **17**, 535–538.

Schilling, E. E., Panero, J. L. and Bohm, B. A. (1988). Flavonoids of *Viguiera* Section *Maculatae*. *Biochemical Systematics and Ecology* **16**, 413–416.

Seaman, F. C. (1982). Sesquiterpene lactones as taxonomic characters in Asteraceae. *Botanical Review* **48**, 121–595.

Seaman, F. C., Bohlmann, F., Zdero, C. and Mabry, T. J. (1990). "Diterpenes of flowering plants: Compositae, Asteraceae." Springer Verlag, Berlin.

Smith, D. M. and Levin, D. A. (1963). A chromatographic study of reticulate evolution in the Appalachian *Asplenium* complex. *American Journal of Botany* **50**, 952–958.

Solereder, H. (1899, Erg. Bd. 1908). "Systematische Anatomie der Dicotyledonen." Enke, Stuttgart.

Spring, O. (1989). Microsampling: an alternative approach using sesquiterpene lactones for systematics. *Biochemical Systematics and Ecology* **17**, 509–517.

Spring, O. (1991). Trichome microsampling of sesquiterpene lactones for the use of systematic studies. *In* "Modern Phytochemical Methods" (N. H. Fischer, ed.). Recent Advances in Phytochemistry, Vol. 25, pp. 319–345. Plenum Press, New York.

Spring, O. and Bienert, U. (1987). Capitate glandular hairs from sunflower leaves: development, distribution and sesquiterpene lactone content. *Journal of Plant Physiology* **130**, 441–448.

Spring, O. and Buschmann, H. (1996). A chemotaxonomic survey of sesquiterpene lactones in the Helianthinae (Compositae). *In* "Compositae: Systematics. Proceedings of the International Compositae Conference, Kew, 1994" (D. J. N. Hind and H. J. Beentje, eds), pp. 307–316. Royal Botanic Gardens, Kew.

Spring, O., Bienert, U. and Klemt, V. (1987). Sesquiterpene lactones in glandular trichomes of sunflower leaves. *Journal of Plant Physiology* **130**, 433–439.

Spring, O., Vargas, D. and Fischer, N. H. (1991). Sesquiterpene lactones and benzofurans in glandular trichomes of three *Pappobolus* species. *Phytochemistry* **30**, 1861–1867.

Spring, O., Panero, J., L. and Schilling, E. E. (1992a). Chemotaxonomic analysis of *Pappobolus* (Asteraceae: Helianthinae). *Biochemical Systematics and Ecology* **20**, 671–684.

Spring, O., Rodon, U. and Macias, F. A. (1992b). Sesquiterpenes from noncapitate glandular trichomes of *Helianthus annuus*. *Phytochemistry* **31**, 1541–1544.

Spring, O., Zitterell-Haid, B., Bierner, M. W. and Mabry, T. J. (1994). Chemistry of glandular trichomes in *Hymenoxys* and related genera. *Biochemical Systematics and Ecology* **22**, 171–195.

Spring, O., Buschmann, H., Vogler, B., Schilling, E. E., Spraul, M. and Hoffmann, M.

(1995). Sesquiterpene lactone chemistry of *Zaluzania grayana* from on-line LC-NMR measurements. *Phytochemistry* **39**, 609–612.

Spring, O., Heil, N. and Eliasson, U. (1999). Chemosystematic studies on the genus *Scalesia* (Asteraceae). *Biochemical Systematics and Ecology* **27**, 277–288.

Stahl, E. (1956). Über Vorgänge in den Drüsenhaaren der Schafgarbe. *Zeitschrift für Botanik* **45**, 297–315.

Staudermann, W. (1924). Die Haare der Monokotylen. *Botanisches Archiv* **8**, 105–184.

Steffens, J. C. and Walters, D. S. (1991). Biochemical aspects of glandular trichome-mediated insect resistance in the Solanaceae. *In* "Naturally Occurring Pest Bioregulators" (P. A. Hedin, ed.), pp. 136–149, ACS Symposium Series 449. American Chemical Society, Washington, DC.

Sticher, V. O. and Flück, H. (1968). Die Zusammensetzung von genuinen, extrahierten und destillierten ätherischen Ölen einiger *Mentha*-Arten. *Pharmaceutica Acta Helvetica* **43**, 411–420.

Stöckigt, D. (1998) How to analyse alkaloids in crude mixtures? By CE/MS! *In* "Natural Product Analysis" (P. Schreier, M. Herderich, H.-U. Humpf and W. Schwab, eds), pp. 177–186. Vieeg Verlag, Braunschweig.

Swain, T. (1977). Secondary compounds as protective agents. *Annual Review of Plant Physiology* **28**, 479–501.

Thompson, W. W., Platt-Aloia, K. and Koller, D. (1979). Ultrastructure and development of the trichomes of *Larrea tridentata* (Creosote bush). *Botanical Gazette* **140**, 249–254.

Thurston, R., Smith, W. T. and Cooper, B. P. (1966). Alkaloid secretion by trichomes of *Nicotiana* species and resistance to aphids. *Entomologia Experimentalis et Applicata* **9**, 428.

Tomas-Barberan, F. A. (1995). Capillary electrophoresis: a new technique in the analysis of plant secondary metabolites. *Phytochemical Analysis* **6**, 177–192.

Uphof, J. C. (1962). Plant hairs. *In* "Encyclopaedia of Plant Anatomy IV" (W. Zimmermann and P. G. Ozenda, eds), pp. 1–206. Gebr. Bornträger, Berlin.

Veit, M., Beckert, C., Höhne, C., Bauer, K. and Geider, H. (1995). Interspecific and intraspecific variation of phenolics in the genus *Equisetum* subgenus *Equisetum*. *Phytochemistry* **38**, 881–891.

Vogler, B., Klaiber, I., Roos, G., Walter, C. U., Hiller, W., Sandor, P. and Kraus, W. (1998). Combination of LC-MS and LC-NMR as tool for the structure determination of natural products. *Journal of Natural Products* **61**, 175–178.

Wagner, G. J. (1991). Secreting glandular trichomes: more than just hairs. *Plant Physiology* **96**, 675–679.

Weber, U. and Deufel, J. (1951). Zur Cytologie der Drüsenhaare von *Achillea millefolium*. *Archiv für Pharmakologie* **284**, 318–323.

Willinsky, M. D. (1973). Analytical aspects of *Cannabis* chemistry. *In* "Chemistry, Pharmacology, Metabolism and Clinical Effects" (R. Mechoulam, ed.), pp. 137–165. Academic Press, New York and London.

Wollenweber, E. (1984). The systematic implication of flavonoids secreted by plants. *In* "Biology and Chemistry of Plant Trichomes" (E. Rodriguez, P. L. Healey and I. Mehta, eds), pp. 53–68. Plenum Press, New York.

Yamaura, T., Tanaka, S. and Tabata, M. (1991). Localization of the biosynthesis and accumulation of monoterpenoids in glandular trichomes of thyme. *Planta Medica* **58**, 153–158.

Yerger, E. H., Grazzini, R. A., Hesk, D., Cox-Foster, D. L., Craig, R. and Mumma, R. O. (1992). A rapid method for isolating glandular trichomes. *Plant Physiology* **99**, 1–7.

Anacardic Acids in Trichomes of *Pelargonium*: Biosynthesis, Molecular Biology and Ecological Effects

D. J. SCHULTZ,[1] J. I. MEDFORD,[2] D. COX-FOSTER,[3] R. A. GRAZZINI,[4] R. CRAIG[5] and R. O. MUMMA[3]

[1]Department of Botany and Plant Pathology, Michigan State University, East Lansing, MI, USA
[2]Department of Biology, Colorado State University, Fort Collins, CO, USA
[3]Department of Entomology, The Pennsylvania State University, University Park, PA, USA
[4]Center Analytical Laboratories, State College, PA, USA
[5]Department of Horticulture, The Pennsylvania State University, University Park, PA, USA

Advances in Botanical Research Vol. 31
ISBN 0-12-005931-2

I. INTRODUCTION

For the past 38 years, we have been examining the interactions of three organisms: the zonal geranium *Pelargonium* × *hortorum* L. H. Bailey and two arthropod pests, as a multidisciplinary research team of horticulturists, botanists, geneticists, entomologists, biochemists, chemists and molecular biologists.

Pelargonium is a genus of approximately 250 species; most are endemic to the Republic of South Africa. This genus should be differentiated from the *Geranium* genus, which also has 250 species endemic to many temperate climates, including Europe and North America. *Geranium* includes many temperate wild flowers and a few cultivated winter hardy perennials. Pelargoniums, however, are often referred to by the common name geranium. Pelargoniums consist of a few major horticultural groups: zonal geraniums, ivy-leaved geraniums, regal pelargoniums or Martha Washingtons, and several scented-leaved species.

We initiated this project in order to understand better a specific host–plant resistance phenomenon between *Pelargonium* and the two-spotted spider mite, *Tetranychus urticae* Koch, and the foxglove aphid, *Acyrthosiphon solani* Kaltenbach. Here we define host–plant resistance as: "Genes controlling a biological function which enables a plant to resist attack of a pathogen or pest", and our efforts to characterize this phenomenon are related in this review.

In 1960, we observed a group of abnormal plants in our research greenhouse. These geraniums were chlorotic and obviously unhealthy, whilst adjacent plants were perfectly normal, green and healthy. The distressed plants were found to be heavily infested with two-spotted spider mites. Additional plants infested with spider mites were located apparently at random throughout the greenhouse. Upon closer examination, it was apparent that all the susceptible plants had resulted from a single hybrid: G10 ("King Midas") × G71 (a dark-scarlet-flowered seedling from "Florist's Mix"). "King Midas" had been developed by X-irradiation and had been selected for its unique citrus orange flower colour. It apparently also contained a mutation that conferred mite susceptibility.

Our curiosity led us to investigate this phenomenon further. Although our original investigations were conducted with "King Midas", it was not sufficiently fertile or vigorous to be used in further genetic analyses. By the mid-1960s, based on bioassays of our extensive germplasm, we chose two model genotypes – one mite-resistant genotype, which had red flowers and which was originally released from the Pennsylvania State University in 1963 as "Nittany Lion", and a pink-flowered mite-susceptible genotype originally known as G60. Both of these genotypes originated from a single seed source called "Florist's Mix". Both are true breeding, fertile and diploid. Almost all of the plants that we have subsequently studied have been derived from these two inbred lines or their progeny.

II. ANATOMY AND MORPHOLOGY

After discovering the existence of mite-resistant and mite-susceptible geraniums, we began to examine the bases of this resistance. The first approach adopted was to investigate leaf morphology and, in particular, that of the abaxial surface of the leaf, which is the area where mites attack the geranium (Eyo *et al.*, 1974).

One of the major features of leaves that has been associated with host plant resistance is the presence of trichomes. Within the Geraniaceae, several studies have been conducted on trichomes. In 1980, Peterson characterized the histochemistry of all trichomes using *Pelargonim graveolens* as a model. Pedro *et al.* (1990) studied *Geranium robertianum* L. and determined that three types of trichomes were present in this species. The research was continued to classify the ultrastructure of Type I trichomes further (Pedro *et al.*, 1991). Navasero and Ramaswamy (1991) attempted to link trichome density in *P. graveolens* to fecundity in *Heliothis virescens*; however, no relationship was observed. Tahir *et al.* (1994) utilized scanning electron microscopy (SEM) to determine if trichomes could be useful in the systematics of *Pelargonium*.

The initial morphological investigations at the Pennsylvania State University determined that the resistant genotype ("Nittany Lion") differed from the susceptible genotype ("King Midas") in that it had: (1) a thicker cuticle, (2) thicker epidermal cells, (3) more trichomes, and (4) fewer arthropod eggs were produced on it than on the susceptible plant (Chang *et al.*, 1972). Stark (1975) conducted a SEM investigation and observed that the geranium leaf had four different types of trichomes: (1) long and (2) short spine-shaped trichomes, (3) short glandular-like trichomes, and (4) tall glandular trichomes. He was able to determine that the tall glandular trichome was critical in mite resistance and, for the first time, he implicated the presence of an orange viscous exudate on the tall glandular trichome as a primary resistance mechanism. Susceptible plants did not produce the viscous exudate; instead, he observed a tan, dry, flaky substance.

Herbert Grossman (Walters *et al.*, 1989a), using both SEM and transmission electron microscopy (TEM), was able to characterize the distribution of all four trichome types on both leaf surfaces of three susceptible and three resistant genotypes. He was also able to study the fine structure of individual trichomes. Again, the exudate produced on the tall glandular trichomes was uniquely associated with resistance. These studies led to the conclusion that tall glandular trichomes were present on almost all chlorophyllous surfaces of the plant and were particularly abundant on sepals and pedicels (stems of individual florets).

III. BIOCHEMISTRY

In 1983, David Gerhold gently touched a microscope slide to the leaf surface of our resistant geranium and noticed tiny droplets of exudate on the slide (Gerhold *et al.*, 1984). If he rinsed that area of the slide with a few drops of methylene chloride, the exudate was completely removed. This technique allowed the collection of milligram quantities of trichome exudate. We had also noticed that pedicels of individual florets possessed abundant tall glandular trichomes and were almost devoid of spiny trichomes; thus, most large-scale collections of exudate were made from pedicels. Unlike the viscous exudate of the resistant plant, the exudate on the tall glandular trichomes from susceptible plants was not sticky, but was hard and amorphous. In order to obtain exudate from susceptible plants, the whole leaf had to be rinsed with methylene chloride. Later, Donald Walters extracted 500 individual glandular trichomes by placing a capillary tube containing methylene chloride over the trichome apical cell, one at a time, from both resistant and susceptible plants. This confirmed that the trichomes were the source of the exudate. Later, we dipped florets into methylene chloride to obtain an extract from which we isolated gram quantities of exudate following silicic acid column separations.

Initial analysis of the exudate by thin layer chromatography (TLC) indicated the composition of the trichome exudate to be primarily one class of components. The exudate was volatile enough to be analysed by gas chromatography (GC), but these analyses were complicated by thermal breakdown with the loss of CO_2. However, after derivitization with diazomethane, the exudate could be analysed by GLC and was thermally stable. It was noted that the presence of methanol was required during derivitization to effect complete methylation and that two methylations occurred with diazomethane, suggesting the presence of phenolic and carboxyl constituents. GLC analysis of the resistant exudate indicated a mixture of four related components. Upon hydrogenation with platinum dioxide and hydrogen, the mixture contained only two related compounds, indicating conversion of two components to the related remaining two products.

Nuclear magnetic resonance, ultraviolet and mass spectrometric (MS) analyses indicated that the components of the trichome exudate are alkyl phenolic acids known as anacardic acids. These salicylic acid analogues possessed C_{15} or C_{17} alkyl chains of which 90% contained a double bond in the side chain with a *cis* configuration in the exudate from resistant plants. Subsequent formation of the dimethylthio adducts and GC/MS analysis

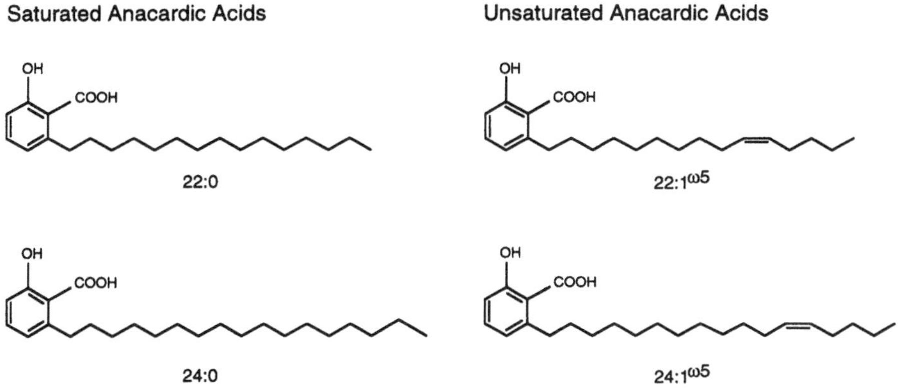

Fig. 1. Molecular structures of the four major anacardic acids from glandular trichome exudate of *Pelargonium* inbreds.

confirmed that the double bond was between the fifth and sixth carbon from the methyl end of the chain [referred to as the omega 5 (ω5) position]. Thus, it was concluded that the exudate of the resistant genotype contained four anacardic acid components (22:0, 24:0, 22:$1^{\omega 5}$, and 24:$1^{\omega 5}$) (Fig. 1). The resistant genotype contained approximately 90% of the ω5 unsaturated anacardic acid, whereas in the susceptible genotype, the saturated 22:0 and 24:0 anacardic acids predominated (Gerhold *et al.*, 1984; Walters *et al.*, 1988).

GLC analysis of the susceptible exudate revealed numerous unresolved minor components. Subsequently, these components were identified with a high-performance liquid chromatography (HPLC) procedure which resolved 17 different anacardic acids, of which only the ω5 unsaturated anacardic acids were present in the resistant genotype (Hesk *et al.*, 1991).

The chemical structure of the anacardic acids led us to surmise that they were biosynthetically derived from fatty acids. However, another important question was whether the unusual double bond was placed in the side chain, before or after anacardic acid formation. To answer these questions, ^{14}C-labelled fatty acid methyl esters were applied to the surface of the mite-resistant geranium leaf. Stearic acid (18:0) was incorporated into primarily the saturated 24:0 anacardic acid, whereas palmitic acid (16:0) was incorporated primarily into the saturated 22:0 anacardic acid. Unsaturated fatty acids, such as oleic and linoleic, were converted into anacardic acids possessing double bonds in the alkyl chain at the same position respective to the fatty acid substrate. It is worthwhile to note that trichomes from resistant plants do not contain anacardic acid species derived from oleic or linoleic acid, yet the trichomes are capable of synthesizing these anacardic acids at high levels when supplied with the appropriate substrates. In summary, these results indicated that: (1) fatty acids are precursors of anacardic acids; (2) the unusual $\omega 5$ double bond must be placed in the fatty acid prior to anacardic acid biosynthesis; (3) the resistant plant must possess an unusual fatty acid desaturase; (4) the resistant plant may contain detectable levels of $\omega 5$ fatty acids; and (5) the resistant plant, under natural conditions, regulated which fatty acid it accepted as a precursor of anacardic acid biosynthesis. Subsequent detailed examination of the fatty acids revealed that resistant plants contained both C_{16} and C_{18} $\omega 5$ fatty acids, albeit in less than 1% concentration, whereas neither was detected in susceptible plants (Walters *et al.*, 1990b; Hesk *et al.*, 1992). A mechanism of biosynthesis of anacardic acids was proposed that involved the addition of six carbon atoms to a preformed fatty acid (C_{16} or C_{18}) via the addition of three malonyl-coenzymeA molecules forming the aromatic ring containing a carboxyl and a phenol. This biosynthetic model parallels that known to exist in fungi producing salicylic acid and similar to other polyketide syntheses.

Critical to the progress of our work was the development of an efficient trichome collection method by Ellen Yerger. Pedicels from individual florets were placed in vials, frozen in liquid nitrogen, then finely ground dry ice was added. The tubes were then vortex mixed for approximately 1 minute at maximum speed. This resulted in the dry ice particles shearing trichomes from the pedicels. By controlling the time of mixing, the purity of the tall glandular trichome preparation could be controlled (Yerger *et al.*, 1992). This procedure proved effective in removing trichomes from numerous additional species (Yerger *et al.*, 1992).

To address the issue of fatty acid precursors to the $\omega 5$ anacardic acids, Grazzini (1993) examined the fatty acid composition of pure glandular trichomes of leaves, of pedicels and of leaf and pedicel tissue after glandular trichomes had been removed. In resistant genotypes, he observed $\Delta 11$ ($\omega 5$) *cis*-hexadecenoic acid ($16{:}1^{\Delta 11}$) and $\Delta 13$ ($\omega 5$) *cis*-octadecenoic acid ($18{:}1^{\Delta 13}$) to be present in the tall glandular trichomes from pedicels at 17.3% and 6.1% relative composition, respectively. Analysis of pedicel tissue, following removal of the trichomes, showed that it did not contain any of these acids. Thus, the $\omega 5$ fatty acids, the precursor for $\omega 5$ anacardic acids, are localized only in the tall glandular trichomes. The presence of 23.4% $\omega 5$ fatty acids in the tall glandular trichome of resistant genotypes, whilst none was detected in those of susceptible genotypes, suggested that there must be a pool of these acids within the trichomes of resistant plants. Evidently an unusual fatty acid desaturase must also be active only in the tall glandular trichome of resistant genotypes. It is important to note that the $\omega 5$ fatty acids were present in less than 1% of the total fatty acids of gross pedicel tissue and could have been easily disregarded as unimportant, but that trichomes were found to be rich in these fatty acids, suggesting that they are a good source of these unusual natural products.

To gain a better understanding of the lipid biosynthetic pathway in trichomes of the resistant genotype, glandular trichome lipid extracts were separated into lipid classes (neutral lipids, glycolipids and polar lipids) by silica gel column chromatography, and each class was further separated into specific lipid compounds by thin layer chromatography (Paul, 1995). The $\omega5$ fatty acids ($16{:}1^{\Delta 11}$ and $18{:}1^{\Delta 13}$) were present in all three lipid classes, the majority (67%) occurring in the polar lipids (phospholipids). Prominent amounts of $\omega5$ fatty acids occurred in lipid bands identified as digalactosyl diglyceride, monogalactosyl diglyceride, phosphatidyl ethanolamine, phosphatidyl inositol and phosphatidyl choline, the latter two being most abundant in $\omega5$ fatty acids.

Using immunochemical analysis, Hage (1995) identified major differences between proteins of trichomes and other plant tissues, and selective protein differences between trichomes of resistant and susceptible genotypes. He was able to demonstrate that an antistearoyl-acyl carrier protein desaturase antibody reacted with two protein bands (36 kDa and 39 kDa) on gel electrophoretic separation of trichome proteins from the resistant genotype, but with a single protein band (39 kDa) from trichomes of the susceptible genotype. Genetic analysis indicated that only the resistant inbred, and its F_1 and backcross progeny with resistant chemistry, possessed this immunoreactive 36 kDa protein. Also the South African *Pelargonium* species, *P. inquinans*, which has similar anacardic acid chemistry to the resistant genotype, exhibited the 36 kDa immunoreactive protein. Conversely, *P. zonale*, the other major progenitor of *P.* × *hortorum*, did not have this immunoreactive protein band. These data suggested that a unique fatty acid desaturase exists in the trichomes of the resistant genotype.

IV. GENETICS

A. CLASSICAL GENETICS

In the earliest days of the project, Balderston (1964) crossed mite-resistant with susceptible plants, and noted that the resulting F_1 generation was resistant. The nine F_2 progeny segregated for resistance and susceptibility. Winner (1975) conducted an extensive genetic investigation involving both mite-resistant and susceptible inbreds. He evaluated both parents, reciprocal F_1s, F_2s and both backcross generations. He bioassayed more than 9000 leaf discs, using a technique described by MacDonald *et al.* (1971) and Chang *et al.* (1972). He evaluated both mite mortality and fecundity, and classified progeny by a resistance index, which included both parameters. Winner's results suggested that the resistance phenomenon was simply inherited, and he suggested a two-gene complementary dominant epistatic model. In this model, he postulated that two dominant genes were responsible for the differences in mite resistance that he observed between the specific genotypes that were investigated. Subsequently, we have determined that the specific genotype that Winner used is temperature sensitive for mite resistance, and that one of the genes that he suggested may in fact be related to this phenomenon (Walters *et al.*, 1991).

In 1987, we initiated an examination of the inheritance of ω5-anacardic acids and tall glandular trichomes (Grazzini *et al.*, 1997). True-breeding resistant and susceptible inbred parents were self-pollinated repeatedly, and the progeny were analysed to verify uniformity and non-segregation. Reciprocal F_1 hybrids were produced, self-pollinated and backcrossed to both parents in all combinations. All plants were bioassayed for evaluation of mite resistance, and the anacardic acid composition of all plants was analysed by gas chromatography. These data indicated that the presence of the ω5-anacardic acid (the mite-resistance compound) is inherited as a single Mendelian dominant gene.

In order to confirm the association between chemistry and resistance, mite bioassays were conducted on 20 F_2 progeny, which had been chemically analysed and selected from four distinct groups of plants: possessing ω5-anacardic acids and a high number of tall glandular trichomes; possessing ω5-anacardic acids and a low number of tall glandular trichomes; and those possessing low/no anacardic acids with both high and low trichome numbers. Those plants with ω5-anacardic acids, regardless of trichome numbers, were shown to be resistant in the bioassays, and those plants which had low/no ω5-anacardic acids were susceptible regardless of the number of trichomes. Hesk *et al.* (1992) used HPLC to analyse these same plants and also to analyse a full testcross population; their results confirmed the single dominant gene hypothesis.

B. MOLECULAR GENETICS

In the early 1990s, biochemical and molecular genetic studies of fatty acid synthesis (FAS) in plants were well under way (see Ohlrogge and Browse, 1995, for a comprehensive review). Production of unsaturated fatty acids is accomplished by two distinct classes of fatty acid desaturase gene products, a soluble acyl-ACP desaturase and a membrane-associated desaturase. The first acyl-ACP desaturase gene to be isolated was that encoding the ubiquitous Δ9 18:0-ACP desaturase, which was isolated from cucumber (*Cucumis sativus*), castor bean (*Ricinus communis*) (Shanklin and Somerville, 1991) and safflower (*Carthamus tinctorius*) (Thompson *et al.*, 1991). Shortly after the isolation of the Δ9 18:0-ACP desaturase gene, it was proposed that an antigenically related protein was responsible for the production of the unique monoenoic fatty acid $18{:}1^{\Delta 6}$ (petroselinic acid) in coriander (*Coriandrum sativum*). Subsequently, the gene for the coriander desaturase was shown to encode a soluble Δ4 16:0-ACP desaturase (Cahoon *et al.*, 1992). The immediate product of this enzyme is $16{:}1^{\Delta 4}$, which is elongated in coriander to $18{:}1^{\Delta 6}$. The second class of plant desaturases are the membrane-associated desaturases that encode the plastid and cytosolic forms of $\omega 6$ and $\omega 3$ fatty acid desaturases. These desaturases function to introduce double bonds into unsaturated fatty acids, which are esterified to lipid backbones (e.g. phosphatidyl choline). Initially, two distinct $\omega 3$ desaturase genes were identified, one encoding a plastid $\omega 3$ desaturase (Iba *et al.*, 1993; Yadav *et al.*, 1993), the other encoding a cytosolic $\omega 3$ desaturase (Arondel *et al.*, 1992). An additional plastid $\omega 3$ desaturase gene that is influenced by temperature has also been identified (Gibson *et al.*, 1994). Similarly to the $\omega 3$ desaturase genes, it was shown that two genes encode $\omega 6$ desaturases, one encoding the plastid enzyme (Hitz *et al.*, 1994), and the second encoding the cytosolic protein (Okuley *et al.*, 1994). These studies provided a foundation for the work on anacardic acids in geranium glandular trichomes.

As was stated earlier, Grazzini *et al.* (1997) determined that the unique $22{:}1^{\omega 5}$ and $24{:}1^{\omega 5}$ anacardic acids found in pest-resistant geraniums are inherited as a single dominant Mendelian trait. Two other findings mentioned earlier were key to the molecular genetic analysis of anacardic acid biosynthesis in glandular trichomes. Firstly, fatty acids were found to be precursors of anacardic acids (Walters *et al.*, 1990b; Hesk *et al.*, 1992). Secondly, unique fatty acids ($16{:}1^{\Delta 11}$ and $18{:}1^{\Delta 13}$) found only in glandular trichomes were identified (Walters *et al.*, 1990b; Yerger *et al.*, 1992; Grazzini, 1993). From these findings, it was postulated that the pathway for the production of the $\omega 5$ anacardic acids involves a unique fatty acid desaturase which could produce both $16{:}1^{\Delta 11}$ and $18{:}1^{\Delta 13}$ (Hesk *et al.*, 1992; Grazzini, 1993). In addition, Hesk *et al.* (1992) showed the double-bond placement in the precursor fatty acids was most likely positioned from the carboxyl (Δ) rather than methyl (ω) end of the acyl chain. Thus, Hesk *et al.* (1992) and Grazzini *et al.* (1997) proposed that a novel desaturase could function to place a double bond at the $\Delta 11$ position of a 16:0 fatty acid. The proposed $16{:}1^{\Delta 11}$ product could then be elongated to $18{:}1^{\Delta 13}$ by the FAS system.

Because two types of desaturase genes exist, in initial studies we tried to provide evidence for the novel desaturase being either membrane-associated or the soluble type. In plants, membrane-associated desaturases typically function on unsaturated substrates, whereas soluble desaturases typically function on saturated substrates, to produce monoenoic fatty acids. Labelling studies suggested that the unique geranium desaturase was a soluble acyl-ACP desaturase (Walters *et al.*, 1990b; Hesk *et al.*, 1992).

Initial attempts to isolate the unique geranium desaturase gene were targeted towards identifying a soluble desaturase gene(s). Research on other desaturases indicated that variant substrate specificity and double-bond placement in coriander ($\Delta 4$ 16:0-ACP desaturase, Cahoon *et al.*, 1992) and *Thunbergia alata* ($\Delta 6$ 16:0-ACP desaturase, Cahoon *et al.*, 1994) could be accomplished through related members of an acyl-ACP desaturase family.

Thus, Hesk *et al.* (1992) and Grazzini *et al.* (1997) proposed that a novel desaturase, perhaps a member of the acyl-ACP desaturase family, could function to place a double bond at the Δ11 position of a 16:0 fatty acid leading to the $16{:}1^{\Delta 11}$ and $18{:}1^{\Delta 13}$ fatty acids (anacardic acid precursors).

Members of the acyl-ACP desaturase gene family encode proteins with greater than 70% amino acid similarity (Schultz, 1996). In order to test if geranium acyl-ACP desaturases were similarly related, a castor bean (*Ricinus communis*) Δ9 18:0-ACP desaturase cDNA was used to analyse expression of corresponding acyl-ACP desaturases in geranium. RNA was isolated from trichomes, pedicels and leaves of pest-resistant and pest-susceptible genotypes as described in Schultz *et al.* (1994). RNA gel blots of these geranium samples were probed under low stringency conditions with the ^{32}P-labelled castor bean cDNA. The preliminary results showed strong expression of Δ9 18:0-ACP desaturases and related genes in the trichomes of both pest-resistant and pest-susceptible genotypes. In order to isolate geranium genes that might encode a unique acyl-ACP desaturase gene, a geranium cDNA library rich in trichome RNA (Clark, 1997) was screened under low-stringency conditions using the labelled Δ9 18:0-ACP desaturase cDNA from castor bean. Two distinct acyl-ACP desaturase cDNAs, named pPXH-A and pPXH-B, were isolated from this screen (Schultz *et al.*, 1996). Sequence analysis of the proteins encoded by the cDNAs showed high homology to the predicted castor bean Δ9 18:0-ACP desaturase (89% with pPXH-A and 79% with pPXH-B). Because our studies suggested a unique desaturase could be involved and because pPXH-B was more divergent, subsequent studies focused on this cDNA.

Since the novel fatty acids $16{:}1^{\Delta 11}$ and $18{:}1^{\Delta 13}$, as well as the ω5 anacardic acids involved in pest resistance, were found to have trichome-specific distribution, we predicted that expression of a novel desaturase gene involved in pest resistance should also show trichome-specific expression. To test this theory, Schultz *et al.* (1996) screened a gel blot containing RNA from leaves, pedicels and trichomes from plants that were genetically and biochemically classified as pest-resistant and pest-susceptible geranium genotypes. In addition, trichome RNA from pooled populations of pest-susceptible and pest-resistant plants, which segregated one to one for the presence of ω5 anacardic acids (pest-resistant) and the absence of ω5 anacardic acids (pest-susceptible), were similarly analysed. Inclusion of this segregating population in the RNA gel blot analysis allowed us to test if expression of the novel desaturase was directly correlated with the pest-resistance phenotype. As predicted, the gene represented in pPXH-B was expressed only in trichome RNA from the pest-resistant plants (homozygous and backcross populations). This RNA gel blot analysis clearly established that expression of the novel desaturase encoded by pPXH-B correlated with production of $16{:}1^{\Delta 11}$ and $18{:}1^{\Delta 13}$ fatty acids, ω5 anacardic acids, and the pest-resistant phenotype (Schultz *et al.*, 1996). As further predicted, expression in plants homozygous for the pest-resistant phenotype was twofold higher than in plants heterozygous for the pest-resistance phenotype (back-cross populations). This indicates that the desaturase expression shows a gene-dosage effect.

To determine the enzymatic function of the desaturase encoded by pPXH-B, the coding region for the desaturase was expressed in *Escherichia coli*. The *E. coli* and plant fatty acid synthesis systems are quite similar, allowing plant enzymes to function in *E. coli* (Cahoon *et al.*, 1996). Two separate assays were conducted to determine the activity, substrate specificity and function of the novel geranium desaturase. First, an *in vivo* assay was conducted to determine if pPXH-B encoded desaturase activity and, if so, the types of fatty acids that are produced. Fatty acids produced following induction of expression of the geranium desaturase in *E. coli* were analysed by gas chromatography. Two new fatty acids, $16{:}1^{\Delta 11}$ and $18{:}1^{\Delta 13}$, were detected in *E. coli* cells expressing the geranium desaturase (Schultz *et al.*, 1996). These fatty acids are the novel fatty acid precursors to the $\omega 5$ anacardic acids found in the pest-resistant line. The *E. coli*-expressed enzyme was also used to evaluate substrate specificity of the geranium desaturase. After induction of expression, a crude protein extract was made, and the soluble protein fraction collected for *in vitro* analysis of activity. The protein fraction was supplied with acyl-ACP substrates (12:0-ACP, 14:0-ACP, 16:0-ACP and 18:0-ACP) and the products analysed. From our previous analyses we predicted the pPXH-B desaturase would act on the 16:0-ACP substrate (Hesk *et al.*, 1992; Grazzini *et al.*, 1997). Surprisingly, the novel geranium desaturase had the highest activity when the 14:0-ACP substrate was supplied (Schultz *et al.*, 1996). In addition, the 14:0 and 16:0 unsaturated products were identified as $14{:}1^{\Delta 9}$ and $16{:}1^{\Delta 9}$ (Schultz *et al.*, 1996). We verified the "Δ9" function of the novel desaturase by GC-mass spectroscopy analysis of unsaturated products produced from a 15:0-ACP substrate (Schultz *et al.*, 1996). Collectively, these results indicated that the novel geranium desaturase encodes a unique Δ9 14:0-ACP desaturase.

Based on the presence of $16{:}1^{\Delta 11}$ and $18{:}1^{\Delta 13}$ in geranium trichomes, we had previously proposed that a 16:0 specific desaturase is likely to be present in the pest-resistant geranium trichomes (Hesk *et al.*, 1992; Grazzini *et al.*, 1997). However, our molecular genetic and biochemical evaluation of the novel geranium desaturase showed that it functions as a Δ9 myristoyl (14:0)-ACP desaturase (Schultz *et al.*, 1996). These results are not inconsistent, but rather indicated an additional step in the process. The primary product of the unique geranium desaturase gene is $14{:}1^{\Delta 9}$, which could then be elongated by the FAS system to $16{:}1^{\Delta 11}$ and $18{:}1^{\Delta 13}$, the direct precursors to the $\omega 5$ anacardic acids. Based on the genetic and biochemical studies, it was concluded that expression of the Δ9 14:0-ACP desaturase gene is required for the production of $\omega 5$ anacardic acids that have been shown to be necessary for pest resistance in geranium (Schultz *et al.*, 1996).

V. MECHANISMS OF ACTION

We have examined a number of different resistance mechanisms during the course of this project (Walters *et al.*, 1989a,b, 1990a,c). One of the primary mechanisms conferring resistance might be that unsaturated anacardic acids act as a sticky trap for insects, whilst the saturated anacardic acids in susceptible plants are not adhesive. This sticky-trap mechanism was confirmed by microscopic observations of pests physically trapped in the exudate. In some detailed studies of mites on resistant and susceptible leaf surfaces, Harmon *et al.* (1996) characterized the behavioural patterns of mites and aphids. On susceptible leaves, they wandered and fed unimpeded, while on resistant leaf surfaces most were immobilized, did not feed and laid a reduced number of eggs.

A secondary pharmacological effect of the anacardic acids also exists. Stark (1975) topically applied trichome exudate to mites, mite eggs and larval whiteflies, and found that the resistant trichome exudate was toxic. Stark's data suggested that at least a portion of the mechanism of action was more than a simple physical sticky trap. An interesting, albeit unclear, result is that resistant trichome exudate also appeared to decrease arthropod fecundity.

Prostaglandins have been implicated in the reproduction of at least a dozen insect species, with potential roles in both male and female arthropods. The metabolism of prostaglandins is linked to the enzyme prostaglandin endoperoxide synthase (PES). Prostaglandins influence insect fecundity and maturation of eggs. Conversely, in the primitive insect species *Thermobia domestica* (silver fish), which lacks endogenous prostaglandins, products mediated by the enzyme lipoxygenase (LOX) play a prostaglandin-like role. LOX has also been implicated in some microorganism infections in plants.

PES activity is inhibited by aspirin (acetyl-salicylic acid) and similar biological inhibitors, whereas LOX activity is not known to be inhibited by aspirin. Anacardic acids are 6-alkyl salicylic acids in which the 6-alkyl chain may vary in length as well as in the position and degree of unsaturation. Thus, the effects of geranium anacardic acids on these enzymes became logical targets for investigation.

Anacardic acids proved to be good inhibitors of PES with an IC_{50} of 27 μM (Grazzini *et al.*, 1991). They were not as good inhibitors as aspirin, which in our test system had an IC_{50} of 3 μM, but were strong enough such that the anacardic acids from one single trichome could produce a significant effect on the biosynthesis of prostaglandins in aphids. Interestingly, anacardic acids strongly inhibited LOX (IC_{50} of 5.6 μM). The length of the alkyl chain (C_{22} versus C_{24}), or the presence or absence of the ω^5 double bond seemed to have little influence on the inhibition of either PES or LOX.

To put all of this information into perspective, the aphid has a body volume of around 1 μl. We have calculated that each tall glandular trichome on a resistant plant has about 100 doses of ω^5-anacardic acid, or enough anacardic acid to create 100 1-μl volumes at a 1 μM concentration. Trichome density on the leaf is about 15 cm^{-2}; on a pedicel, about 75 cm^{-2}. Even though saturated anacardic acids were capable of inhibiting PES and LOX, it should be noted that the lack of viscosity of the saturated compounds would prevent their deposition on arthropod pests. Clearly, the geranium has evolved a multifaceted, elegantly targeted defensive mechanism based on ω^5-anacardic acids and tall glandular trichomes.

These data suggest that we should consider more than a sticky-trap function for the anacardic acids. If an arthropod receives less than a lethal dose of anacardic acid, the acids may inhibit PES, LOX or other untested enzymes, resulting in deleterious viability effects. If these acids do have a physiological effect, most probably they are delayed effects, reducing fecundity in these arthropod pests.

VI. ECOLOGICAL EFFECTS ON RESISTANCE

We have often wondered how resistance is affected by natural environmental factors. During our research we noted that resistant and susceptible geranium plants grown under low irradiance and short day length possessed low levels of water-extractable anacardic acids. The effect of irradiance on mite-resistant geranium leaves was examined biologically by Paul (1995), using bioassays with mites, and chemically, by the quantification of anacardic acids. Increased irradiance resulted in an increase in the concentration of anacardic acids of the leaf but it did not enhance the leaf's resistance to spider mites. Apparently it takes very little anacardic acid to make the plant resistant; however, we expect that there are environmental limits.

It became apparent that the previous bioassay, published in 1971 by MacDonald *et al.* and in 1972 by Chang *et al.*, was both time consuming and, in some cases, difficult to quantify. A new bioassay protocol was designed using small Plexiglas cylinders. This new technique was easy to prepare, required only 24 hours versus 72 hours, accounted for all mites used and provided statistically reliable data (Harmon *et al.*, 1996).

Our first use of this new protocol allowed us to address a question that we had pondered for years: how is mite resistance affected in an outdoor environment, say by a heavy rainstorm? Washing of resistant geranium leaves with water removed sufficient anacardic acids to render the leaves susceptible to the two-spotted spider mite. If resistant plants become susceptible due to removal of anacardic acids, then how long does it take for the regeneration of anacardic acids and concomitant resistance? The Plexiglas cylinder bioassay technique was used to determine how long a period of time was required for regeneration of normal spider-mite resistance after washing resistant geranium leaves (still attached to plants) with water.

It was found that washed leaves regained full resistance after 14 days (Harmon *et al.*, 1996).

VII. EVOLUTIONARY SIGNIFICANCE

The major epicentre of the genus *Pelargonium* is South Africa and most species in the Section Ciconium, which includes the ancestors of *Pelargonium* × *hortorum*, are endemic to the Great Karoo region. Other species are located throughout Southern Africa and in several other geographic locations.

Snetsinger *et al.* (1966) attempted to assess the variability of mite resistance by bioassaying seven species of *Pelargonium* and five cultivars of *Pelargonium* × *hortorum*. They concluded that differential levels of resistance were found among species and among diploid and tetraploid cultivars of the zonal geranium.

Anacardic acid composition and glandular trichome distribution of 21 *Pelargonium* species, 39 diploid and 33 tetraploid cultivars of *P.* × *hortorum* were examined (Grazzini *et al.*, 1995a,b). Only three species of Section Ciconium (*P. acetosum*, *P. frutetorum* and *P. inquinans*) produced anacardic acids. Only *P. inquinans* and *P. frutetorum* contained ω5 anacardic acids and only *P. inquinans* possessed an anacardic acid composition similar to our resistant *P.* × *hortorum*; thus, *P. inquinans* is most probably the original source of the Δ9 14:0-ACP desaturase gene. All 72 *P.* × *hortorum* cultivars contained ω5-anacardic acids; however, wide variation in composition was observed.

Potter and Anderson (1982) conducted mite-resistance studies on *Pelargonium peltatum* (L.) L'Her ex Ait. They found that, unlike *P.* × *hortorum*, mite resistance in *P. peltatum* was not related to the density of simple or glandular trichomes. It is interesting to pose a question about the transmission of the Δ9 14:0-ACP desaturase gene through interspecific hybridization. In a cross of "Yale", tetraploid *P. peltatum* (without glandular trichomes and anacardic acids), and "Veronica", a tetraploid *P.* × *hortorum* (with tall glandular trichomes and ω^5 anacardic acids), the interspecific hybrid produced ω^5 anacardic acids. In backcrosses of the interspecific hybrid to "Veronica", the progeny also contained ω^5 anacardic acids. Thus it appears that we have the ability to produce mite-resistant *P. peltatum* in the future (Laughner, 1985; Grazzini *et al.*, 1995a).

VIII. CONCLUSIONS AND FUTURE PERSPECTIVES

We have presented a story about host–plant resistance in the geranium. What started in 1960 as a chance observation of mite susceptibility is still being vigorously pursued in 1998, using all of the tools of molecular biology.

In J. J. Adri van der Walt's monograph, *Pelargoniums of Southern Africa*, he states that "stems and leaves [of *P. inquinans*] are pounded and used as a headache and cold remedy by tribespeople". Thus anacardic acids and their physiological effects were known many centuries before our work (van der Walt, 1979).

A single Mendelian gene is responsible for the production of a Δ9 14:0-ACP desaturase, which places a single double-bond in a unique location in a simple fatty acid. Subsequently, this results in the production of ω^5 anacardic acids, which when presented to arthropod pests via tall glandular trichomes, causes both physical and chemical effects which limit their herbivory. During almost 400 years of plant breeding, the zonal geranium has conserved this gene for the production of ω5 anacardic acids. When the system is perfect, the exudate produced on the tall glandular trichomes has the proper viscosity to be efficiently presented to mites and aphids and in this way to protect the geranium from attack.

Our group of multidisciplinary scientists has used sophisticated equipment, innovative protocols, and collaborative thinking to investigate host–plant resistance aggressively and strategically. Although such research is obviously important for geranium improvement, it is much more critical as an example of what is possible in other agricultural crops.

ACKNOWLEDGEMENTS

The authors acknowledge the dozens of undergraduate and graduate students, support scientists and faculty who have been involved in this research during the past 38 years. We gratefully note the assistance of Laurel Anderson and Laura Wakefield during the preparation of this manuscript. We thank our respective departments for encouraging these collaborations. We note the financial support from the College of Agricultural Science at the Pennsylvania State University for an intercollege grant which allowed us to complete our investigations.

REFERENCES

Arondel, V., Lemieux, B., Hwang, I., Gibson, S., Goodman, H. M. and Somerville, C. R. (1992). Map-based cloning of a gene controlling omega-3 fatty acid desaturation in Arabidopsis. *Science* **258**, 1353–1355.

Balderston, C. P. (1964). Investigations regarding the natural resistance to the two-spotted spider mite *Tetranychus telarius* (L) (Acarina: Tetranachidae) exhibited by selected species of Pelargonium. M.S. thesis, The Pennsylvania State University, University Park, PA, 81 pp.

Cahoon, E. B., Shanklin, J. and Ohlrogge, J. B. (1992). Expression of a coriander desaturase results in petroselinic acid production in transgenic tobacco. *Proceedings of the National Academy of Sciences USA* **89**, 11 184–11 188.

Cahoon, E. B., Cranmer, A. M., Shanklin, J. and Ohlrogge, J. B. (1994). Δ^6 Hexadecenoic acid is synthesized by the activity of a soluble Δ^6 palmitoyl-acyl carrier protein desaturase in *Thunbergia alata* endosperm. *Journal of Biological Chemistry* **269**, 27 519–27 526.

Cahoon, E. B., Mills, L. A. and Shanklin, J. (1996). Modification of the fatty acid composition of *Escherichia coli* by coexpression of a plant acyl–acyl carrier protein desaturase and ferredoxin. *Journal of Bacteriology* **178**, 936–939.

Chang, K.-K., Snetsinger, R. and Craig, R. (1972). Leaf characteristics of spider mite resistant and susceptible cultivars of *Pelargonium* × *hortorum*. *Entomology News* **83**, 191–197.

Clark, D. G., Richards, C., Hilioti, Z., Lind-Iverson, S. and Brown, K. (1997). Effect of pollination on accumulation of ACC synthase and ACC oxidase transcripts, ethylene production and flower petal abcission in geranium (*Pelargonium* × *hortorum*) L. H. Bailey. *Plant Molecular Biology* **34**, 855–865.

Eyo, B., Snetsinger, R. and Craig, R. (1974). Leaf surface preference of spider mites. *Melsheimer Entomological Series* **14**, 1–11.

Gerhold, D. L., Craig, R. and Mumma, R. O. (1984). Analysis of trichome exudate from mite-resistant geraniums. *Journal of Chemical Ecology* **10**, 713–722.

Gibson, S., Arondel, V., Iba, K. and Somerville, C. (1994). Cloning of a temperature-regulated gene encoding a chloroplast ω-3 desaturase from *Arabidopsis thaliana*. *Plant Physiology* **106**, 1615–1621

Grazzini, R. A. (1993). A biochemical, evolutionary, and genetic model of glandular trichome mediated small pest resistance in *Pelargonium* × *hortorum*. Ph.D. thesis. Pennsylvania State University, University Park, PA, 198 pp.

Grazzini, R. A., Hesk, D., Heininger, E., Hildenbrandt, G., Reddy, C. C., Cox-Foster, D., Medford, J., Craig, R. and Mumma, R. O. (1991). Inhibition of lipoxygenase and prostaglandin endoperoxide synthase by anacardic acids. *Biochemical and Biophysical Research Communications* **176**, 775–780.

Grazzini, R. A., Hesk, D., Yerger, E., Cox-Foster, D., Medford, J., Craig, R. and Mumma, R. O. (1995a). Distribution of biochemical and morphological characters associated with small pest resistance among cultivars of *Pelargonium* × *hortorum*. *Journal of the American Society of Horticultural Science* **120**, 343–346.

Grazzini, R. A., Hesk, D., Yerger, E., Cox-Foster, D., Medford, J., Craig, R. and Mumma, R. O. (1995b). Species distribution of biochemical and morphological characters associated with small pest resistance in *Pelargonium* × *hortorum*. *Journal of the American Society of Horticultural Science* **120**, 336–342.

Grazzini, R. A., Walter, D., Harmon, J., Hesk, D. J., Cox-Foster, D., Medford, J., Craig, R. and Mumma, R. O. (1997). Inheritance of biochemical and morphological characters associated with two-spotted spider mite resistance in *Pelargonium* × *hortorum*. *Journal of the American Society of Horticultural Science* **122**, 373–379.

Hage, T. G. (1995). Characterization of proteins and enzymes associated with resistance to small arthropod pests in *Pelargonium* × *hortorum*. Ph.D. thesis. Pennsylvania State University, University Park, PA, 107 pp.

Harmon, J., Paul, P., Craig, R., Cox-Foster, D., Medford, J. and Mumma, R. O. (1996). Development of a mite bioassay to evaluate plant resistance and its use in determining regeneration of spider mite resistance. *Entomologia Experimentalis et Applicata* **81**, 301–305.

Hesk, D., Craig, R., Collins, L. and Mumma, R.O. (1991). Arthropod-resistant and

-susceptible geraniums: comparison of chemistry. *In* "Naturally Occurring Pest Bioregulators" (P. A. Hedin, ed.), pp. 224–250. American Chemical Society Symposium Series, Vol. 449. American Chemical Society, Washington, DC.

Hesk, D., Craig, R. and Mumma, R. O. (1992). Comparison of anacardic acid biosynthetic capability between insect-resistant and -susceptible geraniums. *Journal of Chemical Ecology* **18**, 1349–1364.

Hitz, W. D., Carlson, R. J., Booth, R., Kinney, A. J., Stecca, K. L. and Yadav, N. S. (1994). Cloning of a higher plant plastid ω-6 fatty acid desaturase cDNA and its expression in a cyanobacterium. *Plant Physiology* **105**, 635–641.

Iba, K., Gibson, S., Nishiuchi, T., Fuse, T., Nishimura, M., Arondel, V., Hughly, S. and Somerville, C. (1993). A gene encoding a chloroplast omega-3 fatty acid desaturase complements alterations in fatty acid desaturation and chloroplast copy number of the fad7 mutant of *Arabidopsis thaliana*. *Journal of Biological Chemistry* **268**, 24 099–24 105.

Laughner, L. J. (1985). Breeding and evaluation of Pelargonium for resistance to geranium rust, *Puccinia pelargonii-zonalis* Doidge. M.S. thesis. The Pennsylvania State University, University Park, PA, 42 pp.

MacDonald, A. J., Root, J., Snetsinger, R. and Craig, R. (1971). Techniques for evaluating host resistance to the two-spotted mite *Tetranychus urticae*. *Melsheimer Entomol. Ser.* **8**, 1–4.

Navasero, R. C. and Ramaswamy, S. B. (1991). Morphology of leaf surface trichomes and its influence on egg laying by *Heliothis virescens*. *Crop Science* **31**, 342–353.

Ohlrogge, J. and Browse, J. (1995). Lipid biosynthesis. *Plant Cell* **7**, 957–970.

Okuley, J., Lightner, J., Feldmann, K., Yadav, N., Lark, E. and Browse, J. (1994). Arabidopsis FAD2 gene encodes the enzyme that is essential for polyunsaturated lipid synthesis. *Plant Cell* **6**, 147–158.

Paul, P. R. (1995). New insights into the chemistry of arthropod resistance in *Pelargonium* × *hortorum*. M.S. thesis. The Pennsylvania State University, University Park, PA, 81 pp.

Pedro, L., Campos, P. and Pais, M. S. (1990). Morphology, ontogeny and histochemistry of secretory trichomes of *Geranium robertianum* L. (Geraniaceae). *Nordic Journal of Botany* **10**, 501–509.

Pedro, L., Campos, P. and Pais, S. (1991). Ultrastructure of the apical cell of procumbent (type I) trichomes in *Geranium robertianum* L. (Geraniaceae). *Israel Journal of Botany* **40**, 209–217.

Peterson, R. L. and Vermeer, J. (1980). Histochemistry of trichomes. *In* "Biology and Chemistry of Plant Trichomes" (E. Rodriguez, P. L. Healey and I. Mehta, eds), pp. 71–94. Plenum Press, New York, London.

Potter, D. A. and Anderson, R. G. (1982). Resistance of ivy geraniums to the two-spotted spider mite. *Journal of the American Society of Horticultural Science* **107**, 1089–1092.

Schultz, D. J. (1996). Molecular and biochemical evaluation of fatty acid desaturase genes that influence the production of anacardic acids in geranium. Ph.D. thesis, 137 pp. The Pennsylvania State University, University Park, PA, USA.

Schultz, D. J., Craig, R., Cox-Foster, D. L., Mumma, R. O. and Medford, J. I. (1994). RNA isolation from recalcitrant plant tissue. *Plant Molecular Biology Reporter* **12**, 193–199.

Schultz, D. J., Cahoon, E. B., Shanklin, J., Craig, R., Cox-Foster, D. L., Mumma, R. O. and Medford, J. I. (1996). Expression of a $\Delta 9$ 14:0-acyl carrier protein fatty acid desaturase gene is necessary for the production of ω5-anacardic acids found in pest-resistant geranium (*Pelargonium* × *hortorum*). *Proceedings of the National*

Academy of Sciences USA **93**, 8771–8775.
Shanklin, J. and Somerville, C. (1991). Stearoyl-acyl-carrier-protein desaturase from higher plants is structurally unrelated to the animal and fungal homologs. *Proceedings of the National Academy of Sciences USA* **88**, 2510–2514.
Snetsinger, R., Balderston, C. P. and Craig, R. (1966). Resistance to the two-spotted spider mite in Pelargonium. *Journal of Economic Entomology* **59**, 76–78.
Stark, R. S. (1975). Morphological and biochemical factors relating to spider mite resistance in the geranium. Ph.D. thesis. Pennsylvania State University, University Park, PA, 83 pp.
Tahir, S. S., Liz-Balchin, M., Husain, S. Z. and Rajput, M. T. M. (1994). SEM studies of trichomes types in representative species of the sect. Polyactium and sect. Ligularia in the genus Pelargonium L'Herit. (Geraniaceae). *Pakistan Journal of Botany* **26**, 397–407.
Thompson, G. A., Scherer, D. E., Foxall-Van Aken, S., Kenny, J. W., Young, J. L., Shintani, D. K., Kridl, J. C. and Knauf, V. C. (1991). Primary structures of the precursor and mature forms of stearoyl-acyl carrier protein desaturase from safflower embryos and requirement of ferredoxin for enzyme activity. *Proceedings of the National Academy of Sciences USA* **88**, 2578–2582.
van der Walt, J. J. A. (1979). "Pelargoniums of Southern Africa", 2nd edn. Pallottinerdruck, Limburg, Germany.
Walters, D. S., Minard, R., Craig, R. and Mumma, R. O. (1988). Geranium defensive agents III. Structural determination and biosynthetic considerations of the anacardic acids on geraniums. *Journal of Chemical Ecology* **14**, 743–751.
Walters, D. S., Grossman, H., Craig, R. and Mumma, R. O. (1989a). Geranium defensive agents IV. Chemical and morphological bases of resistance. *Journal of Chemical Ecology* **15**, 357–372.
Walters, D. S., Craig, R. and Mumma, R. O. (1989b). Glandular trichome exudate is the critical factor in geranium pest resistance to foxglove aphid. *Entomologia Experimentalis et Applicata* **53**, 105–109.
Walters, D. S., Craig, R. and Mumma, R. O. (1990a). Effects of the mite resistance mechanism of geraniums on the mortality and behavior of the foxglove aphid (*Acyrthosiphon solani* Kaltenbach). *Journal of Chemical Ecology* **16**, 877–886.
Walters, D. S., Craig, R. and Mumma, R. O. (1990b). Fatty acid incorporation in the biosynthesis of anacardic acids of geraniums. *Phytochemistry* **29**, 1815–1822.
Walters, D. S., Craig, R. and Mumma, R. O. (1990c). Heritable trichome exudate differences of resistant and susceptible geranium. *In* "Pesticides and Alternatives" (J. Casida, ed.), pp. 317–327. Elsevier Science Publishers, B.V., Amsterdam.
Walters, D. S., Harmon, J., Craig, R. and Mumma, R. O. (1991). Effect of temperature on glandular trichome exudate composition and pest resistance in geraniums. *Entomologia Experimentalis et Applicata* **60**, 61–69.
Winner, B. (1975). Inheritance of resistance to the two-spotted spider mite *Tetranychus urticae* Koch in the geranium *Pelargonium* × *hortorum* Bailey. M.S. thesis. The Pennsylvania State University, University Park, PA, 55 pp.
Yadav, N. S., Wierzbick, A., Aegerter, M., Casterr, C. S., Peréz-Grau, L., Kinney, A. J., Hitz, W. D., Booth, R., Schweiger, B., Stecca, K. L., Allen, S. M., Blackwell, M., Reiter, R. S., Carlson, R. J., Russell, S. H., Feldmann, K. A., Pierce, J. and Browse, J. (1993). Cloning of higher plant omega-3 fatty acid desaturases. *Plant Physiology* **103**, 467–476.
Yerger, E., Grazzini, R. A., Hesk, D., Cox-Foster, D., Medford, J., Craig, R. and Mumma, R. O. (1992). A rapid method for isolating glandular trichomes. *Plant Physiology* **99**, 1–7.

Specification of Epidermal Cell Morphology

B. J. GLOVER[1] and C. MARTIN[2]

[1]*Department of Plant Sciences, University of Cambridge, Downing Street, Cambridge, CB2 3EA, UK*
[2]*Department of Genetics, John Innes Centre, Colney Lane, Norwich, NR4 7UH, UK*

I. PLANT CELL DETERMINATION

The processes of development and morphogenesis, including the commitment of cells to specialized fates, occur in both plants and animals. However, there are fundamental differences in the ways in which spatial patterning is defined in the two groups. Animal development is a highly deterministic process, with little or no environmental input. Organs and tissues develop in

Advances in Botanical Research Vol. 31
ISBN 0-12-005931-2

response to positional cues received by cell lineages during embryogenesis. Plant development, on the other hand, is a much more plastic process and morphogenesis is strongly influenced by external environmental signals, often mediated by the action of phytohormones and their signal transduction pathways, interacting with the internal genetic mechanisms of development. The combination of external and internal controls on plant development can result in an "exploratory" form of morphogenesis, where cells respond initially to internal signals, but then may interrupt and modify their differentiation in response to messages received from other cells. A classic example is the development of the mesophyll tissue of the stem of a young seedling. Cells that have begun their development as mesophyll, as a result of their position in the stem, may later differentiate into vascular cells in response to auxin signals received from the developing vasculature of a nearby leaf. Only cells in the region of a developing leaf make this switch in differentiation, in order to join stem vasculature and leaf vasculature together. Because the vertical position of leaves on the growing stem is determined by environmental cues, the location of those cells that will differentiate into vasculature is determined largely by environmental signals and not by internal gradients of positional information.

It has recently become apparent that the processes of epidermal cell specification in plants are similarly plastic and interactive (Berger *et al.*, 1998; Glover *et al.*, 1998), and in this chapter we will focus on the interactions between developing trichomes and other epidermal cells. Recent advances in identifying genes involved in epidermal cell specification, combined with an appreciation of the influences of plant growth regulators and the environment on cellular patterning, has led to a complex picture of plant cell specification. It also appears that cellular specification is intimately linked to switches between the different growth phases that make up the life history of the plant. The understanding of epidermal patterning requires an appreciation of the changes in the behaviour of all cell types in relation to the adoption of a specialized form and function (such as the formation of trichomes) in one.

The main types of epidermal cell will be described in this chapter, and the mechanics of their development reviewed. The diversity of trichome types will be discussed, and the development of four main categories: unicellular (thale cress, *Arabidopsis thaliana*), fibre (cotton, *Gossypium hirsutum*), multicellular and conical–papillate, will be analysed. Finally, the evidence of interaction between the development of trichomes and other epidermal cells will be reviewed.

II. EPIDERMAL CELL SPECIFICATION

Plant cells differentiate into a relatively limited number of morphologically distinct cell types (Goldberg, 1988). Epidermal cells may adopt several different specialized forms, which are determined by their location in participant organs. Furthermore, differentiation of one cell type may be influenced by the differentiation of other cells, particularly within the cell layer, to the same or different functions. Such interactions may, theoretically, be either positive or negative (competitive). Some distinct cell types may be related by sharing specific parts of a single developmental programme. In such cases, specification also involves the control of timing of initiation and arrest of the differentiation programme in any particular cell. There are therefore several dimensions to the control of cellular specification in plants, and it is necessary to begin any discussion of such specification with an overview of the main types of cells within the epidermis.

A. PAVEMENT CELLS

The epidermal cell layer of all plants is composed primarily of pavement cells, relatively unspecialized morphologically, which fulfil a protective function for the tissue layers underneath and a spacing function for morphologically more specialized cells in the epidermis. Because of their relatively unspecialized morphology, pavement cells have traditionally been regarded as insignificant players in the patterning of the epidermis, filling a default role rather than arising as the result of an active patterning process. However, pavement cells of different organs frequently adopt different shapes in relation to their function. On the leaves and petals of a dicot, pavement cells are usually shaped like the interlocking pieces of a jigsaw puzzle (Fig. 1A). This form gives the leaf a great deal of strength, which is necessary when one considers that the underlying palisade and spongy mesophyll layers are composed primarily of cells with large air spaces between them. Disruption of this interlocking pattern of leaf epidermal cells results in an organ that is less able to withstand tearing stresses, as when the epidermal cells of the tobacco leaf are converted to a petaloid form with hexagonal bases by ectopic expression of the antirrhinum (*Antirrhinum majus*) gene *MIXTA* (Glover *et al.*, 1998; B. Glover, unpublished observations). The pavement cells of stems are usually elongated, with their long axis parallel to the direction of stem elongation. This form is similarly adopted by the pavement cells of other elongated organs, such as the root, staminal filament and style, although it is interesting to note that in many species the stylar pavement cells have elaborate interlocking edges, perhaps to lend additional strength to the structure, which may be subject to considerable mechanical stress during insect pollination (Fig. 1B). Pavement cells also differ in the extent to which they develop intricate cuticular patterning. Root pavement cells are not covered in cuticular material at all, while, as a general rule, the thickness and complexity of the cuticle increase from the vegetative to reproductive areas of the plant. Leaf pavement cells may have some cuticular

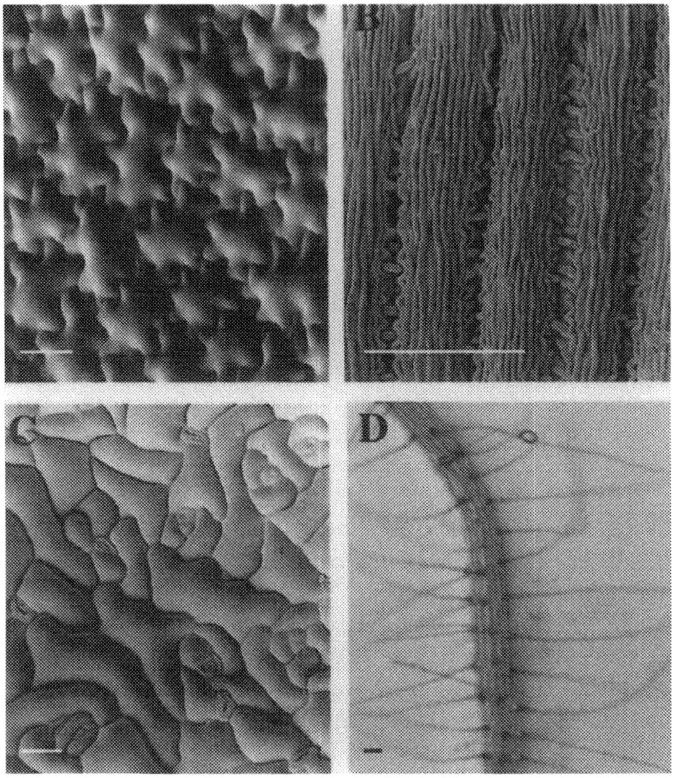

Fig. 1. Epidermal Cell Types. Scale bars = 20 μm. (A) Pavement cells of outer epidermis of antirrhinum petal, showing the interlocking form, which provides material strength. (B) Pavement cells of tomato style, showing the elongated but still interlocking form. (C) *Arabidopsis* leaf epidermis with developing stomatal complexes. (D) *Arabidopsis* root with elongated hair cells in files.

ridges, while petals and reproductive organs differentiate pavement cells with thick cuticular striations, arrayed in random, linear or star-like patterns.

Genetic evidence from mutations affecting root hair formation has suggested that pavement cells are not the default state for root epidermal cells. Analysis of *Arabidopsis* mutants such as *ttg* and *gl2* has demonstrated both genes to be negative regulators of root hair formation. *GL2*, which, when mutated, promotes ectopic root hair formation in non-hair files, is normally expressed in non-hair cells only. Depending on the mechanism of action of genes such as *GL2* and *TTG*, the root pavement cell could therefore be viewed as the specialized form and the root hair cell as the default state (Galway *et al.*, 1994; Masucci *et al.*, 1996). Of course, it could also be argued that neither hair nor non-hair cells are default, but that both represent differently differentiated forms.

B. STOMATAL GUARD CELLS

The specification and morphogenesis of stomata are essential for the photosynthetic activity and water balance of the plant. However, because of the essential nature of these cells, mutants in their development and patterning are rare, and our understanding of the ways in which their characteristics are determined relies more on cellular observations than on genetics. The optimal spatial patterning of stomatal guard cells is closely linked to the size of the cells, and the size and shape of the air space in the mesophyll underneath. Through precise control of these parameters, it is possible for a leaf to develop an epidermis which is as permeable to gases as no epidermis at all, and yet which can become virtually impermeable to water when the environment demands it. The differentiation of a stomatal complex involves both the development of guard cells in the epidermis and the development of a cell-free space in the mesophyll immediately below the guard cells. In the monocot *Tradescantia* this empty space is produced by the localized expansion of the cells of the stomatal complex, which fail to expand towards the mesophyll in the way that pavement cells do (Boetsch *et al.*, 1995). Stomatal patterning in dicots is cell lineage dependent, with cells becoming committed to differentiate as meristemoids, and then, through subsequent ordered cell divisions, producing a pair of crescent-shaped guard cells and a suite of surrounding subsidiary cells, which may serve a role in the ion-channel-mediated opening and closing of the stomatal pore (Fig. 1C). The wave of cell division that results in the differentiation of these cell types from the meristemoid usually occurs late in epidermal development, often after cytokinesis has virtually finished and the pavement cells are undergoing cell expansion. Mutants such as *tmm* (*too many mouths*), which affect the number of meristemoids produced, have been described in arabidopsis and it has been suggested that this gene product limits the timing of specification of stomatal initials relative to the development of the rest of the epidermis (Yang and Sack, 1995). The *FLP* (*FOUR LIPS*) locus also affects stomatal production, with mutation at this locus resulting in paired stomata. The *FLP* gene product is believed to affect guard mother cell production (Yang and Sack, 1995). It is suggested that this locus acts at the stage of development where guard mother cells are differentiated from meristemoids, but that *TMM* can act both at this stage and at entry into the pathway by determining precursor cell number (Yang and Sack, 1995). Stomatal initials can also arrest later in development, possibly in response to signals from neighbouring initials. In the monocot *Tradescantia*, 10% of stomatal initials do not complete their development (Boetsch *et al.*, 1995). The density of developing stomatal initials is under environmental control, as an adaptation to maximize favourable water and gas exchange regimes. Atmospheric carbon dioxide levels, for example, affect stomatal numbers in a highly species-specific manner. In *Tradescantia* the number of guard cell pairs does not change in response to elevated CO_2 but the number of subsidiary cells increases. This results in a consequent decrease in the number of pavement cells and emphasizes that in monocots specification of subsidiary cells may be under independent control from specification of guard cells, unlike the situation in dicots where both types of cells are linked by lineage (Boetsch *et al.*, 1996). Environmental cues affecting stomatal formation may be mediated by the hormones ethylene and cytokinin, for both of which there is evidence to suggest that increased concentrations may increase stomatal density (Serna and Fenoll, 1997; Wang *et al.*, 1997). The later stages of guard cell morphogenesis involve the arrangement of microtubules and cellulose microfibrils, and the subsequent deposition of thickened wall material around the stomatal pore, which allows the opening and closing of the pore in response to changes in guard cell turgor.

C. ROOT HAIRS

Newly formed root epidermal cells may differentiate into root pavement cells or into root hair cells. The root hairs of arabidopsis are single cells that expand to increase the surface area of the root available for nutrient and water uptake (Fig. 1D). This extension grows by tip-growth, the addition of cell wall material to the growing tip by microfilament- and actin-driven vesicle deposition. In *Arabidopsis* the developmental decision is made on the basis of positional cues, with root hairs always located over the join of adjacent cortical cells below them (Dolan *et al.*, 1993). Root hairs can thus develop only in certain files. This pattern is typical of the Brassicaceae, whereas the more common root hair pattern in dicotyledonous plants is one in which any root epidermal cell can become a hair (Dolan, 1996).

Mutation at the *TTG* (*TRANSPARENT TESTA GLABRA*) locus, which also serves a role in trichome development, results in a loss of patterning in the arabidopsis root, with epidermal cells in all positions able to differentiate into root hairs (Galway *et al.*, 1994), and mutation at the *GL2* (*GLABRA 2*) locus produces a similar phenotype (Masucci *et al.*, 1996), although the two mutants show some differences in the cytological characteristics of the cells. *TTG* encodes a WD40 repeat protein, likely to be involved in signal transduction (A. Walker and J. Gray, personal communication), while *GL2* encodes a homeodomain transcription factor (Rerie *et al.*, 1994) and is preferentially expressed in the non-hair cells of the developing root (Masucci *et al.*, 1996). Another locus involved in the early stages of root hair differentiation is *CAPRICE* (Wada *et al.*, 1997), which encodes a MYB-related DNA-binding protein lacking a distinctive transcriptional activation domain. Mutation at the *CPC* locus results in roots with few hairs. Wada *et al.* (1997) proposed that *CPC* acts by failing to activate transcription of the hair fate repressor, *GL2*, in hair cell files, allowing the development of hairs in these cells. CPC may be regulating *GL2* expression competitively with positive transcriptional regulators. The *WEREWOLF* locus is a good candidate for a gene encoding such a product, since *WEREWOLF* mutants have ectopic hairs in non-hair files (J. Schiefelbein, personal communication). However, the molecular basis of action of this gene has not yet been established and awaits its molecular cloning. The *ERH1* locus is also believed to play a role in specifying cell identity in the root epidermis (Schneider *et al.*, 1997). *erh1* mutant plants produce ectopic root hairs on all epidermal cell files. Like *WEREWOLF*, *ERH1* remains to be cloned, but it is likely to play a role in several root-specific processes, as the mutant plants also have significantly thicker roots as a result of increased radial expansion of cells in all layers (Schneider *et al.*, 1997).

Ethylene also plays a very important role in the specification of root hairs. The *CTR1* (*CONSTITUTIVE TRIPLE RESPONSE 1*) locus, for instance, encodes a negative regulator of root hair development and a negative regulator of the ethylene response (Kieber *et al.*, 1993; Dolan *et al.*, 1994). Ethylene treatment of *Arabidopsis* roots results in the ectopic development of root hairs in non-hair files as well as their formation in hair files. It is possible that TTG and GL2 regulate the pattern of ethylene biosynthesis.

Following the determination of root hair fate, the developing hair cell forms a bulge at the apical end. Mutations at the *RHD1* (*ROOT HAIR DEVELOPMENT 1*) and *REB1* (*ROOT EPIDERMAL BULGER 1*) loci result in an altered bulge, possibly swollen as a result of a defect in regulated degradation of the cell wall (Schiefelbein and Somerville, 1990; Baskin *et al.*, 1992). The later tip growth of this hair is regulated by a number of loci, controlling the extent and direction of growth (Schiefelbein and Somerville, 1990).

III. DIVERSITY OF PLANT TRICHOMES

While the morphology of stomata, pavement cells and root hairs is largely consistent between unrelated plant species, trichome morphology can vary widely. In fact, the definition of a trichome is far from clear, with some authors including leaf glands lying directly on the epidermal surfaces, others including petal and stomatal papillae (Fig. 2A and B), and others limiting the term only to the long hairs that limit predation. These definitions are discussed in detail in the chapter by Werker in this volume. For the purpose of this chapter, any structure that grows in a direction perpendicular to the plane of the aerial epidermis will be considered a trichome. The diversity of these structures is astonishing. They may be unicellular or multicellular, and vary in size from a few micrometres in length to structures over a centimetre long. Nor is this size related to the number of cells they contain; the longest trichomes are the fibres of cotton ovules, at 28–52 mm in length, and yet

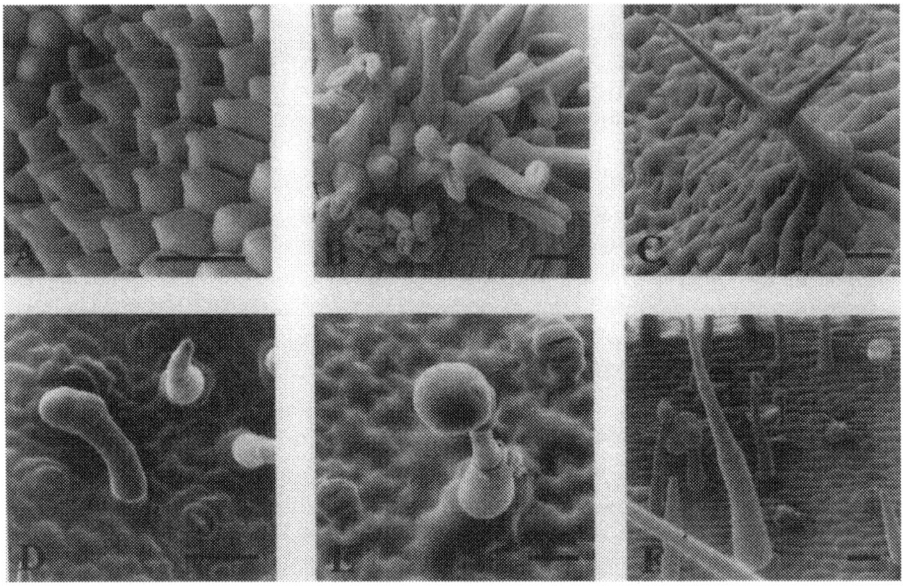

Fig. 2. Types of trichome. Scale bars = 30 μm. (A) Conical–papillate cells on the inner epidermis of an antirrhinum petal. (B) Stigmatic papillae of arabidopsis, with germinating pollen grains. (C) Unicellular arabidopsis trichome. (D) Short glandular trichome of tobacco, which secretes alkaloids. (E) Long glandular trichome of tobacco, with a large drop of terpenoid-rich secretion. (F) Tomato stem with four types of trichomes present; pendant glandular, four-lobed, long glandular and Luckwill's Type 1, the very large trichome with specialized basal cells.

these are unicellular. The long trichomes of tobacco may contain up to 32 cells, and yet, at 200 μm in length, are a very similar size to the single-celled arabidopsis trichome. Furthermore, the morphological differences between trichomes extend to the production of glands, branching and the elaboration of the cuticle. The arabidopsis trichome, for example, is stellate, with two branches from the primary stem. It has elaborate cuticular markings (Fig. 2C). It does not, however, produce a gland. The trichomes of tobacco (*Nicotiana tabacum*) generally do not branch, but they do produce glandular heads. The short glandular trichomes are thought to exude nicotine and other alkaloids transported from the roots from this head (Fig. 2D), while the longer trichomes secrete terpenoids, which are synthesized in the glandular head (Fig. 2E) (Akers *et al.*, 1978; Meyberg *et al.*, 1991; Nielsen *et al.*, 1991). *Nicotiana tomentosa* produces trichomes very similar to those of *N. tabacum*, with the exception that they do branch, often at several points. *Quercus* (oak) species produce apparently stellate trichomes that actually consist of groups of trichomes clustered at a single point on the epidermis. Many species, including tobacco, develop more than one morphologically distinct type of trichome, presumably specified by multiple developmental programmes. For example, the trichomes of tomato were classified by Luckwill (1943), who found at least four types on the stem alone (some shown in Fig. 2F).

Evidence is accumulating that these many different types of trichome are specified by different developmental programmes. Constitutive expression of the *Lc* gene from maize results in the production of ectopic trichomes in *Arabidopsis* but not in tobacco (Lloyd *et al.*, 1992). It is also not required for trichome specification in maize. In contrast, constitutive expression of the *MIXTA* gene from antirrhinum results in the production of ectopic trichomes in tobacco but not in *Arabidopsis* (Glover *et al.*, 1998). Following these observations, several attempts have been made to identify orthologues of the *Arabidopsis* trichome genes *GL1* and *GL2* in tobacco, thus far without success (B. Glover, unpublished results) and there is evidence that the petunia (*Petunia hybrida*) homologue of the *Arabidopsis* trichome-determining gene *TTG* is not involved in trichome development (de Vetten *et al.*, 1997). Similarly, attempts to identify orthologues of *GL1* in developing cotton fibres have also failed (T. Wilkins, personal communication). However, it is possible that, although developmental programmes may vary, the functions of different trichomes may be very similar, particularly with regard to the mechanisms of exudate secretion and the processes of heavy metal and salt accumulation.

IV. TRICHOME DEVELOPMENT

A. *ARABIDOPSIS* TRICHOME DEVELOPMENT

The understanding of epidermal cell differentiation is greatest for the unicellular trichomes of arabidopsis. However, since the patterning and shaping of these trichomes are covered in detail elsewhere in this volume in the chapter by Hülskamp and Kirik, they will be described only briefly here for purposes of comparison to other trichome types. The *Arabidopsis* trichome is believed to function primarily as a deterrent to predation by herbivores and consequent infection by pathogenic agents. It acts as a simple barrier to animals trying to cross the leaf surface. It has recently been demonstrated that natural enemies do impose selection for increased densities of trichomes in *Arabidopsis* (Mauricio and Rausher, 1997). However, there is also some evidence that trichomes may play a role in stress responses (Yamagushi-Shinozaki and Shinozaki, 1993; Wang and Cutler, 1995; Gotor *et al.*, 1997).

Hülskamp *et al.* (1994) divided trichome differentiation into three distinct stages: pattern formation, cellular specification and morphogenesis. Trichome pattern formation is primarily under the control of three genes, *TTG*, *GL1* and *TRY* (*TRIPTYCHON*). *TTG* encodes a WD40 repeat protein, potentially involved in signal transduction (A. Walker and J. Gray, personal communication), and appears to interact with, or activate another gene whose product interacts with, *GL1*. *GL1* encodes a MYB-related transcription factor (Oppenheimer *et al*., 1991; Larkin *et al*., 1994). The *TRY* gene product appears to act by modifying activity of *GL1*, negatively regulating trichome development. It has been proposed that this is achieved through negative regulation of endoreduplication or promotion of cytokinesis, since the *Arabidopsis* trichome is a unicell with an elevated C value (Hülskamp *et al*., 1998; Martin and Glover, 1998). These three genes work together to specify trichome patterning via induction of the developmental programme in some cells with subsequent lateral inhibition of those surrounding it. The *RTN* locus is thought to regulate the length of time cells are competent to respond to these patterning signals (Larkin *et al*., 1996). Clonal analysis has demonstrated that the trichome subsidiary cells are not related to the main cell through a single cell lineage, suggesting that it is not lineage that determines final pattern (Larkin *et al*., 1996). The *COT1* locus also acts to regulate negatively the specification of trichome cells (Szymanski *et al*., 1998). The *cot1* allele has no effect on trichome initiation in a wild-type background but combined with ectopic expression of *GL1* acts in a dosage-dependent manner to increase trichome density (Szymanski *et al*., 1998). Molecular cloning of this gene should provide more information about its mode of action.

Trichome cell specification involves enlargement of the cell and three rounds of endoreduplication. These processes may be direct targets of TTG/GL1 activity. Activation of *GL2* is also essential, and the promoter of this gene contains a motif suitable for binding by GL1. *GL2* encodes a homeodomain transcription factor whose function is the promotion of cellular outgrowth. *gl2* mutants produce cells that have become committed to trichome development but fail to grow out from the blade of the leaf, remaining as large cells with the same high C content as trichomes (Rerie *et al*., 1994).

The later morphogenesis of the *Arabidopsis* trichome is primarily concerned with branching and cuticle elaboration. GL2 may also play a role in these later stages (Hülskamp *et al*., 1998). Primary and secondary branching of the cell have recently been shown to be under separate genetic control (Folkers *et al*., 1997) with *AN* (*ANGUSTIFOLIA*) and *STA* (*STACHEL*) required for secondary and primary branches, respectively, *ZWI* (*ZWICHEL*), which encodes a kinesin-like microtubule motor protein, required for maintenance of the stalk (Oppenheimer *et al*., 1997), and *NOK* (*NOECK*) necessary for repression of further branches.

Although our understanding of the genetic control of trichome development in other species is much less than in *Arabidopsis*, the *Arabidopsis* trichome is somewhat specialized and unique to the Cruciferae, and therefore may not be a relevant model for all trichome differentiation.

B. COTTON FIBRE DEVELOPMENT

The trichomes, or fibres, of cotton (*Gossypium hirsutum*) seeds develop from many of the outer integument epidermal cells of ovules and grow to lengths of several centimetres. Despite the remarkable length (up to 52 mm) and rapid growth of cotton fibres, their development appears to proceed through diffuse growth (the deposition of cell wall material all over the aerial surface of the growing cell) like other trichomes, rather than the tip growth associated with similarly sized structures such as pollen tubes (Tiwari and Wilkins, 1995). Their outgrowth from the ovule is first visible on the day of cotton flower anthesis, but there is evidence for ultrastructural changes in fibre initials before expansion is visible (Wilkins and Jernstedt, 1999), presumably reflecting the process of commitment to the fibre developmental programme. Following specification the cells undergo a period of rapid expansion, lasting approximately 15 days, after which they deposit a thick secondary cellulosic cell wall. Finally, 50 days after they first became visible, the fibres desiccate as the seed matures, ready for dispersal.

In comparison to the *Arabidopsis* trichome, relatively little is known about the molecular mechanisms involved in the specification and morphogenesis of the cotton trichome. Attempts to isolate *GL1* homologues resulted in the isolation of six cDNAs encoding MYB-related transcription factors from cotton ovules, although none shows significant homology to *GL1* and none can be conclusively linked with fibre differentiation (Wilkins and Jernstedt, 1999). Differential levels of expression of these genes at different stages of fibre development may indicate a role for them in cellular morphogenesis, but their non-fibre-specific spatial expression patterns suggest that they may be involved in cellular processes not restricted to the differentiation of fibres.

Expansion of the fibres post-specification requires a complex interplay between cell turgor and cell wall extensibility, combined with regulation of the cytoskeleton and cell wall production. Many genes have been isolated that show enhanced mRNA levels during the rapid fibre-elongation stage, and these genes have been shown to encode products required for the cellular processes of expansion. One of the first such genes to be expressed in the developing initials is *ACP*, which encodes an acyl carrier protein believed to function in the synthesis of membrane lipids (Song and Allen, 1997). The maintenance and enhancement of cell turgor requires the activity of two membrane-bound proton pumps, the vacuolar H^+-ATPase (V-ATPase) and the plasma membrane H^+-ATPase (PM-ATPase). The V-ATPase is a multimeric enzyme composed of three subunits. The gene encoding one of these subunits, *CVA69.24*, has been shown to be differentially expressed during fibre development, with transcript levels increasing by several orders of magnitude during the rapid expansion phase (Wilkins and Jernstedt, 1999). Two genes encoding cell wall enzymes involved in regulating the intramolecular and intermolecular bonds of the cellulose-xyloglucan framework during cell wall loosening have been reported to be induced during fibre expansion (Shimizu *et al.*, 1997), and the genes encoding two isoforms of a lipid transfer protein, LTP3 and LTP6, have also been shown to be preferentially expressed in rapidly expanding fibres, 10–15 days after anthesis (Ma *et al.*, 1997). The presence of these two sets of transcripts together demonstrates that the cell is undergoing both the processes of wall loosening and wall deposition (since LTPs are believed to synthesize cutins) simultaneously. At the same time, populations of the subunits of tubulin also increases in the developing fibre (Wilkins and Jernstedt, 1999).

Between 16 and 21 days post-anthesis, the developing fibres switch from producing primary cell wall to synthesizing secondary cell wall, as expansion is completed. The cell's entry into the stage of secondary wall production correlates with a sharp increase in the expression of cellulose synthase catalytic subunit genes (Pear *et al.*, 1996). Sucrose synthase mRNA levels also peak during this phase of secondary cell wall synthesis (Shimizu *et al.*, 1997), perhaps reflecting a utilization of sucrose as a carbon source for cell-wall polysaccharide biosynthesis.

The final maturation of the cotton fibre is a relatively uncharacterized process. Cell death may result from desiccation as the cotton boll opens, or occurs as an active process, but these ideas are, as yet, speculative. It is clear that the cotton fibre, as a morphologically and developmentally unique trichome of considerable economic importance, deserves further study, in particular with respect to the molecular and cellular mechanisms of fibre specification.

C. MULTICELLULAR TRICHOME DEVELOPMENT

The majority of flowering plants produce trichomes, which, in contrast to *Arabidopsis* and cotton, are multicellular (Werker, this volume). Although these trichomes must undergo processes of commitment, expansion and morphogenesis, which may in principle be similar to those of unicellular trichomes, they must also do this in the context of continuing cell division. In *Arabidopsis*, trichomes are specified while the other cells of the leaf epidermis continue to divide, but they do not undertake further cytokinesis so that the DNA is duplicated in line with the surrounding cells to generate cells with a high C value. In contrast, cotton trichomes extend from the epidermis of the integument after the phase of cell division, so cytokinesis is not repressed in the same way. Multicellular trichomes may develop as the surrounding cells continue to divide, changing their orientation of growth, but continuing to divide in line with the rest of the epidermis. Multicellularity allows trichomes to function as miniature organs, with differentiation within the trichome with regard to cell function. Many multicellular trichomes are glandular, developing a single or multicelled gland at the tip. The gland may secrete a variety of compounds, including excess salts and chemicals to deter or poison predators.

The trichomes of tobacco are the most extensively studied multicellular trichomes, but little is known of the molecular control of their development. Tobacco leaves develop two discrete types of multicellular trichome, described as short- and long-stalked. The short-stalked trichomes have an eight-celled glandular head, and exude nicotine, while the long-stalked trichomes also have a gland, which is believed to produce and secrete diterpenoids (Akers *et al.*, 1978; Meyberg *et al.*, 1991; Nielsen *et al.*, 1991). The glands of the long-stalked trichomes contain many chloroplasts, while the glands of the short-stalked form contain undifferentiated plastids and a large extraplasmic space between the cell wall and the plasmalemma (Akers *et al.*, 1978). There is substantial evidence that the two types of trichome are under the control of separate developmental programmes, as mutations which affect the development of the gland on the long-stalked trichomes do not affect the short-stalked trichome (Burk *et al.*, 1982). The mutant line TI-1112 fails to develop glands on the long-stalked trichomes, while line NC-744 develops non-secretory glands on the long-stalked trichomes, a result of the failure to differentiate the chloroplasts presumably necessary for diterpenoid production (Nielsen *et al.*, 1991). Diterpenoid formation is inherited as a single locus, designated *TE*, as is gland formation (Nielsen *et al.*, 1982; B. Glover, unpublished observations). There has also been some discussion as to whether tobacco produces a third variety of multicellular trichome, identical to the long-stalked glandular trichome but without a gland at the tip. However, a number of lines of evidence suggest that simple (glandless) trichomes represent immature glandular trichomes, as suggested by Cannon (1909). Very occasionally the long-stalked trichomes of tobacco branch and it is then possible to find branched trichomes in which one branch point has a gland and the other does not (Roberts *et al.*, 1981). Furthermore, ectopic expression of the *MIXTA* gene from antirrhinum in tobacco increases the frequency of both glandular and non-glandular long-stalked trichomes, suggesting a common developmental route, but does not affect the development of the short-stalked trichomes (Glover *et al.*, 1998).

The development of multicellular, glandular trichomes from many species has been described, and the process usually proceeds through the enlargement of single epidermal cells, followed by several cell divisions to generate a structure perpendicular to the epidermal surface (Vermeer and Peterson, 1979). Analysis of mutant lines has provided more information about the regulation of the specification and morphogenesis of the multicellular trichome. In tomato (*Lycopersicon esculentum*), different mutants in the production of several of the trichome types have been described. The *Woolly* (*Wo*^{m}) mutant (Rick and Butler, 1956) produces many more type 1 trichomes (Luckwill, 1943) than wild-type, and they are longer and branch more frequently (B. Glover, unpublished observations). No other trichome type is affected. The mutant allele of this gene is dominant. The recessive *dialytic* mutation (*dl*) of tomato prevents the formation of multicellular trichomes on the anthers, causing the anther cone to fall apart, as it is normally held together by interlocking hairs (Rick, 1947). The *DL* gene is involved in the regulation of the plane of cell division within the trichomes on all organs. In contrast to the strictly anticlinal division of cells in the developing trichomes of wild-type plants, trichomes of types 1 and 2 on the mutant plants undergo periclinal cell division, or between anticlinal and periclinal division, resulting in trichomes of abnormal shape, which often appear as blocks of cells. Cell division appears to be normal in all other structures in this line (B. Glover, unpublished observations). The *hairless* (*hl*) mutant of tomato produces trichomes of all types with abnormally shaped cells. The base cell of the trichome may swell and cells further up the stalk may also swell. Trichomes also appear in which the direction of cell expansion has reorientated by 90 degrees, so that the final trichome has a helical appearance (Reeves, 1977; B. Glover, unpublished observations). These tomato mutants demonstrate the activity of genes, which regulate the processes of cell division and subsequent expansion in multicellular trichomes, as well as the activity of genes which regulate the total number of epidermal cells specifying trichomes.

Mutants have also been described in *Datura wrightii* and antirrhinum, both of which suggest that multicellular trichome development may be under the control of different regulatory genes at different stages of development. The *velvety* phenotype in *Datura wrightii* is associated with the loss of the type 4 glandular trichome but, notably, the mutant plants do produce these trichomes on their first few true leaves, indicating that the structural genes required for their development are present (van Dam *et al.*, 1999). The *angus* (*antirrhinum glabrous stem*) mutant of antirrhinum fails to produce trichomes on the inflorescence stem, but develops normal numbers of trichomes on the hypocotyl and the flowers, again indicating that more than one regulatory gene is involved in specifying trichome patterning (B. Glover, unpublished observations). These two mutants may reflect the importance of trichomes as markers for changes in the phase of plant growth and development. Trichome distribution and density on the leaves of arabidopsis change with the age of the plant, and particularly with switches between vegetative and reproductive growth phases (Telfer *et al.*, 1997). The juvenile and adult leaves of maize show similar differences in trichome production, with the juvenile leaves developing hairs only on the leaf margins. When the switch between growth phases is disturbed, for instance in the maize mutant *glossy15*, which undergoes the transition to the adult phase early, the production of trichomes is also altered (Moose and Sisco, 1994). It is therefore possible that the *velvety* and *angus* mutants of *Datura wrightii* and antirrhinum represent a disruption of the growth stages of the plant rather than simply a disruption in trichome development. Molecular cloning of the genes involved may help to address this point.

A small number of data concerning the molecular control of multicellular trichome development has recently become available. Ectopic expression of the *MIXTA* gene from antirrhinum, required for the production of conical–papillate cells on the petal epidermis (see Section IV.D), in tobacco, resulted in the development of excess trichomes, both on epidermal surfaces that normally produce trichomes and on non-trichome-producing surfaces (Glover *et al.*, 1998). This indicates a relationship between the developmental pathways leading to conical–papillate cell and multicellular trichome production. In antirrhinum, *MIXTA* is not required for trichome development (Noda *et al.*, 1994), but the expression pattern of another antirrhinum gene with strong sequence identity to *MIXTA*, *ML1* (*MIXTA-LIKE 1*), indicates that it might play a role in the development of the trichomes of antirrhinum, which are very similar to the trichomes of tobacco (Glover *et al.*, 1998). Until more molecular data become available, it is possible to conclude only that the developmental steps leading to multicellular trichome development may be significantly different to those of unicellular trichomes, and it is likely that the molecular mechanisms may also be distinct.

D. CONICAL–PAPILLATE CELL DEVELOPMENT

The conical cells that form on the petals of 80% of flowering plants (Kay *et al.*, 1981) have also been classed as trichomes (Esau, 1953) and provide a cellular morphology distinct enough to be used as a marker for petal identity (Simon *et al.*, 1994; Goodrich *et al.*, 1997; Sablowski *et al.*, 1998). These cells are involved in enhancing light absorption by the petal and focusing the light into its pigment-containing regions (Gorton and Vogelmann, 1996). Petal epidermal cells may have very elaborate morphologies in some species, for example, multipapillate, with several cones per cell, or reverse papillate, with a lower cell wall extending into the mesophyll layer. Further to their role in the absorption and reflectance of light, there is some evidence to suggest that the shape of petal epidermal cells may be used as a tactile cue by bees in their discrimination between particular flowers (Kevan and Lane, 1985). It is possible that the texture of a flower may provide tactile signals about the position of a pollinator on a petal, through the sensilla trichodea on the tips of the antennae of the animal.

The establishment of conical cell form in antirrhinum requires the action of a MYB-related transcription factor, MIXTA (Noda *et al.*, 1994). Mutation at the *MIXTA* locus results in uniformly pigmented but pale red flowers, lacking the sheen of darker wild-type petals. This phenotypic difference is observed only on the inner epidermis of the flower in the region of the corolla lobes where conical–papillate cells normally develop. The perceived change in colour is due to the change in cell shape, and the subsequent change in ability to absorb and focus light. The mutant phenotype is recessive and the gene is expressed only in the inner petal epidermis (Noda *et al.*, 1994; Glover *et al.*, 1998). The *PHMYB1* locus in petunia is similarly required for the establishment of conical cell form (Avila *et al.*, 1993; van Houwelingen *et al.*, 1998) and a gene encoding a similar protein is believed to be expressed in tobacco flowers (Glover *et al.*, 1998).

Ectopic expression of *MIXTA* in tobacco results in the development of conical–papillate cells on all aerial organs (Glover *et al.*, 1998). This demonstrates that the *MIXTA* gene product is both necessary and sufficient to induce conical cell formation, implying that it is a master gene controlling this developmental programme. Therefore, to understand how conical cells are specified, it is necessary to understand the functions of the targets of transcriptional activation by MIXTA. There are a number of possible ways in which conical–papillate cells could be formed. A simple expansion in cell size could result from an increase in turgor pressure, causing the cell wall to loosen to allow expansion. Alternatively, directed extension of the cell wall could be the primary driving force behind conical–papillate cell formation. Ultrastructural analysis of conical and mutant flat cells has shown that the primary difference between the two cell shapes is an excess of cuticular cell-wall material in the conical cells (Fig. 3). This cell-wall material greatly increases the surface area of the conical cells but their volume is not significantly altered as the wall material is produced in a directed fashion (Noda *et al.*, 1994). Morphologically, the directional formation of buds and projections in the budding yeast *Saccharomyces cerevisiae* is similar to the growth of conical cells. In yeast, bud formation depends in part on the interactions of the fast-growing ends of microfilaments and actin filaments with the cell cortex (Drubin, 1991). The processes of polarity establishment and bud growth are regulated by a large number of genes, the products of which function together to control polarity and morphogenesis (Drubin, 1991). Many of these gene products, such as CDC24, CDC42 and CDC43, are required for the establishment of cytoskeletal polarity and thus for locating cell growth to the chosen bud site (Chant and Pringle, 1991). It is likely that the directional outgrowth of conical cells in petals also involves alterations to the cytoskeleton to maintain polarity, analogous to the situation in yeast. However, since the growth of conical–papillate cells and trichomes appears to be diffuse, and therefore associated with microtubule-driven processes, rather than with the microfilament-driven growth of the bud in budding yeast, it may be that the molecular mechanisms by which cell polarity is maintained through microtubules in fission yeast, *Schizosaccharomyces pombe*, represent better models for the morphogenesis of these cells (Martin and Glover, 1998). However, it is necessary to wait for an analysis of the targets of transcriptional activation by *MIXTA* before any further conclusions about the mechanisms of conical cell morphogenesis can be drawn.

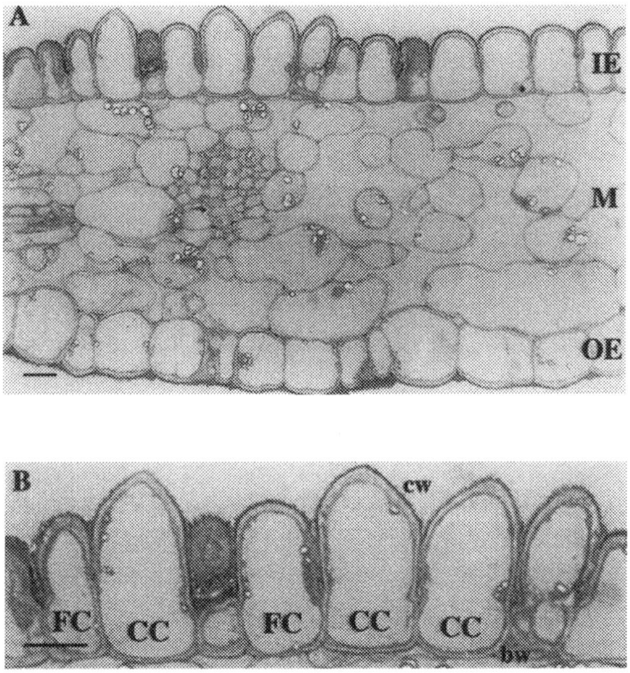

Fig. 3. Conical–papillate and mutant flat cells of the petal of the antirrhinum unstable *mixta* mutant line. Scale bars = 20 μm. (A) Conical–papillate and flat cells on the inner epidermis (IE) are both morphologically distinct from the less regular cells of the outer epidermis (OE). Mesophyll cells (M) have less regular shapes than epidermal cells. (B) The conical cells (CC) have the same volume and the same amount of basal wall (bw) as the flat cells (FC). However, they do produce more cuticular cell wall (cw) in a directional manner.

V. INTERACTIONS BETWEEN TRICHOMES AND OTHER EPIDERMAL CELLS

The flexibility inherent in plant development, required for environmental entrainment, makes overlap between developmental pathways highly probable. As discussed in Sections I and II, this overlap is likely to extend to the developmental programmes specifying epidermal cell form and function. Such overlap may be expected to result in multiple cell types being produced by elements of the same developmental programme and, conversely, by competitive or antagonistic interactions between developmental programmes. This section of this review focuses on the evidence for such interactions between developmental programmes specifying the various types of trichomes described above and those specifying the differentiation of the other epidermal cell types.

Selection of traits modifying trichome density can be demonstrated under changing environmental conditions (Mauricio and Rausher, 1997). It is thus likely that the development of trichomes is also responsive to signals from the environment. Changes in trichome density as a result of environmental factors will also affect the number and distribution of other cell types, particularly pavement cells. It has been observed many times that cells of the stomatal complex of *Arabidopsis*, including subsidiary cells, do not differentiate into trichomes. A competitive interaction may therefore exist between the developmental programmes specifying the two cell types. If this is the case, selection for alteration in the distribution of trichomes will be expected to influence the distribution of stomata as well.

Two of the *Arabidopsis* trichome mutants also affect the development of other epidermal cell types. The *ttg* mutant, which does not develop trichomes, has a greatly increased number of root hairs that form ectopically in what would normally be non-hair cell files. The *gl2* mutant plant also shows ectopic root hair formation. None of the other genes involved in the morphogenesis of the *Arabidopsis* trichome affect the development of any other epidermal cell type. It is interesting that in these two examples the gene products appear to be required in the root epidermis for the cell commitment and patterning step, but in the shoot epidermis they are required at different stages of trichome development. TTG is required for the initiation of trichome patterning, while GL2 is required for later morphogenesis, specifically the outgrowth of the trichome cell.

A more complex interaction between trichome and root hair differentiation was described by Wada *et al.* (1997). The *Arabidopsis* mutant *caprice* fails to produce root hairs, and the *CAPRICE* gene product encodes a MYB-related protein without an activation domain. Wada *et al.* (1997) therefore proposed that CPC acts to repress *GL2* activity, by binding the promoter region of the gene but failing to activate transcription. In this way, *GL2* expression is prevented in certain cells which may then differentiate into root hairs. In the *cpc* mutant, *GL2* activity is much higher and few cells are able to differentiate. It is predicted that, in the non-hair cells of the root epidermis, a similar MYB-related protein with an activation domain is required to activate *GL2* expression. However, constitutive expression of *CPC* in *Arabidopsis* results in a loss of trichomes on all aerial organs. Wada *et al.* (1997) argued that this phenotype may result from CPC competing with GL1, another MYB-related protein, to bind the promoter region of *GL2* in committed trichome cells. In the absence of an activation domain, CPC is unable to activate *GL2* expression and the plant fails to differentiate trichomes. This is an example where, in the natural situation, *GL1* and *CPC* do not interact in any way and trichome and root hair development are not normally interlinked. However, the genes may be evolutionarily related and the two developmental pathways may therefore have ancestrally shared common components.

Further evidence for the connections between different developmental programmes has recently arisen in *Arabidopsis*, where it has been demonstrated that stomatal distribution on the hypocotyl is governed by the same patterning process as governs root hair distribution. Berger *et al.* (1998) showed that, on the hypocotyl, guard cells developed only on the cell files corresponding to the root epidermal cell files capable of generating root hairs, that is in those cell files overlying a cortical anticlinal cell wall. This was the first demonstration that stomatal development may be linked to the development of other cell fates. Berger *et al.* (1998) further showed that hypocotyl stomatal patterning is disrupted in both the *ttg* and *gl2* mutants, with ectopic stomata arising in other cell files, mirroring the effects of these two mutations on root hair patterning. This correlation between root hair and stomatal distribution is specific to the hypocotyl, and is not seen on any other aerial organ, reflecting the common derivation of the two structures from the basal pole of the embryo (Scheres *et al.*, 1994). The *Arabidopsis* stomatal patterning mutants *flp* and *tmm* (Yang and Sack, 1995) were both generated in hairless lines, so no comparison of stomata and trichome distribution was possible, and they were not found to be affected in their root hair development.

The interaction between the differentiation of different cell types in the shoot epidermis has also recently been described in tobacco (Glover *et al.*, 1998). Constitutive expression of the *MIXTA* gene from antirrhinum in tobacco results in effects on several developmental pathways. Some of the transgenic lines produced conical cells, resembling petal epidermal cells, on their leaves as might be predicted from a gene required for petal conical cell formation. Other lines produced an excess of long multicellular trichomes. The formation of either conical cells or trichomes was strongly correlated to the timing of expression of *MIXTA* in relation to the developmental stage of the tissue and particularly its competence for further cell divisions. Leaf tissue expressing *MIXTA* relatively late (maximum expression when the cells had finished dividing) differentiated cones, whereas epidermal leaf tissue that had high *MIXTA* expression while the cells were still mitotically active, differentiated predominantly trichomes. Further to this, a gene product, AmMYBML1, with 89% identity to the MIXTA protein across the MYB DNA-binding domain and significant homology in the C-terminal activation domain, was identified in floral tissues. The *AmMYBML1* gene is expressed early relative to cell division in developing petals and may represent a gene controlling trichome development in antirrhinum flowers. Antirrhinum trichomes are morphologically very similar to the long trichomes of tobacco. Therefore trichome and conical cell development in antirrhinum and tobacco appear to share parts of a common developmental programme. Separation of cellular differentiation may have arisen by gene duplication. Differentiation of expression of two genes with a common function, particularly in relation to the progress of cell division, may result in different forms of cellular morphogenesis.

Development of either conical cells or excess trichomes in the transgenic tobacco leaves was also correlated with a dramatic reduction in stomatal density. This suggests that the conical cell/trichome developmental programme acts competitively with the guard cell developmental programme. This idea is supported by the observation that the petal epidermis, a tissue composed entirely of conical cells, contains no stomata and on cotton ovules the cells surrounding guard cells remain free of the late-forming trichomes. It is clear that cells that have been recruited to a conical or trichome cell fate cannot then differentiate into guard cells. Observation of both the 35S-*MIXTA* lines and of wild-type plants indicates that the converse is also true – the cells associated with the stomatal complex do not differentiate into trichomes or cones.

Analysis of the development of other epidermal cell types in the trichome mutants of tomato has also demonstrated an interaction between trichome and guard cell developmental pathways, and one between trichomes and root hairs (B. Glover, unpublished observations). The Wo^m mutant of tomato produces excess numbers of long multicellular trichomes on all shoot surfaces. Production of stomata on the leaves of these plants is severely reduced, with an almost complete absence of stomata on the adaxial epidermis of the leaf (Fig. 4), and a less striking reduction on the abaxial epidermis (B. Glover, unpublished observations). The *Hirsute* (*Hr*) mutant of tomato produces type 1 trichomes, which are considerably longer than wild-type, on only the adaxial epidermal surfaces of the leaf. The roots of this line show a dramatic increase in root hair density and length, with almost no non-hair cells visible (B. Glover, unpublished observations). This interaction could be explained by the presence of a protein required for limiting extension growth, but it is intriguing that its effects are limited to only one of the types of tomato trichomes, and to root hairs. It is possible that these two cell types first developed at the same time, and therefore share components of their developmental programmes.

VI. CONCLUSIONS

The determination of epidermal cell shape in plants is a complex process. The environmentally responsive nature of plant development means that cell lineage does not play the same role in plant cell patterning as it does in animals. Instead, signalling between differentiating cells is more important. Variations on the basic plant cell type, developed in response to different environmental inputs, may mean that discrete cell types share components of the same developmental programme, responding to that programme differently as the plant grows. The final shape of a plant epidermal cell is therefore determined by an extremely complex set of potentially conflicting signals, all interpreted with regard to the growth phase of the tissue and the competence of the cells for further division and differentiation.

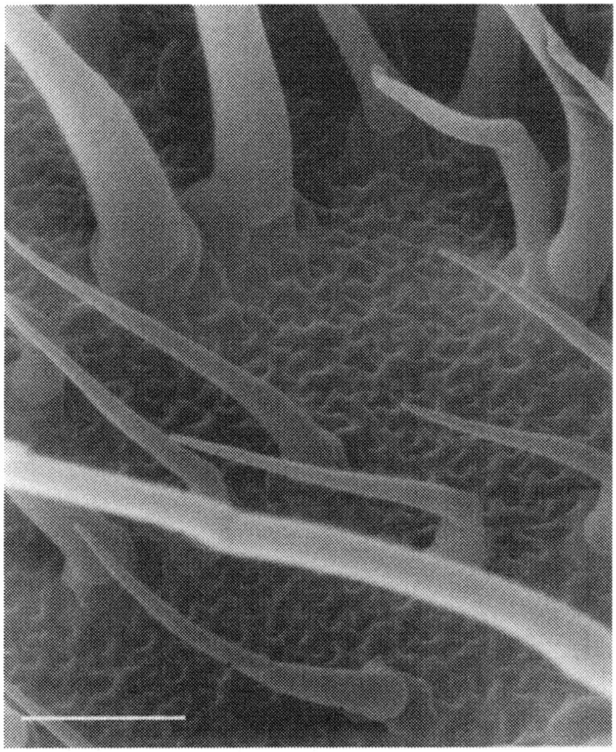

Fig. 4. Epidermal surface of the leaf of the tomato mutant *Wo*^{m}. Scale bar = 50 μm. The increased density of trichomes is correlated with a virtual absence of stomata from the adaxial epidermis.

ACKNOWLEDGEMENTS

We would like to thank David Hanke for helpful discussions. Work in our laboratories on this subject is funded by the BBSRC CAD Initiative, the John Innes Foundation, the Royal Society and the Gatsby Charitable Trust.

REFERENCES

Akers, C. P., Weybrew, J. A. and Long, R. C. (1978). Ultrastructure of glandular trichomes of leaves of *Nicotiana tabacum* L., cv Xanthi. *American Journal of Botany* **65**, 282–292.

Avila, J., Nieto, C., Canas, L., Benito, M. J. and Paz-Ares, J. (1993). Petunia hybrida genes related to the maize regulatory C1 gene and to animal myb proto-oncogenes. *Plant Journal* **3**, 553–562.

Baskin, T. I., Betzner, A. S., Hoggart, R., Cork, A. and Williamson, R. E. (1992).

Root morphology mutants in *Arabidopsis thaliana*. *Australian Journal of Plant Physiology* **19**, 427–437.

Berger, F., Linstead, P., Dolan, L. and Haseloff, J. (1998). Stomata patterning on the hypocotyl of *Arabidopsis thaliana* is controlled by the genes involved in the control of root epidermis patterning. *Developmental Biology* **194**, 226–234.

Boetsch, J., Chin, J. and Croxdale, J. (1995). Arrest of stomatal initials in *Tradescantia* is linked to the proximity of neighbouring stomata and results in the arrested initials acquiring properties of epidermal cells. *Developmental Biology* **168**, 28–38.

Boetsch, J., Chin, J., Ling, M. and Croxdale, J. (1996). Elevated carbon dioxide affects the patterning of subsidiary cells in *Tradescantia* stomatal complexes. *Journal of Experimental Botany* **47**, 925–931.

Burk, L. G., Chaplin, J. F. and Jackson, D. M. (1982). Inheritance of the glandless leaf trichome trait in *Nicotiana tabacum* L. *Tobacco International* **184**, 80–82.

Cannon, W. A. (1909). "Studies in Heredity as Illustrated by the Trichomes of Species and Hybrids of *Juglans*, *Oenothera*, *Papaver* and *Solanum*." Carnegie Institute of Washington Publication No. 17. Carnman Printing Co., Carlisle, PA.

Chant, J. and Pringle, J. R. (1991). Budding and cell polarity in *Saccharomyces cerevisiae*. *Current Opinion in Genetics and Development* **1**, 342–350.

de Vetten, N., Quattrocchio, F., Mol, J. and Koes, R. (1997). The an11 locus controlling flower pigmentation in petunia encodes a novel WD-40 repeat protein conserved in yeast, plants and animals. *Genes and Development* **11**, 1422–1434.

Dolan, L. (1996). Pattern in the root epidermis: An interplay of diffusible signals and cellular geometry. *Annals of Botany* **77**, 547–553.

Dolan, L., Janmaat, K., Willemsen, V., Linstead, P., Poethig, S., Roberts, K. and Scheres, B. (1993). Cellular organisation of the *Arabidopsis thaliana* root. *Development* **119**, 71–84.

Dolan, L., Duckett, C. M., Grierson, C., Linstead, P., Schneider, K., Lawson, E., Dean, C. and Roberts, K. (1994). Clonal relationships and cell patterning in the root epidermis of Arabidopsis. *Development* **120**, 2465–2474.

Drubin, D. G. (1991). Development of cell polarity in budding yeast. *Cell* **65**, 1093–1096.

Esau, K. (1953). "Plant Anatomy." John Wiley & Sons, Inc., New York.

Folkers, U., Berger, J. and Hülskamp, M. (1997). Cell morphogenesis of trichomes in Arabidopsis: differential control of primary and secondary branching by branch initiation regulators and cell growth. *Development* **124**, 3779–3786.

Galway, M. E., Masucci, J. D., Lloyd, A. M., Walbot, V., Davis, R. W. and Schiefelbein, J. W. (1994). The *TTG* gene is required to specify epidermal cell fate and cell patterning in the Arabidopsis root. *Developmental Biology* **166**, 740–754.

Glover, B. J., Perez-Rodriguez, M. and Martin, C. (1998). Development of several epidermal cell types can be specified by the same MYB-related plant transcription factor. *Development* **125**.

Goldberg, R. B. (1988). Plants: novel developmental processes. *Science* **240**, 1460–1467.

Goodrich, J., Puangsomiee, P., Martin, M., Long, D., Meyerowitz, E. and Coupland, G. (1997). A polycomb-group gene regulates homeotic gene expression in Arabidopsis. *Nature* **386**, 44–51.

Gorton, H. L. and Vogelmann, T. C. (1996). Effects of epidermal cell shape and pigmentation on optical properties of Antirrhinum petals at visible and ultraviolet wavelengths. *Plant Physiology* **112**, 879–888.

Gotor, C., Cejudo, F. J., Barroso, C. and Vega, J. M. (1997). Tissue specific expression of ATCYS-3A, a gene encoding the cytosolic isoform of O-acetylserine(thiol)lyase in *Arabidopsis*. *Plant Journal* **11**, 347–352.

Hülskamp, M., Miséra, S. and Jürgens, G. (1994). Genetic dissection of trichome cell development in Arabidopsis. *Cell* **76**, 555–566.

Hülskamp, M., Folkers, U. and Grini, P. E. (1998). Cell morphogenesis in Arabidopsis. *Bioessays* **20**, 20–29.

Kay, Q. O. N., Daoud, H. S. and Stirton, C. H. (1981). Pigment distribution, light reflection and cell structure in petals. *Botanical Journal of the Linnean Society* **83**, 57–84.

Kevan, P. G. and Lane, M. A. (1985). Flower petal microtexture is a tactile cue for bees. *Proceedings of the National Academy of Sciences (USA)* **82**, 4750–4752.

Kieber, J. J., Rotheberg, M., Roman, G., Feldmann, K. A. and Ecker, J. (1993). CTR1, a negative regulator of the ethylene response pathway in Arabidopsis, encodes a member of the Raf family of protein kinases. *Cell* **72**, 427–441.

Larkin, J. C., Oppenheimer, D. G., Lloyd, A. M., Paparozzi, E. T. and Marks, M. D. (1994). Roles of the *GLABROUS1* and *TRANSPARENT TESTA GLABROUS* genes in Arabidopsis trichome development. *Plant Cell* **6**, 1065–1076.

Larkin, J. C., Young, N., Prigge, M. and Marks, M. D. (1996). The control of trichome spacing and number in Arabidopsis. *Development* **122**, 997–1005.

Lloyd, A. M., Walbot, V. and Davis, R. W. (1992). Arabidopsis and *Nicotiana* anthocyanin production activated by maize regulators *R* and *C1*. *Science* **258**, 1773–1775.

Luckwill, L. C. (1943). "The Genus *Lycopersicon*: An Historical, Biological, and Taxonomic Survey of the Wild and Cultivated Tomatoes." Aberdeen University Press, Aberdeen.

Ma, D. P., Liu, H. C., Tan, H., Creech, R. G. and Jenkins, J. N. (1997). Cloning and characterisation of a cotton lipid transfer protein gene specifically expressed in fiber cells. *Biochimica et Biophysica Acta* **1344**, 111–114.

Martin, C. and Glover, B. J. (1998). Cellular differentiation in the shoot epidermis. *Current Opinion in Plant Biology* **1**, 511–519.

Masucci, J. D., Rerie, W. G., Foreman, D. R., Zhang, M., Galway, M. E., Marks, M. D. and Schiefelbein, J. W. (1996). The homeobox gene GLABRA2 is required for position-dependent cell differentiation in the root epidermis of *Arabidopsis thaliana*. *Development* **122**, 1253–1260.

Mauricio, R. and Rausher, M. D. (1997). Experimental manipulation of putative selective agents provides evidence for the role of natural enemies in the evolution of plant defense. *Evolution* **51**, 1435–1444.

Meyberg, M., Krohn, S., Brummer, B. and Kristen, U. (1991). Ultrastructure and secretion of glandular trichomes of tobacco leaves. *Flora* **185**, 357–363.

Moose, S. P. and Sisco, P. H. (1994). *Glossy15* controls the epidermal juvenile-to-adult phase transition in maize. *Plant Cell* **6**, 1343–1355.

Nielsen, M. T., Jones, G. A. and Collins, G. B. (1982). Inheritance pattern for secreting and nonsecreting glandular trichomes in tobacco. *Crop Science* **22**, 1051–1053.

Nielsen, M. T., Akers, C. P., Jarlfars, U. E., Wagner, G. J. and Berger, S. (1991). Comparative ultrastructural features of secreting and non-secreting glandular trichomes of two genotypes of *Nicotiana tabacum* L. *Botanical Gazette* **152**, 13–22.

Noda, K.-I., Glover, B. J., Linstead, P. and Martin, C. (1994). Flower colour intensity depends on specialized cell shape controlled by a Myb-related transcription factor. *Nature* **369**, 661–664.

Oppenheimer, D. G., Herman, P. L., Sivakumaran, S., Esch, J. and Marks, M. D. (1991). A *MYB* gene required for leaf trichome differentiation in Arabidopsis is expressed in stipules. *Cell* **67**, 483–493.

Oppenheimer, D. G., Pollock, M. A., Vacik, J., Szymanski, D. B., Ericson, B., Feldmann, K. and Marks, M. D. (1997). Essential role of a kinesin-like protein in Arabidopsis trichome morphogenesis. *Proceedings of the National Academy of Sciences (USA)* **94**, 6261–6266.

Pear, J. R., Kawagoe, Y., Schreckengost, W. E., Delmer, D. P. and Stalker, D. M. (1996). Higher plants contain homologs of the bacterial *celA* genes encoding the catalytic subunit of cellulose synthase. *Proceedings of the National Academy of Sciences (USA)* **93**, 12 637–12 642.

Reeves, A. F. (1977). Tomato trichomes and mutations affecting their development. *American Journal of Botany* **64**, 186–189.

Rerie, W. G., Feldmann, K. A. and Marks, M. D. (1994). The *GLABRA2* gene encodes a homeo domain protein required for normal trichome development in Arabidopsis. *Genes and Development* **8**, 1388–1399.

Rick, C. M. (1947). Partial suppression of hair development indirectly affecting fruitfulness and the proportion of cross-pollination in a tomato mutant. *American Naturalist* **81**, 185–202.

Rick, C. M. and Butler, L. (1956). Cytogenetics of the tomato. *Advances in Genetics* **8**, 267–382.

Roberts, K. L., Thurston, R. and Jones, G. A. (1981). Density of types of trichomes on leaf lamina of various tobacco cultivars. *Tobacco Science* **25**, 68–69.

Sablowski, R. W. M and Meyerowitz, E. M. (1998). A homolog of *NO APICAL MERISTEM* is an immediate target of the floral homeotic genes *APETALA3/PISTILLATA*. *Cell* **92**, 93–103.

Scheres, B., Wolkenfelt, H., Willemsen, V., Terlouw, M., Lawson, E., Dean, C. and Weisbeek, P. (1994). Embryonic origin of the Arabidopsis primary root and root-meristem initials. *Development* **120**, 2475–2487.

Schiefelbein, J. W. and Somerville, C. (1990). Genetic control of root hair development in *Arabidopsis thaliana*. *Plant Cell* **2**, 235–243.

Schneider, K., Wells, B., Dolan, L. and Roberts, K. (1997). Structural and genetic analysis of epidermal cell differentiation in *Arabidopsis* primary roots. *Development* **124**, 1789–1798.

Serna, L. and Fenoll, C. (1997). Tracing the ontogeny of stomatal clusters in *Arabidopsis* with molecular markers. *Plant Journal* **12**, 747–755.

Shimizu, Y., Aotsuka, S., Hasegawa, O., Kawada, T., Sakuno, T., Sakai, F. and Hayashi, T. (1997). Changes in levels of mRNAs for cell wall-related enzymes in growing cotton fiber cells. *Plant and Cell Physiology* **38**, 375–378.

Simon, R., Carpenter, R., Doyle, S. and Coen, E. (1994). *Fimbriata* controls flower development by mediating between meristem and organ identity genes. *Cell* **78**, 99–107.

Song, P. and Allen, R. D. (1997). Identification of a cotton fiber-specific acyl carrier protein cDNA by differential display. *Biochimica et Biophysica Acta* **1351**, 305–312.

Szymanski, D. B., Klis, D. A., Larkin, J. C. and Marks, M. D. (1998). *cot1*: a regulator of Arabidopsis trichome initiation. *Genetics* **149**, 565–577.

Telfer, A., Bollman, K. M. and Poethig, R. S. (1997). Phase change and the regulation of trichome distribution in *Arabidopsis thaliana*. *Development* **124**, 645–654.

Tiwari, S. C. and Wilkins, T. A. (1995). Cotton (*Gossypium hirsutum*) seed trichomes expand via diffuse growing mechanism. *Canadian Journal of Botany* **73**, 746–757.

van Dam, N. M., Hare, J. D. and Elle, E. (1999). Inheritance and distribution of trichome phenotypes in *Datura wrightii*. *Journal of Heredity* **90**, 220–227.

van Houwelingen, A., Souer, E., Spelt, K., Kloos, D., Mol, J. and Koes, R. (1998). Analysis of flower pigmentation mutants generated by random transposon mutagenesis in *Petunia hybrida*. *Plant Journal* **13**, 39–50.

Vermeer, J. and Peterson, R. L. (1979). Glandular trichomes on the inflorescence of *Chrysanthemum morifolium* cv. Dramatic (Compositae). 2. Ultrastructure and histochemistry. *Canadian Journal of Botany* **57**, 714–729.

Wada, T., Tachibana, T., Shimura, Y. and Okada, K. (1997). Epidermal cell differentiation in *Arabidopsis* determined by a *myb* homolog, *CPC*. *Science* **277**, 1113–1116.

Walker, A. R., Davison, P. A., Bolognesi-Winfield, A. C., James, C. M., Srinivasan, N., Blundell, T. L., Esch, J. J., Marks, M. D. and Gray, J. C. (1999). The *TRANSPARENT TESTA GLABRA1* locus, which regulates trichome differentiation and anthocyanin biosynthesis in Arabidopsis, encodes a WD40-repeat protein. *Plant Cell* **11**, 1337–1350.

Wang, H. and Cutler, A. J. (1995). Promoters from kin1 and cor6-6, two *Arabidopsis thaliana* low-temperature and ABA-inducible genes, direct strong β-glucuronidase expression in guard cells, pollen and young developing seeds. *Plant Molecular Biology* **28**, 619–634.

Wang, J., Letham, D. S., Cornish, E. and Stevenson, K. R. (1997). Studies of cytokinin action and metabolism using tobacco plants expressing either the *ipt* or the *GUS* gene controlled by a chalcone synthase promoter 1. Developmental features of the transgenic plants. *Australian Journal of Plant Physiology* **24**, 661–672.

Wilkins, T. A. and Jernstedt, J. A. (1999). Molecular genetics of developing cotton fibers. *In* "Cotton Fibers" (A. M. Basra, ed.). Food Products Press, New York (in press).

Yamagushi-Shinozaki, K. and Shinozaki, K. (1993). Characterization of the expression of a desiccation-responsive rd-29 gene of *Arabidopsis thaliana* and analysis of its promoter in transgenic plants. *Molecular and General Genetics* **236**, 331–340.

Yang, M. and Sack, F. D. (1995). The too many mouths and four lips mutations affect stomatal production in Arabidopsis. *Plant Cell* **7**, 2227–2239.

Trichome Initiation in *Arabidopsis*

A. R. WALKER[1] and M. D. MARKS[2]

[1]*Department of Plant Sciences, University of Cambridge, Downing Street, Cambridge CB2 3EA, UK*
[2]*Department of Genetics and Cell Biology, University of Minnesota, St Paul, Minnesota 55108, USA*

I. INTRODUCTION

Arabidopsis possesses a single type of aerial trichome. The simplicity and visibility of the trichomes in this species have made them the perfect

Advances in Botanical Research Vol. 31
ISBN 0-12-005931-2

candidates for studying a single developmental process in plants. Many genes (Marks, 1997) and some environmental processes affect trichome formation and development. In this chapter, genes and factors affecting the initiation of trichomes in *Arabidopsis* will be discussed.

Mature *Arabidopsis* trichomes are very large single cells that are present on the surface of leaves, stems and sepals. On leaves, they are normally branched with a single stalk supporting three spikes (Fig. 1A), whereas on stems they are unbranched (Fig. 1B). Despite this morphological difference, they are classified as the same type of simple trichome (unicellular and non-glandular). Since mutations that alter the formation of leaf trichomes also affect stem trichomes, the steps involved in initiating the formation of both stem and leaf trichomes appear to be the same processes. However, there have been more studies conducted on leaf trichomes because the developing leaf goes through specific stages defined by length and has a more determinate structure than a stem. As *Arabidopsis* trichomes are not glandular and do not secrete substances, their main role is to provide a physical barrier against insects (Mauricio, 1998), and possibly to protect the meristem and developing leaf primordia from damage by ultraviolet light. However, under laboratory conditions, trichomes are not essential to the normal growth and development of the plant.

Root hairs are sometimes referred to as trichomes, as the word "trichome" means simply "an epidermal hair". The formation of root hairs is linked to the formation of aerial hairs in *Arabidopsis* by the action of several genes. Some of these genes appear to have an opposite effect by, for example, negatively regulating root hair initiation while positively regulating the initiation of leaf hairs. We shall mention root hairs in connection with trichome formation, but a detailed account is beyond the remit of this chapter. Excellent reviews have been published recently by Dolan and Scheres (1998) and van der Berg *et al.* (1998).

II. NORMAL TRICHOME INITIATION IN *ARABIDOPSIS*

A. PATTERN OF TRICHOME INITIATION ON A LEAF

As a developing leaf primordium approaches 100 μm in length, trichome precursor cells become visually distinct from the surrounding cells of the protoepidermis on the adaxial or upper side of the leaf. At this stage the initiating trichomes are about four epidermal cells apart (Hülskamp *et al.*, 1994; Larkin *et al.*, 1996). As the leaflet enlarges, the epidermal cells continue to divide. New trichome initials, sometimes called intercalary trichomes, form between the developing trichomes until the leaf primordium reaches a length of about 600 μm. After this stage, trichomes are no longer initiated (also suggested by the work of Lloyd *et al.*, 1994).

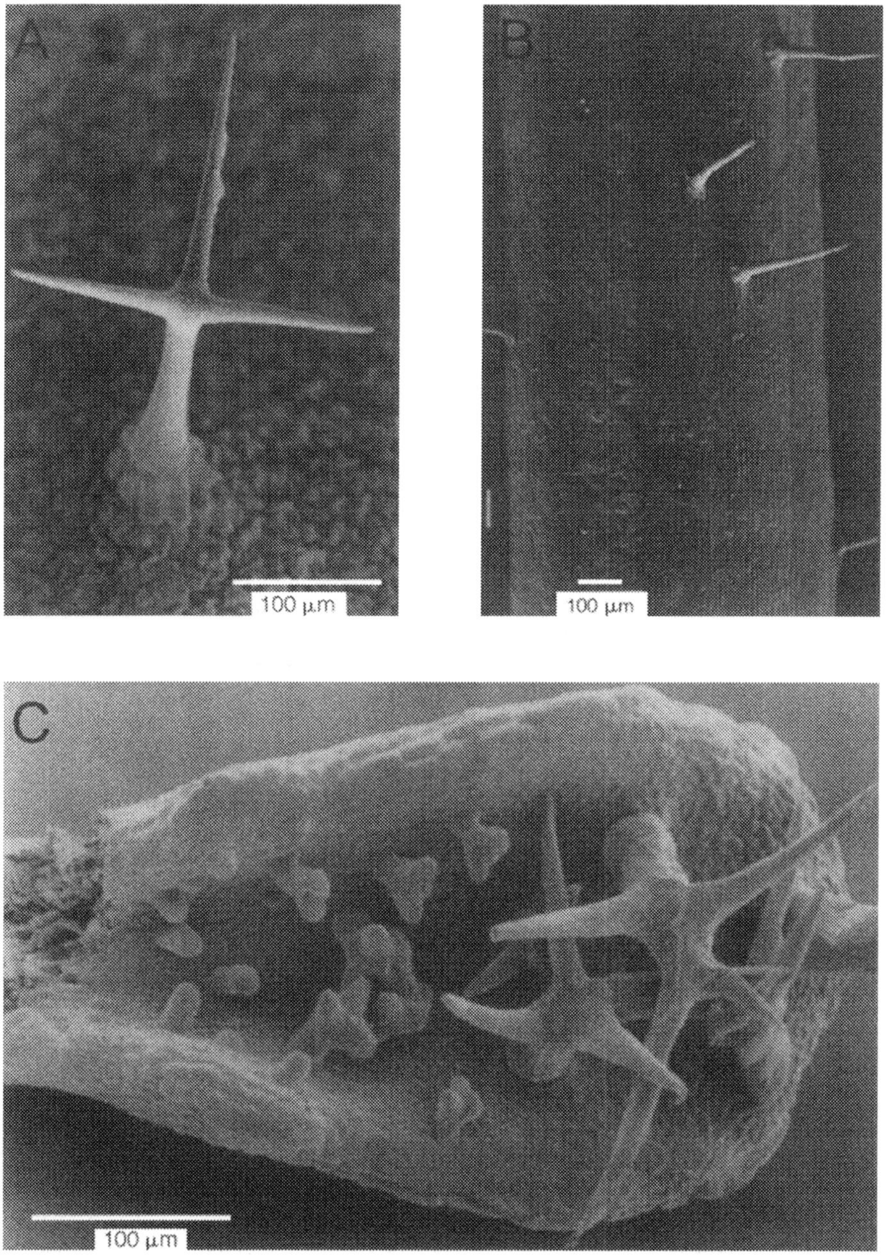

Fig. 1. *Arabidopsis* trichomes (scanning electron micrographs). (A) Single mature leaf trichome. (B) Stem with trichomes. (C) Developing leaf primordium showing gradient initiation (photograph by Vava Grbic).

The trichomes do not belong to the same cell lineage and are spaced regularly (Larkin *et al.*, 1996). This suggests a model of inhibition by developing trichomes, which prevents neighbouring cells from becoming trichomes. Trichomes do not initiate side by side, and support cells do not become trichomes. A field of inhibition originating within each developing trichome and extending 2–3 cells beyond the trichome (suggested by Hülskamp *et al.*, 1994 and Larkin *et al.*, 1996) would achieve this even spacing.

Trichomes initiate in a gradient of development starting at the tip of the leaf and extending down to the base near the petiole (Fig. 1C). Epidermal cells continue to divide to give rise eventually to small guard cells and larger pavement cells. The final spacing between trichomes is about 30 pavement cells apart on the first set of leaves (Hülskamp *et al.*, 1994). The development of guard cells, which comprise the stomatal complex, is an independent process from that of the formation of trichomes and occurs later in leaf development (for a recent review, see Larkin *et al.*, 1997). The trichome precursor cell undergoes several rounds of endoreplication of the nuclear DNA, then expands to form a larger cell. The cell expansion continues outwards away from the leaf surface. Another round of endoreplication of the nuclear genetic material is followed by branch formation (Folkers *et al.*, 1997). Trichome morphology is discussed in the chapter by Hülskamp and Kirik in this volume. Pavement cells can also undergo several rounds of endoreplication with the larger cells containing more nuclear DNA. However, the nuclei of guard cells remain at 2*c* (Melaragno *et al.*, 1993).

About 8–10 columnar support or socket cells in the epidermis surround each mature trichome. These accessory cells are not derived from a single cell lineage (Larkin *et al.*, 1996), suggesting that proximity to a developing trichome provides the necessary signal(s) to initiate the cellular programme in proto-accessory cells that would otherwise become pavement cells. Further evidence for this possibility comes from the observation that glabrous or hairless mutants do not develop support cells.

B. POSITION OF TRICHOMES ON THE PLANT

Only some epidermal surfaces of *Arabidopsis* are covered with hairs. The seed coat and embryo lack hairs. As the seedling emerges on germination, all aerial parts of the young plant, including the hypocotyl and cotyledons, are glabrous (hairless). The first trichomes appear on the adaxial or upper side of the first pair of leaves, which lack trichomes on the abaxial or underside of the leaves. As the plant matures and more rosette leaves are formed, trichomes appear on the abaxial side of later leaves. The first abaxial trichomes appear on leaves near the base and usually on the midrib. On later leaves abaxial trichomes appear with greater frequency and further towards the tip of the leaf. This is very pronounced in the cauline leaves on the flowering stem, where, as trichomes are less frequent on the basal end of the adaxial side, they increase in the basal region of the abaxial surface. On later leaves, more trichomes appear on the abaxial side towards the leaf tip as they are lost from the adaxial side towards the tip of the leaf (Chien and Sussex, 1996).

The primary flowering stem develops unbranched trichomes near the base of the stem, as do some lateral flowering stems. Further up the stems, the number of trichomes decreases until the stems become glabrous. If flowers develop where the stems have trichomes, then the sepals on the outside of the flower have some trichomes. It appears that once the signal to initiate trichomes is lost in the flowering stem, then fewer trichomes appear on the sepals of the flowers. Whether the sepals have trichomes or not, other flower parts are glabrous. Different ecotypes exhibit differences in the number, density and pattern of trichome formation on leaves and stems, suggesting that many genes are involved in these processes.

C. ROOT HAIR FORMATION

Both the primary and lateral roots have similar developmental patterns of root hair differentiation. A root hair forms as an extension of an epidermal cell in the cell elongation zone after the cells have completed cell division. Files of cells (trichoblasts) that lie over the boundary of two neighbouring cortical cells can give rise to root hairs. These cells differ in vacuolation, size and surface structure from the atrichoblast cells, which do not have root hair extensions (Dolan *et al*., 1994). During the development of the root hair, the nucleus is translocated into the extended part of the cell. This shows similarity to the movement of the nucleus up the stalk of the developing trichome.

III. GENETIC AND MOLECULAR ASPECTS OF THE INITIATION OF TRICHOMES

A. TRANSCRIPTION FACTORS CONTROL INITIATION OF TRICHOMES

1. A MYB Transcription Factor Initiates Trichome Formation

Many mutants involved in initiation and development of trichomes have been identified over the years (Hülskamp *et al*., 1994; Marks, 1997; Oppenheimer, 1998). However, only one gene affects initiation of trichomes alone. Plants with mutations at the *GLABROUS1* (*GL1*) locus lack mature trichomes and trichome precursor cells cannot be differentiated from other epidermal pavement cells. Plants with a deletion of the *GL1* gene can still produce a few peripheral trichomes, indicating that the function of this gene is not an essential requirement for the formation of every trichome. The severe allele causes no other identified phenotypic effects, suggesting that the action of this gene is confined to trichome initiation and development. *GL1*, one of the first T-DNA-tagged genes to be isolated (Herman and Marks, 1990), encodes a MYB transcription factor (Oppenheimer *et al.*, 1991). Since *GL1* expression persists as trichomes develop and there is genetic evidence that GL1 might regulate later aspects of trichome morphology (Esch *et al.*, 1994), this protein is thought to play a role in processes after trichome initiation. Overexpression studies using cauliflower mosaic virus *35S:GL1* constructs in transgenic *Arabidopsis* have shown that the gene product is not sufficient to activate trichome formation because few ectopic trichomes are formed on parts of the plant normally lacking trichomes, such as cotyledons or flower parts. In fact, there are fewer trichomes on the adaxial side of leaves in plants carrying the *35S:GL1* construct (Larkin *et al.*, 1996). GL1 is necessary for the formation of most trichomes, but it is not sufficient to induce more than normal numbers of trichomes. GL1 controls the first committed step so far identified in trichome initiation, independent of other processes.

2. MYC Transcription Factors may Regulate Trichome Initiation

Anthocyanins are purple, red and blue pigments found in the vacuoles of flowers, leaves, stems and seed coats of plants. The regulation of the anthocyanin biosynthetic pathway has been studied extensively in maize, petunia and *Antirrhinum majus* (snapdragon) (reviewed in Mol *et al.*, 1996). In maize, MYC (e.g. C1) and MYB (e.g. R) transcription factors act together to regulate the transcription of genes encoding enzymes involved in anthocyanin biosynthesis (Goff *et al.*, 1992; Lesnick and Chandler, 1998).

By analogy with the regulation of the anthocyanin biosynthetic pathway in maize, the MYB transcription factor GL1 might require a MYC partner protein to activate trichome initiation in *Arabidopsis*. Lloyd *et al.* (1992) showed that the MYC factor R from maize (Ludwig *et al.*, 1989) could induce more leaf trichomes and form ectopic trichomes on flower parts compared to wild-type plants when placed under the control of the nearly constitutive *35S* promoter. This result suggests there might be an endogenous *MYC* gene, which encodes a partner for GL1 to switch on trichome initiation. GL1 and R have been shown to interact *in vitro* (Syzmanski *et al.*, 1998a), so it seems probable that GL1 and a MYC protein from *Arabidopsis* might act together to regulate trichome formation.

Constructs containing the constitutively expressed *Delila* (*Del*) gene [the *MYC* gene from snapdragon, which regulates anthocyanin biosynthesis (Goodrich *et al.*, 1992)], when used to generate transgenic *Arabidopsis*, did not lead to an increase in trichome numbers or a significant increase in pigmentation (Mooney *et al.*, 1995). In transgenic tobacco, constitutive expression of *Del* and *R* can intensify flower colour from pink to purple with no change in trichome number (Mooney *et al.*, 1995; Lloyd *et al.*, 1992). Since

Del RNA was present in *Arabidopsis*, it is probable that the Delila protein was also available. The differences in *Arabidopsis* between the action of the two MYC proteins might be due to their intrinsic level of activity or to their regulation. The maize R protein might escape a form of negative regulation by being less similar to dicot MYC proteins.

Two *MYC* genes with sequence similarity to *Delila* have been identified in *Arabidopsis* (Payne *et al.*, 1999; A. R. Walker, unpublished). However, whether either gene is involved in trichome formation or anthocyanin regulation is still under investigation.

B. REGULATION OF TRANSCRIPTION FACTORS

Mutations at the *TRANSPARENT TESTA GLABRA1* (*TTG1*) locus result in loss of trichomes, anthocyanin and seed mucilage (Koornneef, 1981) and the formation of ectopic root hairs (Galway *et al.*, 1994). The expression of *GL1* is not reduced in *ttg1* mutants indicating that TTG1 does not regulate GL1 at the transcriptional level. It had been suggested that *TTG1* might encode a MYC transcription factor because *35S:R* constructs were able to complement all the defects in the *ttg1-1* mutant (Lloyd *et al.*, 1992). The cloning of the *TTG1* gene has revealed that it encodes a WD40-repeat protein (Walker *et al.*, 1999). WD40 motifs are common to many proteins (Neer *et al.*, 1994) and are involved in protein–protein interactions. These data are consistent with the hypothesis that TTG1 might regulate the activity of the transcription factors directly or be part of a signal transduction pathway that controls the transcription factors.

In roots, it has been proposed that TTG1 could regulate the rate of cell division to produce more trichoblasts compared to atrichoblast cells in the files of the root epidermis (Berger *et al.*, 1998). TTG1 might function to prevent cell division both in atrichoblasts and in trichome precursors. This hypothesis explains the apparent anomaly of opposite effects on root hairs and leaf hairs of the *TTG1* gene. Analysis of coiled-body structures in the nuclei of root epidermal cells of wild-type and *ttg1* mutant plants (Boudonck *et al.*, 1998) supports other observational evidence that all epidermal cells in *ttg1* mutants resemble atrichoblast cells.

C. INTERACTION BETWEEN GL1 AND TTG1

Genetic studies (Larkin *et al.*, 1994, 1999) have shown that there is a specific relationship between GL1 and TTG1. In the first study, when plants overexpressing *GL1* were crossed to a *ttg1* mutant, the next heterozygous generation had trichomes in clumps reminiscent of the effect of weak *ttg1* mutant alleles. This suggests that the excess GL1 protein is removing some TTG1 protein from the available pool.

Larkin *et al.* (1999) showed that there is an allele-specific interaction between GL1 and TTG1 using the mutants *gl1-2* and *ttg1-10*. The *ttg1-10* mutation results in a new initiation codon (out of frame) in the 5′UTR leading to a reduction in the level of message, possibly due to nonsense-mediated message degradation. Plants homozygous for this allele may have very small quantities of wild-type protein (Larkin *et al.*, 1999). Mutants with the genotype *ttg1-10/ttg1-10 gl1-2/+* have an unexpected phenotype. Since *gl1-2/+* appears similar to wild-type with normal trichomes and *ttg1-10* is leaky having some hairs in clumps, one would expect to find some hairs. However, these plants have no trichomes, suggesting that the *gl1-2* protein (a truncated form) is outcompeting the GL1 protein for the small amounts of wild-type TTG1 protein predicted to be present in the *ttg1-10* mutant.

A similar allele-specific interaction was detected using a specific weak allele of *GL1* named *gl1-EM2* and two alleles of *TTG1*, *ttg1-1* (the strong reference allele) and *ttg1-10* (Schnittger *et al.*, 1999). Either of the two *ttg1* alleles together with *gl1-EM2/+*, both in a heterozygous state with wild-type alleles, leads to a trichome-clumping phenotype. The molecular nature of the lesion in *gl1-EM2* is not known. *ttg1-1* results in a truncated protein missing 25 amino acids at the *C*-terminus (Walker *et al.*, 1999). Another *ttg1* allele (*ttg-w* identified as *ttg1-9*, A. R. Walker, unpublished), resulting in an amino-acid substitution of serine 282 to phenylalanine, probably disrupting secondary structure, does not lead to clumping of trichomes in the same conditions (Schnittger *et al.*, 1999).

These results cannot rule out the possibility that TTG1 activates a pathway leading to active GL1 or MYC components but do show a close relationship between TTG1 and GL1. It is not yet known whether these two proteins interact directly, although the genetic evidence suggests they may associate.

D. REGULATION OF TRICHOME INITIATION IN OTHER SPECIES

In other species, *MYB* genes affect trichome development. *MIXTA* from snapdragon and *CotMYBA* from cotton both change trichome differentiation when placed in tobacco plants (Glover *et al.*, 1998; Payne *et al.*, 1999), whereas constitutive *GL1* expression does not. However, since overexpression of *GL1* in *Arabidopsis* does not lead to a marked phenotype, subtle changes in the number of simple trichomes of tobacco could be overlooked. Constitutive *MIXTA* expression does not alter trichome number or shape in *Arabidopsis* (Glover *et al.*, 1998). *MIXTA* is discussed in detail in the chapter by Glover and Martin.

Comparison of the roles of TTG1 from *Arabidopsis* and a similar protein called AN11 from petunia (de Vetten *et al.*, 1997) suggests there are major differences in trichome initiation between these two species. AN11 regulates anthocyanin biosynthesis in the flowers only and does not play a known role in trichome initiation. Trichome differentiation appears to be regulated differently in *Arabidopsis* and petunia. Moreover, complementation experiments using a transient assay system indicate that AN11 might regulate the MYB transcription factor involved in anthocyanin biosynthesis. The results of this complementation are at odds with the results of complementation of the *ttg1-1* mutation by the maize *R* gene, suggesting that TTG1 and AN11 might function differently to regulate anthocyanin biosynthesis. All of these data suggest that the pathways of trichome formation might differ between *Arabidopsis* and other dicots such as tobacco, snapdragon and petunia.

Matthiola incana (10-week stock) is a brassica species that appears to be more similar to *Arabidopsis* in its trichome development. Two non-allelic mutations that affect both trichomes and anthocyanin have been identified (Kappert, 1949). Neither of these loci has yet been isolated, so the nature of these genes is still a mystery. However, the similarity between these mutants and the *ttg1* mutants of *Arabidopsis* leads to the prediction that one of the *M. incana* mutants might be defective in the *TTG1* homologue.

E. DOWNSTREAM TARGETS OF GL1 AND TTG1

Another transcription factor encoded by the *GLABRA2* (*GL2*) locus is a homeodomain protein (Rerie *et al.*, 1994). Plants with mutations at this locus bear trichomes reduced both in size and number, lack seed mucilage (Koornneef *et al.*, 1982) and have ectopic root hairs (Di Cristina *et al.*, 1996; Masucci *et al.*, 1996). Recent studies have shown that, in *ttg1* mutants, the *GL2* transcript is much reduced, although the pattern of expression is maintained in the root (Hung *et al.*, 1998). Szymanski *et al.* (1998a) have also shown that *GL2* expression is initially down-regulated in *gl1* mutants. As the leaf matures, *GL2* expression rises to normal levels in the absence of the *GL1* gene, suggesting activation by another MYB factor or a different mechanism at this point in development. Rudimentary unbranched trichomes on *gl2* mutant plants are formed as partially expanded cells, indicating that GL2 is important in continuing the development of a trichome after the initiation steps.

Another locus, *TTG2*, encodes a protein belonging to a fourth class of transcription factor (C. Johnson and D. Smyth, personal communication). Plants with mutations at this locus are similar but less severe in phenotype than those of *ttg1* mutants. The relationship between this locus and the other genes affecting trichome development is not yet understood. One possibility is that TTG2 acts downstream of TTG1.

F. MUTATIONS AFFECTING TRICHOME SPACING

The *triptychon* (*try*) mutant has clumps or nests of trichomes, which appear to initiate at the same time as though the field of inhibition is no longer operating (Hülskamp *et al.*, 1994; Hülskamp and Schnittger, 1998). Initiation of each trichome in a single clump is an independent event and not through cell division of a trichome precursor (Schnittger *et al.*, 1999). As the *TRY* gene has not been isolated, there are no molecular clues as to how the gene product might interact with the components of the initiation pathway that have already been identified. Trichomes on *try* mutants are larger and have a larger nucleus than the wild type, suggesting that TRY controls the cell cycle in trichomes.

A new phenotype is seen when the *try* mutation is introduced into plants containing the *35S:GL1* construct. These plants have subepidermal trichomes formed in the layer of cells below the epidermis (Schnittger *et al.*, 1998; Szymanski and Marks, 1998). This layer of cells is derived from the L2 of the meristem, and therefore it is not clonally related to the epidermis. These ectopic trichomes sometimes erupt through the epidermis to form trichomes of relatively normal appearance. The socket or support cells are not present. This suggests that TRY has an inhibitory role to prevent trichomes from forming in the subepidermal cell layers.

Plants containing weak *ttg1* alleles such as *ttg1-10* have some trichomes and some anthocyanin. The *ttg1-10* mutant has clumps of trichomes (Larkin *et al.*, 1994) leading to the proposition that TTG1 and TRY together might behave as inhibitors of trichome development in cells adjacent to developing trichomes.

A genetic interaction between TTG1 and TRY has been shown in double mutants of these loci together with *35S:GL1* (Schnittger *et al.*, 1999). Overexpression of *GL1* does not rescue the *ttg1-1* mutant phenotype, although there are more trichomes than in the *ttg1-1* mutant. However, if TRY is removed, trichomes develop normally around the edge of the leaf in the absence of TTG1. The role of TRY as an inhibitor of trichome initiation is clearly seen in plants constitutively expressing *GL1* and *R* in the absence of TRY. These plants are extremely hairy with trichomes covering most plant surfaces.

The *REDUCED TRICHOME NUMBER* (*RTN*) locus prevents the formation of the secondary or intercalary trichomes that develop later in the formation of the leaf (Larkin *et al.*, 1996) and accounts for the observation that there are fewer trichomes on ecotype Landsberg *erecta* (L*er*) plants than on Columbia (Col) plants. New trichomes are not initiated once the leaf primordia reach 400 μm in length, resulting in fewer trichomes with larger spaces between them.

G. OTHER GENES INFLUENCING TRICHOME FORMATION

The *ROOT HAIRLESS1* gene is required for the formation of root hairs and normal trichomes (Schneider *et al.*, 1998). The gene is epistatic to and independent of GL2 and plants mutated at this locus die before reaching maturity. The nuclear protein of unknown function encoded by this gene could be part of a normal cellular function pathway not regulated by GL1 and TTG1.

A mutation called *increased chalcone synthase expression1* (*icx1*) increases transcription of the gene encoding chalcone synthase in response to light, reduces the number of trichomes and alters leaf shape (Jackson *et al.*, 1995). Other mutations increasing or decreasing anthocyanin content often result in increased or decreased trichome number; for example, *cop1* mutants (mentioned in section H) show an increase in both trichomes and anthocyanin, whereas *ttg1* decreases both. It is not known how *icx1* affects other aspects of trichome formation. However, it is clear that the pathways leading to trichome formation are complex.

H. INHIBITORS OF TRICHOME INITIATION

Epidermal cells on leaves, stems and sepals are competent to make trichomes; cells in other organs might also be able to initiate trichome development under the right conditions. If organ identity is altered, trichomes can appear on cotyledons. The *leafy cotyledon* (*lec*) mutants have cotyledons that resemble leaves in shape, veination and surface structure, including the presence of trichomes (Meinke *et al.*, 1994).

In plants related to *Arabidopsis*, there are various patterns of trichome distribution. White mustard (*Sinapis alba*) has trichomes on the cotyledons and the hypocotyl, whereas oil-seed rape (*Brassica napus*) has very few trichomes on the plant. These differences might be due to negative regulators of trichome initiation. The mutants *constitutively photomorphogenic* (*cop*) and *fusca* (*fus*) have many pleiotropic effects, including ectopic trichomes. COP1 is part of a complex of nuclear proteins which negatively regulates photomorphogenesis, and mutants have ectopic trichomes on sepals and carpels (Miséra *et al.*, 1994). Many of the other *cop/fus* mutations are seedling-lethal, so determining the effect on trichome distribution is not possible.

A gene called *COT1* that appears to negatively regulate the initiation of hairs on cotyledons negatively has been identified (Szymanski *et al.*, 1998b). The phenotype of a mutant at this locus was identified by mutating a line carrying *35S:GL1* and screening for increased numbers of cotyledon trichomes. The *cot1* mutation in a wild-type background has no visible phenotype (Szymanski *et al.*, 1998b).

CAPRICE encodes a MYB-related transcription factor involved in root hair development. Overexpression of this gene (*CPC*) in *Arabidopsis* reduces the number of leaf trichomes, although mutants at this locus do not have a marked aerial phenotype (Wada *et al.*, 1997). If this gene is normally expressed in leaf primordia, it could have a role in suppressing trichome formation in conjunction with TRY.

IV. OTHER FACTORS

A. PLANT GROWTH FACTORS

Gibberellins are the main group of plant growth factors that influence the pattern and timing of trichome initiation. Applications of exogenous gibberellic acid (GA) or GA biosynthesis inhibitors, and mutants defective in GA biosynthesis, signalling or response to signalling have been used to elucidate the effects of GA on trichome formation (Chien and Sussex, 1996; Telfer *et al.*, 1997).

The timing of the initiation of the first abaxial trichomes can be brought forward by the addition of exogenous GA, whereas inhibitors of GA biosynthesis increase the leaf number before the appearance of abaxial trichomes. Once the signal to initiate abaxial trichomes has been perceived by the plant, the stimulus can be removed and the plant continues to make abaxial trichomes on new leaves (Chien and Sussex, 1996).

Telfer *et al.* (1997) highlighted the developmental stages that the plant goes through from vegetative to flowering to assess whether abaxial trichome pattern and number are indicative of the juvenile to adult transition. Their results show that trichome initiation on the abaxial side of leaves within the rosette is influenced by the exogenous application of GA at various stages of development. Some mutants defective in GA signalling had reduced numbers of abaxial trichomes while those involved in signal perception had normal trichomes.

A mutant in the perception of GA called *spindly* seems to behave as though it is constitutively receiving a GA signal (Jacobsen *et al.*, 1996). This mutant has more trichomes than normal, showing that GA is important in the regulation of trichome initiation. The severe allele *ga1-3* has a deletion of the gene encoding the first committed step of the GA biosynthetic pathway, ent-kaurene synthase A (Sun *et al.*, 1992). These plants have very few trichomes (Chien and Sussex, 1996). However, this phenotype can be rescued by the exogenous application of GA (Chien and Sussex, 1996) or by the mutation *rga* (*repressor of ga1-3*) (Silverstone *et al.*, 1997). Not all mutations affecting GA biosynthesis are reported to influence trichome formation. A later step in GA biosynthesis, ent-kaurene oxidase encoded by the *GA3* locus (Helliwell *et al.*, 1998), does not greatly affect the number of trichomes.

Until recently, there was no direct evidence for how GA might act. Perazza *et al.* (1998) showed that the transcription of the *GL1* gene was greatly reduced in *ga1-3* plants when GA biosynthesis was at a low level, resulting in the loss of trichome initiation. *GL1* is normally expressed in the epidermis of leaf primordia and it is this expression that is vital to trichome initiation. Constitutive expression of *GL1* and *R* leads to many ectopic trichomes in *Arabidopsis*. In the *ga1-3* background, expression of these two genes results in trichomes, showing that they overcome the *ga1* mutation. This confirms that transcription factors involved in trichome initiation act downstream of GA signalling.

Mutations in pathways affecting the synthesis, transport or perception of other plant growth factors, such as ethylene, auxin, brassinosteroids or abscisic acid, have not been reported to affect trichome initiation in *Arabidopsis*. It is possible that subtle effects have gone unnoticed.

B. LIGHT

Long days or continuous light can alter trichome distribution in *Arabidopsis*. More abaxial trichomes are induced on earlier leaves of the rosette in long days (Chien and Sussex, 1996). This response could be the result of a series of pathways that are not directly related to trichomes and are more general responses to light. White mustard (*Sinapis alba*) has trichomes on and anthocyanin in the hypocotyl when germinated in the light. In the dark, both of these characters are missing (Wagner and Mohr, 1966). Since *Arabidopsis* has a glabrous hypocotyl, there is no easy way to show that light is a requirement for the formation of trichomes. However, mutations in pathways which regulate the plant's response to light (for example, *cop1*) have suggested that light would normally play a role in initiating trichomes in *Arabidopsis*.

C. γ-RADIATION AND ACTIVE OXYGEN SPECIES

In a recent study, Nagata *et al.* (1999) have shown that active oxygen species, generated by γ-radiation, increase the number of trichomes on the adaxial side of developing leaves but do not change the number of trichomes on fully developed leaves. In the Columbia ecotype, the additional trichomes are of normal appearance and are equally spaced. Strong alleles of *ttg1* and *gl1* cannot be suppressed by γ-radiation, whereas irradiated *gl3* mutants had more trichomes than the *gl3* mutant. Anthocyanin accumulation in the irradiated plants suggests that TTG1-activated pathways are up-regulated by the radiation. These observations indicate that the processes that initiate normal trichome development are required for the formation of these additional trichomes.

The ecotype Columbia is more sensitive than Landsberg *erecta* to the radiation treatment. The authors suggested that this difference could be due to higher levels of superoxide dismutase in L*er* plants compared with levels in Col. However, other differences, such as the *RTN* locus, could also be important.

D. RELATIVE HUMIDITY

Relative humidity is known to affect trichome numbers (Chien and Sussex, 1996; Perazza *et al.*, 1998) but no extensive study has been conducted. High relative humidity seems to reduce trichome initiation by an unknown mechanism.

V. CONCLUSION

The pathways involved in trichome formation have been illustrated in Fig. 2. There are many areas where components of these pathways have not yet been identified. The genes that govern the formation of trichomes in *Arabidopsis* are numerous and there may yet be more that have not been

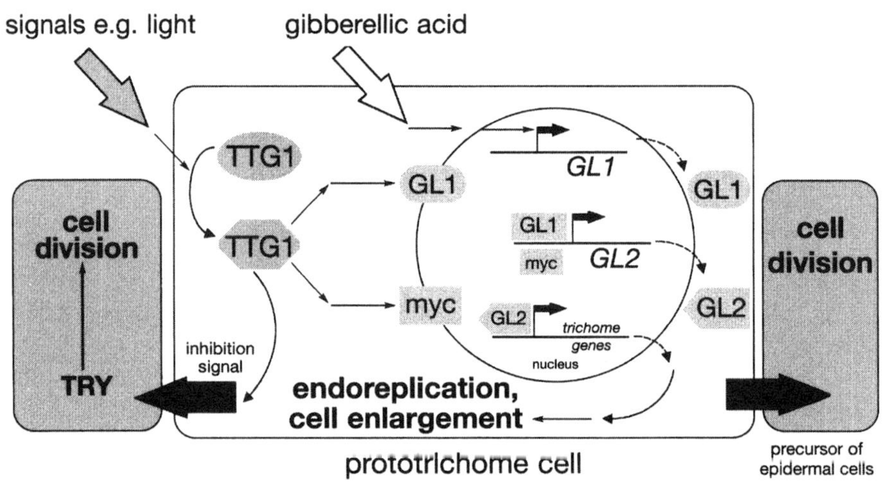

Fig. 2. Scheme of the interaction between the major components of trichome initiation. The *GL1* gene is transcribed in the presence of the GA pathway (large open arrow). To form a trichome precursor cell, a signal (large shaded arrow) (whether exogenous or endogenous) might alter the TTG1 protein, which then modifies the GL1 and/or MYC transcription factors. *GL2* transcription is activated, which in turn activates the transcription of other genes required for trichome development. The activated form of TTG1, together with TRY, produces an inhibition effect (large black arrows) on neighbouring cells, which continue to divide forming pavement cells, guard cells, new trichome precursors and socket or support cells. The small black arrows indicate areas of unknown components or interactions. The nature of the activating and inhibiting signal transduction pathways is not yet understood.

identified. Some gene products act positively to initiate trichome development, while other gene products have an inhibitory effect. The way in which these genes act is becoming clearer as more components of these pathways are uncovered. The isolation of the genes and identification of the gene products will allow the eventual elucidation of the mechanisms behind the decision for a cell to become a trichome. The way in which the trichome precursor communicates with neighbouring cells, firstly, to prevent trichome initiation in those cells, and later in the epidermal formation to induce a ring of socket or support cells, will be important in increasing our understanding of plant development.

REFERENCES

Berger, F., Hung, C.-Y., Dolan, L. and Schiefelbein, J. (1998). Control of cell division in the root epidermis of *Arabidopsis thaliana*. *Developmental Biology* **194**, 235–245.

Boudonck, K., Dolan, L. and Shaw, P. J. (1998). Coiled body numbers in the *Arabidopsis* root epidermis are regulated by cell type, developmental stage and cell cycle parameters. *Journal of Cell Science* **111**, 3687–3694.

Chien, J. C. and Sussex, I. M. (1996). Differential regulation of trichome formation on the adaxial and abaxial leaf surfaces by gibberellins and photoperiod in *Arabidopsis thaliana* (L.) Heynh. *Plant Physiology* **111**, 1321–1328.

de Vetten, N., Quattrocchio, F., Mol, J. and Koes, R. (1997). The *an11* locus controlling flower pigmentation in petunia encodes a novel WD-repeat protein conserved in yeast, plants, and animals. *Genes and Development* **11**, 1422–1434.

Di Cristina, M., Sessa, G., Dolan, L., Linstead, P., Baima, S., Ruberti, I. and Morelli, G. (1996). The Arabidopsis Athb-10 (GLABRA2) is an HD-Zip protein required for regulation of root hair development. *Plant Journal* **10**, 393–402.

Dolan, L. and Scheres, B. (1998). Root pattern: shooting in the dark. *Seminars in Cell and Developmental Biology* **9**, 201–206.

Dolan, L., Duckett, C. M., Grierson, C., Linstead, P., Schneider, K., Lawson, E., Dean, C., Poethig, S. and Roberts, K. (1994). Clonal relationships and cell patterning in the root epidermis of Arabidopsis. *Development* **120**, 2465–2474.

Esch, J. J., Oppenheimer, D. G. and Marks, M. D. (1994). Characterization of a weak allele of the *GL1* gene of Arabidopsis thaliana. *Plant Molecular Biology* **24**, 203–207.

Folkers, U., Berger, J. and Hülskamp, M. (1997). Cell morphogenesis of trichomes in Arabidopsis: differential control of primary and secondary branching by branch initiation regulators and cell growth. *Development* **19**, 3779–3786.

Galway, M. E., Masucci, J. D., Lloyd, A. M., Walbot, V., Davis, R. W. and Schiefelbein, J. W. (1994). The *TTG* gene is required to specify epidermal cell fate and cell patterning in the Arabidopsis root. *Developmental Biology* **166**, 740–754.

Glover, B. J., Perez-Rodriguez, M. and Martin, C. (1998). Development of several epidermal cell types can be specified by the same MYB-related plant transcription factor. *Development* **125**, 3497–3508.

Goff, S. A., Cone, K. L. and Chandler, V. L. (1992). Functional analysis of the transcriptional activator encoded by the maize *B* gene: evidence for a direct

functional interaction between two classes of regulatory proteins. *Genes and Development* **6**, 864–875.
Goodrich, J., Carpenter, R. and Coen, E. S. (1992). A common gene regulates pigmentation pattern in diverse plant species. *Cell* **68**, 955–964.
Helliwell, C. A., Sheldon, C. C., Olive, M. R., Walker, A. R., Zeevaart, J. A. D., Peacock, W. J. and Dennis, E. S. (1998). Cloning of the Arabidopsis ent-kaurene oxidase gene *GA3*. *Proceedings of the National Academy of Sciences (USA)* **95**, 9019–9024.
Herman, P. L. and Marks, M. D. (1990). Trichome development in Arabidopsis thaliana. II. Isolation and complementation of the *GLABROUS1* gene. *Plant Cell* **1**, 1051–1055.
Hülskamp, M., Miséra, S. and Jürgens, G. (1994). Genetic dissection of trichome cell development in Arabidopsis. *Cell* **76**, 555–566.
Hülskamp, M. and Schnittger, A. (1998). Spatial regulation of trichome formation in Arabidopsis thaliana. *Seminars in Cell and Developmental Biology* **9**, 213–220.
Hung, C. Y., Lin, Y., Zhang, M., Pollock, S., Marks, M. D. and Schiefelbein, J. (1998). A common position-dependent mechanism controls cell-type patterning and *GLABRA2* regulation in the root and hypocotyl epidermis of Arabidopsis. *Plant Physiology* **117**, 73–84.
Jackson, J. A., Fuglevand, G., Brown, B. A., Shaw, M. J. and Jenkins, G. I. (1995). Isolation of Arabidopsis mutants altered in the light-regulation of chalcone synthase gene expression using a transgenic screening approach. *Plant Journal* **8**, 369–380.
Jacobsen, S. E., Binkowski, K. A. and Olszewski, N. E. (1996). SPINDLY, a tetratricopeptide repeat protein involved in gibberellin transduction in Arabidopsis. *Proceedings of the National Academy of Sciences (USA)* **93**, 9292–9296.
Kappert, H. (1949). Die Genetik des incana-Charakters und der Anthocyanbildungbeider Levkoje. *Züchter* **19**, 289–297.
Koornneef, M. (1981). The complex syndrome of *ttg* mutants. *Arabidopsis Information Service* **18**, 45–51.
Koornneef, M., Dellaert, S. W. M. and van der Veen, J. H. (1982). EMS- and radiation-induced mutation frequencies at individual loci in *Arabidopsis thaliana* (L.) Heynh. *Mutation Research* **93**, 109–123.
Larkin, J. C., Marks, M. D., Nadeau, J. and Sack, F. (1997). Epidermal cell fate and patterning in leaves. *Plant Cell* **9**, 1109–1120.
Larkin, J. C., Oppenheimer, D. G., Lloyd, A. M., Paparozzi, E. T. and Marks, M. D. (1994). Roles of the *GLABROUS1* and *TRANSPARENT TESTA GLABRA* genes in Arabidopsis trichome development. *Plant Cell* **6**, 1065–1076.
Larkin, J. C., Young, N., Prigge, M. and Marks, M. D. (1996). The control of trichome spacing and number in Arabidopsis. *Development* **122**, 997–1005.
Larkin, J. C., Walker, J. D., Bolognesi-Winfield, A. C., Gray, J. C. and Walker, A. R. (1999). Allele-specific interactions between *ttg* and *gl1* during trichome development in *Arabidopsis thaliana*. *Genetics* **151**, 1591–1604.
Lesnick, M. L. and Chandler, V. L. (1998). Activation of the maize anthocyanin gene *a2* is mediated by an element conserved in many anthocyanin promoters. *Plant Physiology* **117**, 437–445.
Lloyd, A. M., Walbot, V. and Davis, R. W. (1992). Anthocyanin production in dicots activated by maize anthocyanin-specific regulators, R and C1. *Science* **258**, 1773–1775.
Lloyd, A. M., Schena, M., Walbot, V. and Davis, R. W. (1994). Epidermal cell fate determination in Arabidopsis: patterns defined by steroid-inducible regulator.

Science **266**, 436–439.

Ludwig, S. R., Habera, L. F., Dellaporta, S. L. and Wessler, S. R. (1989). *Lc*, a member of the maize *R* gene family responsible for tissue-specific anthocyanin production, encodes a protein similar to transcription activators and contains the myc-homology region. *Proceedings of the National Academy of Sciences (USA)* **86**, 7092–7096.

Marks, M. D. (1997). Molecular genetic analysis of trichome development in Arabidopsis. *Annual Review of Plant Physiology and Plant Molecular Biology* **48**, 137–163.

Masucci, J. D., Rerie, W. G., Foreman, D. R., Zhang, M., Galway, M. E., Marks, M. D. and Schiefelbein, J. W. (1996). The homeobox gene *GLABRA2* is required for position-dependent cell differentiation in the root epidermis of *Arabidopsis thaliana*. *Development* **122**, 1253–1260.

Mauricio, R. (1998). Costs of resistance to natural enemies in field populations of the annual plant *Arabidopsis thaliana*. *American Naturalist* **151**, 20–28.

Meinke, D. W., Franzmann, L. H., Nickle, T. C. and Yeung, E. C. (1994). Leafy cotyledon mutants of Arabidopsis. *Plant Cell* **6**, 1049–1064.

Melaragno, J. E., Mehrota, B. and Coleman, A. W. (1993). Relationship between endoploidy and cell size in epidermal tissue of Arabidopsis. *Plant Cell* **5**, 1661–1668.

Miséra, S., Müller, A. J., Weiland-Heidecker, U. and Jürgens, G. (1994). The *FUSCA* genes of Arabidopsis: negative regulators of light responses. *Molecular and General Genetics* **244**, 242–252.

Mol, J., Jenkins, G., Schafer, E. and Weiss, D. (1996). Signal perception, transduction, and gene expression involved in anthocyanin biosynthesis. *Critical Reviews in Plant Sciences* **15**, 525–557.

Mooney, M., Desnos, T., Harrison, K., Jones, J., Carpenter, R. and Coen, R. (1995). Altered regulation of tomato and tobacco pigmentation genes caused by the *delila* gene of Antirrhinum. *Plant Journal* **7**, 333–339.

Nagata, T., Todoriki, S., Hayashi, T., Shibata, Y., Mori, M., Kanegae, H. and Kikuchi, S. (1999). γ-Radiation induces leaf trichome formation in Arabidopsis. *Plant Physiology* **120**, 113–119.

Neer, E. J., Schmidt, C. J., Nambudripad, R. and Smith, T. F. (1994). The ancient regulatory protein family of WD-repeat proteins. *Nature* **371**, 297–300.

Oppenheimer, D. G. (1998). Genetics of plant cell shape. *Current Opinion in Plant Biology* **1**, 520–524.

Oppenheimer, D. G., Herman, P. L., Sivakumaran, S., Esch, J. and Marks, M. D. (1991). A *myb* gene required for leaf trichome differentiation in Arabidopsis is expressed in stipules. *Cell* **67**, 483–493.

Payne, T., Clement, J., Arnold, D. and Lloyd, A. (1999). Heterologous *myb* genes distinct from *GL1* enhance trichome production when overexpressed in *Nicotiana tabacum*. *Development* **126**, 671–682.

Perazza, D., Vachon, G. and Herzog, M. (1998). Gibberellins promote trichome formation by up-regulating *GLABROUS1* in Arabidopsis. *Plant Physiology* **117**, 375–383.

Rerie, W. G., Feldmann, K. A. and Marks, M. D. (1994). The *GLABRA2* gene encodes a homeodomain protein required for normal trichome development in Arabidopsis. *Genes and Development* **8**, 1388–1389.

Schneider, K., Mathur, J., Boudonck, K., Wells, B., Dolan, L. and Roberts, K. (1998). The *ROOT HAIRLESS 1* gene encodes a nuclear protein required for root hair initiation in Arabidopsis. *Genes and Development* **12**, 2013–2021.

Schnittger, A., Jürgens, G. and Hülskamp, M. (1998). Tissue layer and organ specificity of trichome formation are regulated by *GLABRA1* and *TRIPTYCHON* in Arabidopsis. *Development* **125**, 2283–2289.

Schnittger, A., Folkers, U., Schwab, B., Jürgens, G. and Hülskamp, M. (1999). Generation of a spacing pattern: the role of *TRIPTYCHON* in trichome patterning in Arabidopsis. *Plant Cell* **11**, 1105–1116.

Silverstone, A. L., Mak, P. Y. A., Martínez, E. V. and Sun, T.-P. (1997). The new *RGA* locus encodes a negative regulator of gibberellin response in *Arabidopsis thaliana*. *Genetics* **146**, 1087–1099.

Sun, T. P., Goodman, H. M. and Ausubel, F. M. (1992). Cloning the Arabidopsis *GA1* locus by genomic subtraction. *Plant Cell* **4**, 119–128.

Szymanski, D. B. and Marks, M. D. (1998). *GLABROUS1* overexpression and *TRIPTYCHON* alter the cell cycle and trichome cell fate in Arabidopsis. *Plant Cell* **10**, 2047–2062.

Szymanski, D. B., Jilk, R. A., Pollock, S. M. and Marks, M. D. (1998a). Control of *GL2* expression in Arabidopsis leaves and trichomes. *Development* **125**, 1161–1171.

Szymanski, D. B., Klis, D. A., Larkin, J. C. and Marks, M. D. (1998b). *cot1*: a regulator of Arabidopsis trichome initiation. *Genetics* **149**, 565–577.

Telfer, A., Bollman, K. M. and Poethig, R. S. (1997). Phase change and the regulation of trichome distribution in *Arabidopsis thaliana*. *Development* **124**, 645–654.

van der Berg, C., Weisbeek, P. and Scheres, B. (1998). Cell fate and cell differentiation status in the Arabidopsis root. *Planta* **205**, 483–491.

Wada, T., Tachibana, T., Shimura, Y. and Okada, K. (1997). Epidermal cell differentiation in Arabidopsis determined by a Myb homolog, *CPC*. *Science* **277**, 1113–1116.

Wagner, E. and Mohr, H. (1966). Primary and secondary differentiation in connection with photomorphogenesis of seedling *Sinapis alba*. *Planta* **71**, 204–221.

Walker, A. R., Davison, P. A., Bolognesi-Winfield, A. C., James, C. M., Srinivasan, N., Blundell, T. L., Esch, J. J., Marks, M. D. and Gray, J. C. (1999). The *TTG1* (*Transparent Testa, Glabra1*) locus which regulates trichome differentiation and anthocyanin biosynthesis in Arabidopsis encodes a WD40-repeat protein. *Plant Cell* **11**, 1337–1349.

Trichome Differentiation and Morphogenesis in *Arabidopsis*

M. HÜLSKAMP and V. KIRIK

ZMBP, Entwicklungsgenetik, Universität Tübingen, Auf der Morgenstelle 1, D-72076 Tübingen, Germany

I. INTRODUCTION

During plant development, a large diversity of specialized cell types is generated. As a first step, cells become committed to a certain cell fate depending on their position in a given tissue layer and organ. Then, cells differentiate and adopt specific physiological properties, and may undergo characteristic changes in their cell size and morphology. For most cell types, the analysis of cell differentiation and cell morphogenesis at a single-cell

Advances in Botanical Research Vol. 31
ISBN 0-12-005931-2

level is usually hampered by the inaccessibility to a morphological, genetic and molecular analysis (Hülskamp *et al.*, 1998).

Arabidopsis leaf trichomes (leaf hairs) and root hairs are the best-established model systems in plants to address these questions (Aeschbacher *et al.*, 1994; Marks, 1997; Hülskamp *et al.*, 1999). Being epidermal derivatives, trichomes and root hairs can easily be inspected, and their highly uniform cell shape provides excellent criteria to enable the identification of specific cell morphogenesis mutants. Mutants affecting trichomes or root hairs can easily be recovered and maintained because both cell types are not essential for normal plant growth and development under laboratory conditions.

The aim of this chapter is to summarize the morphological, genetic and molecular data on trichome and root hair differentiation and morphogenesis. Particular emphasis is given to new insights in general developmental or cellular mechanisms, such as the regulation of cell differentiation, endoreduplication, cell elongation and cell polarity.

II. LEAF TRICHOMES

In *Arabidopsis*, trichomes are unicellular, non-secreting hairs. While trichomes on the stem and sepals are usually unbranched, leaf trichomes are stellate. Depending on the ecotype, 2–4 separate branches are formed. Most studies have focused on the development of leaf trichomes in the ecotype *Landsberg erecta*, which typically exhibits three branches.

A. WILD-TYPE TRICHOME MORPHOGENESIS

Trichome initiation takes place in a narrow zone at the base of developing leaves (Hülskamp *et al.*, 1994; Larkin *et al.*, 1996). While protodermal cells continue to divide, trichome initials stop cell divisions and start to endoreduplicate, i.e. they proceed through several DNA replication cycles in the absence of nuclear and cell divisions (Hülskamp *et al.*, 1994). The incipient trichome cell expands out of the leaf surface and undergoes two successive branching events. The orientation of the primary branching event is coaligned with the proximal–distal leaf axis (Fig. 1) (Hülskamp *et al.*, 1994; Folkers *et al.*, 1997). Secondary branching is initiated only on that branch that points to the leaf tip; the new branching plane is perpendicular to the first branching plane. By the time branch initiation is completed, the DNA content of the trichome cell has increased sixteenfold (32C) as compared to normal protodermal cells (2C), suggesting that the trichome cell has executed four endoreduplication cycles (Hülskamp *et al.*, 1994). Subsequent morphogenesis is characterized by the onset of vacuolization and cell elongation. The mature trichome cell is approximately 200–400 μm tall and 40–60 μm wide at the trichome base. During trichome maturation, the cell

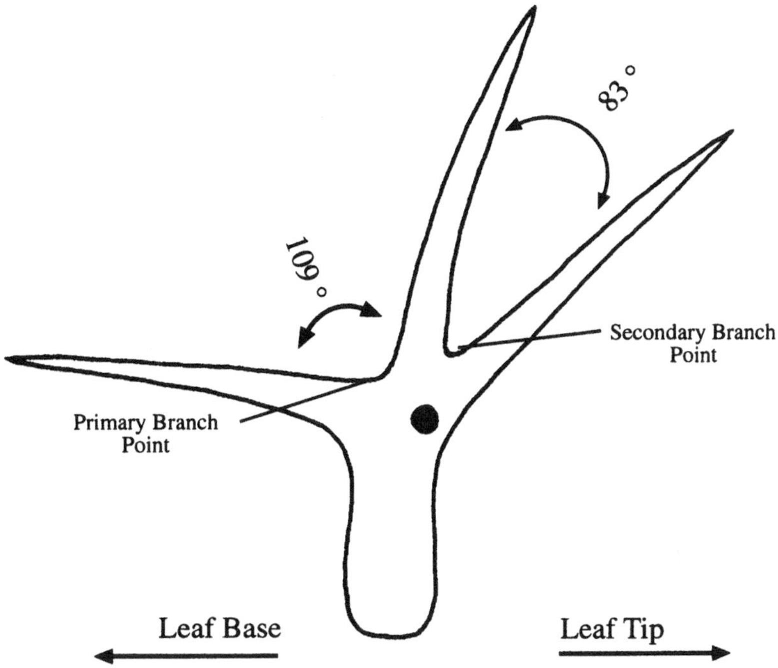

Fig. 1. Spatial organization of a leaf trichome. The spatial arrangement of trichome branches is highly predictable. The branch pointing to the leaf tip branches again perpendicular to the first two branches. The angle between the primary and secondary branches is different.

wall is solidified and small papillae of unknown origin and function are formed.

B. GENETIC DISSECTION OF TRICHOME MORPHOGENESIS

A powerful approach to study trichome morphogenesis is genetic analysis. Based on mutant phenotypes, regulatory steps can be defined and the number of genes and their genetic interactions allows an analysis of the complexity of the process to be made. In addition, the identification of gene function by a mutant phenotype facilitates the cloning of the corresponding gene and thus permits a further biochemical analysis.

To date, 40 trichome mutants have been identified that affect various steps of trichome morphogenesis (Table I). While some genes appear to act in a trichome-specific manner, others are also involved in the control of cell morphogenesis in other cell types. In Table I, genes are grouped into five

TABLE I

Trichome and root hair morphogenesis genes in Arabidopsis

Gene	Trichome phenotype	Root hair phenotype	Molecular nature	Reference
Cell Differentiation				
GLABRA2 (GL2)	reduced trichome expansion	ectopic root hairs	homeo-domain transcription factor	Hülskamp *et al.*, 1994; Masucci *et al.*, 1996; Rerie *et al.*, 1994; Koornneef *et al.*, 1982
ROOT HAIRLESS1 (RHL1)	reduced trichome expansion	no root hairs	n.a.	Schneider *et al.*, 1997
ROOT HAIRLESS2 (RHL2)	reduced trichome expansion	no root hairs	n.a.	Schneider *et al.*, 1997
ROOT HAIRLESS3 (RHL3)	reduced trichome expansion	no root hairs	n.a.	Schneider *et al.*, 1997
ECTOPIC ROOT HAIR3 (ERH3)	reduced trichome expansion	ectopic root hairs	n.a.	Schneider *et al.*, 1997
Endoreduplication				
GLABRA3 (GL3)	one endoreduplication cycle missing, fewer branches		n.a.	Koornneef *et al.*, 1982; Hülskamp *et al.*, 1994
TRIPTYCHON (TRY)	clustered trichomes, one additional endoreduplication cycle, more branches		n.a.	Hülskamp *et al.*, 1994
KAKTUS (KAK)	one additional endoreduplication cycle, more branches		n.a.	Hülskamp *et al.*, 1994; Perazza *et al.*, 1999
POLYCHOME (PYM)	one additional endoreduplication cycle, more branches		n.a.	Hülskamp *et al.*, 1994; Perazza *et al.*, 1999

TABLE I (*continued*)

Gene	Trichome phenotype	Root hair phenotype	Molecular nature	Reference
FURCA1 (FRC1)	reduced branching		n.a.	Luo and Oppenheimer as cited in Oppenheimer *et al.*, 1999
FURCA2 (FRC2)	reduced branching		n.a.	Luo and Oppenheimer as cited in Oppenheimer *et al.*, 1999
FURCA3 (FRC3)	reduced branching		n.a.	Luo and Oppenheimer as cited in Oppenheimer *et al.*, 1999
FURCA4 (FRC4)	reduced branching		n.a.	Luo and Oppenheimer as cited in Oppenheimer *et al.*, 1999
SUPPRESSOR OF ZWICHEL2 (SUZ2)	increased branching		n.a.	Krishnakumar and Oppenheimer as cited in Oppenheimer *et al.*, 1999
ROOT HAIR DEFECTIVE6 (RHD6)		site selection of root hair outgrowth affected	n.a.	Schiefelbein *et al.*, 1990
AUXIN RESISTANT (AXR2)		site selection of root hair outgrowth affected	n.a.	Masucci *et al.*, 1994
ETHYLENE RESISTANT (ETR1)		site selection of root hair outgrowth affected	ethylene-binding receptor protein	Masucci *et al.*, 1994; Chang *et al.*, 1993
Expansion Growth				
ALIEN (ALI)	distorted trichomes		n.a.	Hülskamp *et al.*, 1994
CROOKED (CRK)	distorted trichomes		n.a.	Hülskamp *et al.*, 1994
DISTORTED1 (DIS1)	distorted trichomes		n.a.	Feenstra *et al.*, 1978; Hülskamp *et al.*, 1994

RASTAFARI (RFI)	one additional endoreplication cycle, more branches	n.a.	Hülskamp *et al.*, 1994; Perazza *et al.*, 1999
SPINDLY (SPY)	one additional endoreplication cycle, more branches, abaxial trichome formation	tetratricopeptide repeat protein	Jacobsen *et al.*, 1996; Telfer *et al.*, 1997; Perazza *et al.*, 1999
GUMBY (GMB)	fewer endoreduplication cycles	n.a.	Venezia, Larkin and Oppenheimer as cited in Oppenheimer, 1999
SMALL TRICHOMES1 (SML1)	fewer endoreduplication cycles	n.a.	Venezia, Larkin and Oppenheimer as cited in Oppenheimer, 1999
ELCH (ELC)	multiple nuclei	n.a.	Schellmann and Hülskamp, unpublished
Cell Polarity and Branching			
STICHEL (STI)	no branches	n.a.	Hülskamp *et al.*, 1994; Folkers *et al.*, 1997
ANGUSTIFOLIA (AN)	no secondary branching	n.a.	Hülskamp *et al.*, 1994; Koornneef *et al.*, 1982; Tsuge *et al.*, 1996
STACHEL (STA)	no primary branching	n.a.	Hülskamp *et al.*, 1994; Folkers *et al.*, 1997
ZWICHEL (ZWI)	reduced trichome branching	kinesin-like protein	Hülskamp *et al.*, 1994; Oppenheimer *et al.*, 1997
NOECK (NOK)	more branches	n.a.	Folkers *et al.*, 1997
FASS (FS)	unbranched trichomes	n.a.	Torres-Ruiz *et al.*, 1994; Traas *et al.*, 1995; McClinton *et al.*, 1997

DISTORTED2 (DIS2)	distorted trichomes		n.a.	Feenstra *et al.*, 1978; Hülskamp *et al.*, 1994
GNARLED (GRL)	distorted trichomes		n.a.	Hülskamp *et al.*, 1994
KLUNKER (KLK)	distorted trichomes		n.a.	Hülskamp *et al.*, 1994
SPIRRIG (SPI)	distorted trichomes		n.a.	Hülskamp *et al.*, 1994
WURM (WRM)	distorted trichomes		n.a.	Hülskamp *et al.*, 1994
SINGED (SNE)	distorted trichomes	shorter root hairs	n.a.	Marks *et al.*, 1992
ROOT HAIR DEFECTIVE1 (RHD1)		local outgrowth of root hair affected	n.a.	Schiefelbein *et al.*, 1990
ROOT HAIR DEFECTIVE2 (RHD2)		elongation growth affected	n.a.	Schiefelbein *et al.*, 1990
ROOT HAIR DEFECTIVE3 (RHD3)	slightly distorted trichomes	elongation and directionality of root hair growth affected	GTP-binding protein	Schiefelbein *et al.*, 1990; Wang *et al.*, 1997; Galway *et al.*, 1997
ROOT HAIR DEFECTIVE4 (RHD4)		elongation and directionality of root hair growth affected	n.a.	Schiefelbein *et al.*, 1990
CAN OF WORMS1 (COW1)		elongation and directionality of root hair growth affected	n.a.	Grierson *et al.*, 1997
TIP GROWTH DEFECTIVE1 (TIP1)		elongation and directionality of root hair growth affected	n.a.	Schiefelbein *et al.*, 1993
ROOT EPIDERMAL BLEBBING (REB1)		local outgrowth of root hair affected	n.a.	Baskin *et al.*, 1992
Maturation				
UNDERDEVELOPED TRICHOME (UDL)	defects in papillate surface of trichomes		n.a.	Haughn *et al.*, 1988
CHABLIS (CHA)	glassy trichomes		n.a.	Hülskamp *et al.*, 1994
CHARDONNAY (CDO)	glassy trichomes		n.a.	Hülskamp *et al.*, 1994
RETSINA (RTS)	glassy trichomes		n.a.	Hülskamp *et al.*, 1994
TRICHOME BIREFRINGENCE (TBR)	cell wall defects		n.a.	Potikha *et al.*, 1995

n.a., not available.

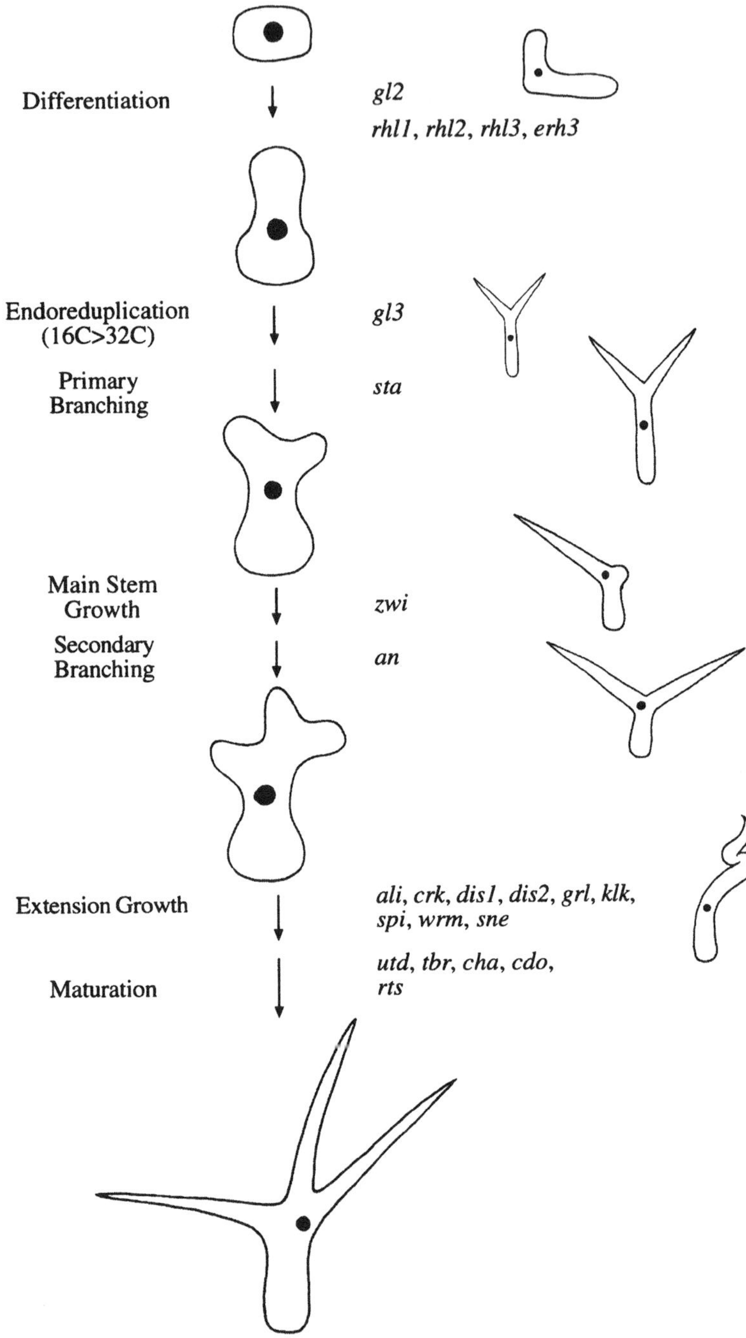

Fig. 2. Development of leaf trichomes. A series of developmental steps is shown in the middle. Mutants affecting these steps are shown on the right.

classes according to their phenotypic trichome and root hair defects. The first class includes genes that appear to be involved in early steps of trichome differentiation, and the corresponding mutants have reduced and not fully developed trichomes (Hülskamp *et al.*, 1994; Schneider *et al.*, 1997). The second class comprises genes important for the regulation of endoreduplication. Five genes of this class function to restrict the number of endoreduplication cycles (Hülskamp *et al.*, 1994; Perazza *et al.*, 1999), three genes appear to be required to undergo the normal number of endoreduplication cycles (Hülskamp *et al.*, 1994; Oppenheimer, 1998) and one seems to be important in preventing nuclear divisions during endoreduplication cycles (S. Schellmann and M. Hülskamp, unpublished observations). Genes grouped in the third class play a role in the regulation of trichome branching: some act as positive regulators, others as negative regulators (Folkers *et al.*, 1997; Oppenheimer, 1998). The fourth class contains nine genes that are involved in controlling the directionality of extension growth (Feenstra, 1978; Marks and Esch, 1992; Hülskamp *et al.*, 1994). Finally, the fifth class contains genes involved in the maturation of the trichome cell and corresponding mutants appear glassy, suggesting that cell wall encrustation is affected (Haughn and Somerville, 1988; Hülskamp *et al.*, 1994; Potikha and Delmer, 1995).

The five phenotypic classes correspond closely to steps occurring during the normal progression of trichome development. Genes required to execute distinct developmental steps have been used to define a series of regulatory steps during trichome development (Fig. 2). In addition, each phenotypic class includes genes that are likely to act as negative regulators or modifiers. In the following paragraphs, the genetic and molecular data on trichome morphogenesis are summarized.

C. REGULATION OF TRICHOME DIFFERENTIATION

Mutations in five genes, *GLABRA2*, *ROOT HAIRLESS1* (*RHL1*), *RHL2*, *RHL3* and *ECTOPIC ROOT HAIR3* (*ERH3*), result in a wide range of trichome phenotypes. Mutant trichomes are generally reduced in size and branching, and often lack the papillae on their surface. Because these phenotypic aspects resemble the single phenotypes of various other trichome morphogenesis mutants, it is conceivable that these mutants act early in trichome development and function to regulate gene expression of later-acting genes. As outlined below, several lines of evidence support this concept.

Trichomes in *gl2* mutants exhibit a wide range of phenotypes depending on the leaf position (Hülskamp *et al.*, 1994; Rerie *et al.*, 1994). While *gl2* trichomes located at the centre of the leaf blade undergo normal cell enlargement but do not extend out of the leaf surface, trichomes at the leaf margin show a much weaker phenotype resembling *sta* or *zwi* mutant trichomes (see below). The finding that the *GL2* gene encodes a protein with sequence similarity to homeodomain transcription factors and that it is expressed early in trichome development is consistent with the idea that *GL2* regulates the expression of later-acting morphogenesis genes (Rerie *et al.*, 1994). A second line of arguments comes from the analysis of *GL2* function during root hair differentiation. During root hair differentiation, two morphologically distinct cell files are specified, the atrichoblasts and the trichoblasts. *GL2* is expressed in atrichoblasts, where it suppresses root hair differentiation (Masucci *et al.*, 1996). Thus, *GL2* acts to maintain a certain cell differentiation state in both the shoot and root epidermis.

In addition to *GL2*, *RHL1*, *RHL2*, *RHL3* and *ERH3* are important in root hair cell differentiation (Schneider *et al.*, 1997). Although no molecular data are available, genetic analysis suggests that these genes act early in root hair differentiation. *ERH3* is thought to act at the same regulatory level as *GL2*, whereas *RHL1*, *RHL2* and *RHL3* are thought to act upstream of *GL2* and *ERH3*. Thus, by analogy, one could speculate that all genes play a role in the regulation of later-acting trichome morphogenesis genes.

D. ENDOREDUPLICATION

In plants, the majority of cells eventually undergo one or several endoreduplication cycles. Endoreduplication begins after cell fate has been committed and can be considered part of the cell differentiation programme (Traas *et al.*, 1998). By contrast, the commitment of a protodermal cell to adopt trichome cell fate appears to be intimately coupled to the switch from mitotic cell divisions to endoreduplication. DNA replication in trichomes is likely to be a cyclic process as wild-type and endoreduplication mutants display a DNA content corresponding to an arithmetic progression, base two (Hülskamp *et al.*, 1994). Furthermore, while other cell types show various degrees of DNA content, the number of endoreduplication cycles in trichomes is fairly constant and strictly controlled by at least eight genes (Hülskamp *et al.*, 1994; Oppenheimer, 1998; Perazza *et al.*, 1999).

It is not known how the initial switch from mitotic cell divisions to endoreduplication is regulated. However, it is noteworthy that two genes important in trichome initiation are also involved in the regulation of endoreduplication in these cells. *TRIPTYCHON* (*TRY*) acts as a negative regulator of trichome initiation and endoreduplication (Hülskamp *et al.*, 1994; Schnittger *et al.*, 1998; Szymanski and Marks, 1998). Corresponding mutants produce trichome clusters of 2–4 trichomes that exhibit double the DNA content in comparison to wild-type trichomes. Conversely, *GLABRA1* (*GL1*) appears to be a positive regulator of trichome initiation and endoreduplication in trichomes (Schnittger *et al.*, 1998). *gl1* mutants lack trichomes and overexpression of *GL1* results in an additional cycle of

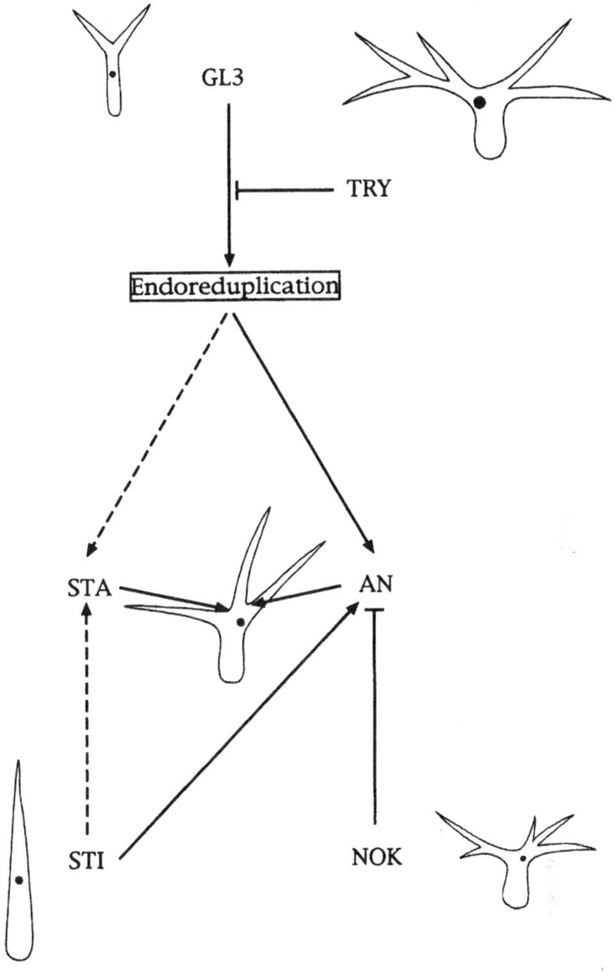

Fig. 3. Regulation of endoreduplication in trichomes. The number of endoreduplication cycles in trichomes is regulated by positive and negative regulators. Positive interactions are shown as arrows. Negative regulation is indicated by blunted lines.

endoreduplication compared to the wild type. These findings have led to the suggestion that the switch from mitosis to endoreduplication is an integral part of the mechanism to bias the decision towards trichome cell fate (Schnittger *et al.*, 1998).

The number of endoreduplication cycles is negatively controlled by four genes, *TRY*, *KAKTUS* (*KAK*), *RASTAFARI* (*RFI*), *POLYCHOME* (*PYM*) and *SPINDLY* (*SPY*) (Hülskamp *et al.*, 1994; Perazza *et al.*, 1999). The corresponding mutant trichomes show twice the DNA content compared to the wild type, suggesting that these genes function to restrict the number of endoreduplication cycles. The *GLABRA3* gene appears to function as a positive regulator of endoreduplication. In *gl3* mutants, one endoreduplication cycle is missing; *gl3* trichomes have only half the DNA content of the wild type. Although the genetic analysis is far from complete, a simplified model suggests several partially independent pathways (Fig. 3) (Perazza *et al.*, 1999).

It is not known exactly how the initial endoreduplication cycles are controlled, but it is conceivable that these depend on the function of *GL1*. *GL3* acts genetically upstream of *TRY* as the *gl3* mutation is epistatic to *try* mutations; *gl3 try* double mutants undergo only three cycles of endoreduplication (Hülskamp *et al.*, 1994). The analogous *gl3* double mutants with *kak*, *rfi*, *pym* and *spy* have not been tested yet. A systematic analysis of double mutants between the four negative regulators of endoreduplication revealed two independent pathways. One pathway is represented by *TRY*, a second pathway is constituted by *KAK*, *RFI*, *PYM* and *SPY* (Fig. 3). The plant hormone gibberellic acid (GA) is likely to be involved in the second pathway because the *SPY* gene has been shown to be an important component in the GA response.

Unlike the mutants described above, one gene, *ELCH* (*ELC*), has been discovered that appears to be required to inhibit nuclear divisions during endoreduplication; *elc* mutant trichomes occasionally contain between up to four nuclei, and cell divisions have not been observed (S. Schellmann and M. Hülskamp, unpublished observations).

E. TRICHOME BRANCHING

Trichome branching is an excellent model system to study cell morphogenesis. Because trichome branching takes place when the cell has already extended from the surface, all morphogenetic processes must be strictly regulated by intracellular mechanisms. In addition, the three-dimensional arrangement of branch initiation provides excellent criteria to classify mutants affecting the branch pattern.

1. Genetic Analysis of Branching

Morphological and genetic data suggest that the trichome cell has two characteristic spatial elements. First, the branch pointing towards the leaf tip seems to dominate in that only this branch undergoes further growth and branching. This branch is called the main stem (Folkers *et al.*, 1997). This morphological notion is supported by the *zwi* mutant phenotype in weak and strong alleles. While weak *zwi* alleles have two equally long branches, stronger alleles exhibit a progressively reduced main stem, whereas the proximal stem is not affected (M. Hülskamp, unpublished observations). The gradual loss of main stem growth in weak and strong *zwi* alleles suggests that *ZWI* is mainly required to support main stem growth and branching. Second, primary and secondary branching appear to be qualitatively different. They differ in two morphological aspects: primary branching is spatially oriented with respect to the proximal–distal leaf axis, and secondary branching is oriented with reference to primary branching. In addition, the angles between branches are different. Primary branches are arranged at an angle of approximately 109 degrees, secondary branches at an angle of 83 degrees (Folkers *et al.*, 1997). Two mutants are known that appear to affect either of the two branching events. Trichomes on *angustifolia* (*an*) mutants have two equally long branches resting on a short stalk and the branches are oriented with the proximal–distal leaf axis. This suggests that primary branching is normal, but secondary branching does not occur. The *STACHEL* (*STA*) gene appears to regulate primary branching. Trichomes on *sta* mutants have a long stalk and two short branches that are oriented at random with respect to the leaf axis. Because the *an sta* double mutant is unbranched, it has been suggested that primary and secondary branching are genetically distinct, with *AN* regulating secondary branching and *STA* regulating primary branching (Folkers *et al.*, 1997). However, the recent identification of five additional branching mutants (unpublished data cited in Oppenheimer, 1998) challenges this model. The genetic analysis of their interactions with *an* and *sta* suggest a more complex model of branch initiation with at least four independent pathways, and put under question that primary and secondary branching are genetically separable. Cloning of the corresponding genes and their molecular and biochemical analysis will help to address this question.

The number of trichome branches is regulated by two independent pathways. One pathway includes the two genes *STICHEL* (*STI*) and *NOECK* (*NOK*). The other pathway regulates branch number via cell growth or cell size (Folkers *et al.*, 1997) (Fig. 4).

STI and *NOK* are thought to act at the same regulatory level and upstream of *STA* and *AN*. *STI* appears to act as a positive regulator. Strong *sti* alleles have unbranched trichomes, whereas in weak alleles, trichomes are branched with a long stalk and two short branches. This observation suggests that branch initiation by *STI* is regulated in a concentration-dependent manner. *NOK* seems to act as a negative regulator of trichome branching and *nok* mutants are overbranched displaying up to seven branches. Double mutants of strong and weak *sti* alleles and *nok* display intermediate phenotypes of the respective parental lines, i.e. a reduction of a positive activity can be compensated by the additional reduction of the counteracting negative regulator.

The second branch initiation pathway depends on DNA endoreduplication such that an increased DNA content in the trichome is correlated with a higher branch number and a decrease in the DNA content results in fewer branches. This correlation was first noticed in trichome mutants that alter

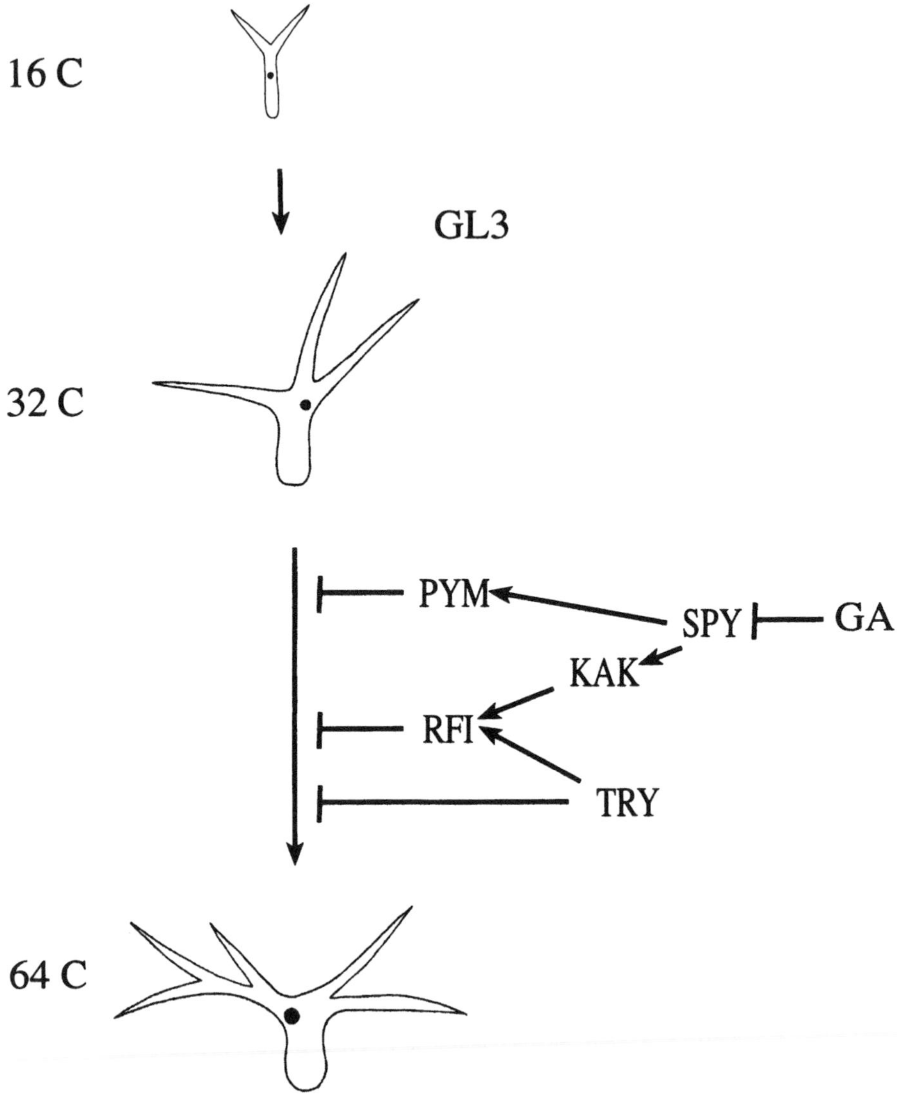

Fig. 4. Regulation of trichome branching. Trichome branching is regulated by two independent pathways. One pathway is defined by genes that regulate the number of endoreduplication cycles and, as a consequence, cell size/growth, which is correlated with the number of branches. The second pathway includes at least two genes, that regulate branch initiation independently of the DNA content. Positive interactions are shown as arrows. Negative regulation is indicated by blunted lines.

the nuclear DNA content. In *gl3* mutants trichomes contain half the DNA content as the wild type, resulting in smaller trichomes that have two branches or are unbranched (Hülskamp *et al.*, 1994). Mutations in *TRY*, *KAK*, *RFI* and *PYM* result in extra cycles of endoreduplication. The corresponding mutants exhibit larger and overbranched trichomes (Hülskamp *et al.*, 1994; Folkers *et al.*, 1997; Perazza *et al.*, 1999). The correlation between nuclear DNA content and branch number also holds true for tetraploid wild-type lines that have been shown to have trichomes with twice the DNA content than diploid wild-type lines and also have supernumerary branches (Perazza *et al.*, 1999). The cellular mechanisms underlying this correlation are unknown, but it is likely that the regulation of trichome morphogenesis by the DNA content is indirectly regulated by either cell growth or cell size. By genetic criteria, branch number regulation via a DNA content-dependent mechanism can be considered as a pathway different to that defined by *STI* and *NOK*. This is evident from *nok try* double mutants, which show an additive phenotype where trichomes form up to 13 branches (Folkers *et al.*, 1997).

The induction of supernumerary branches in *try* and *nok* mutants appears to involve the *AN*-controlled initiation of secondary branches but not *STA*-dependent branch initiation. This is suggested by the double mutant analysis of *an* and *sta* with *nok* and *try*. While *an nok* and *an try* double mutants exhibit the same branching phenotype as *an* mutants, the corresponding double mutants with *sta* result in additional branch formation. This indicates that *AN* is essential for extra branch formation in *nok* and *try* mutants.

2. *Molecular Analysis of Branching*

To date only one gene involved in trichome morphogenesis has been cloned, the *ZWI* gene (Oppenheimer *et al.*, 1997). *ZWI* has been cloned by T-DNA tagging and encodes a member of the kinesin superfamily of motor proteins that contains a calmodulin-binding site. Kinesins are motor proteins that move along microtubules with a characteristic directionality and carry various intracellular components in the cell. *ZWI* is a member of the minus-end directed motor proteins family. The *ZWI* gene had been independently identified biochemically based on its ability to bind to calmodulin and named KCBP, for kinesin-like calmodulin-binding protein (Reddy *et al.*, 1996). *In vitro* studies have shown that binding of KCBP to calmodulin is Ca^{2+} dependent and that binding of KCBP to microtubules is inhibited by Ca^{2+}/calmodulin (Song *et al.*, 1997). *In vivo* studies on cell tissue cultures revealed that KCBP co-localize with microtubule arrays in the preprophase band, spindle and phragmoplast in dividing *Arabidopsis* and tobacco cells (Bowser and Reddy, 1997).

Its biological role in growth and branch formation in trichomes remains unknown. In principle, the present data are consistent with two functions.

ZWI could be involved in the transport of important intracellular components in the main growth direction. As a consequence, *zwi* mutants would fail to accumulate components critical for growth and branching. This would, in particular, explain the reduction of the main stem growth in *zwi* alleles of increased strength. Alternatively, *ZWI* could be involved in the reorganization of microtubules prior to branch initiation. This would be consistent with *KCBP* being a member of the C-type kinesins that are thought to participate in the movements of microtubules (Hirokawa, 1998). More biochemical studies and the analysis of biochemical and/or genetic interactors of *ZWI* should enable distinction between these possibilities.

The genetic and molecular data on *ZWI* strongly suggest that the microtubules play an important role in branching and growth of trichomes. This notion is further supported by genetic and molecular data on the *FASS* (*FS*) gene. The *FASS* gene was identified as an embryonic lethal mutant affecting the cell division pattern during embryogenesis (Mayer *et al.*, 1991). *fs* mutants show irregular cell expansion and abnormal alignments of the cell division planes. These abnormalities are correlated with a strong impairment of the microtubule arrays and the absence of the preprophase band (Traas *et al.*, 1995; McClinton and Sung, 1997). In weak *fs* alleles in which embryonic development is nearly normal, trichomes exhibit a *sti*-like phenotype (Torres-Ruiz and Jürgens, 1994), and hence connect the regulation of trichome branching with the regulation of the microtubules and the orientation of cell division planes. One interesting interpretation is that the mechanisms underlying branch formation in trichomes are derived from already existing general mechanisms employed in regulating the orientation of cell divisions.

F. DIRECTIONALITY OF GROWTH

After the three branches are initiated, the trichome cell starts to expand tremendously, increasing cell size by a factor of 7–10. The directionality of cell expansion requires the function of a group of nine genes, the *DISTORTED* (*DIS*) genes (Feenstra, 1978; Marks and Esch, 1992; Hülskamp *et al.*, 1994). *dis* mutants all share a very similar phenotype, mutant trichomes grow irregular and the mature trichome exhibits a twisted, distorted phenotype. This phenotype suggests that the corresponding genes are involved in the maintenance or the stabilization of growth direction, possibly by organizing the actin or tubulin cytoskeleton. Although the mutant phenotypes of the nine genes are indistinguishable, genetic mosaic data indicate that the eight genes act differently. While *DISTORTED2* (*DIS2*) was shown to act in a cell autonomous manner, *DIS1* appears to act non-autonomously (Hülskamp *et al.*, 1994). This suggests that *DIS2* functions in intracellular morphogenetic processes, whereas *DIS1* seems to be involved in the production of diffusible products that can be transported between cells.

G. TRICHOME MATURATION

During trichome maturation, the cell wall thickens and becomes increasingly solidified. Five mutants, *underdeveloped trichome* (*udt*), *trichome birefringence* (*tbr*), *chablis* (*cha*), *chardonnay* (*cdo*) and *retsina* (*rts*), have been identified in which mature trichomes look glassy and often collapse (Hülskamp *et al.*, 1994; Potikha and Delmer, 1995). Although these mutants have not been characterized in detail, these phenotypes suggest that cell wall strengthening is affected. It should be noted also that *nok* mutants share the glassy phenotype, hence providing a genetic link between branching and cell wall maturation on the basis of at least one gene function (Folkers *et al.*, 1997).

III. ROOT HAIRS

A. WILD-TYPE ROOT HAIR MORPHOGENESIS

Root hairs are specialized epidermal cells that play an important role in nutrient and water uptake and in anchoring the plant to the soil. Root hairs form long tubular projections at the basal end of the epidermal cells. Two aspects of root hair development serve excellent model systems to study the regulation of cell morphogenesis (Hülskamp *et al.*, 1998): first, the origin and interpretation of cell polarity that is required to direct the outgrowth of the hair to the basal end of the cell; and second, the directionality of tip growth, by which the growth of the tubular projection is driven.

B. GENETIC DISSECTION OF ROOT HAIR MORPHOGENESIS

Normal root hair morphogenesis proceeds in a well-described sequence of events and several mutants have been identified that affect *Arabidopsis* root hair cell development at various steps (Fig. 5). Two functionally and morphologically distinct cell files are initially created in the root meristematic zone. Cells that contact only one wall of underlying cortex cells form atrichoblast files. Cells overlying the clefts between underlying cortical cells are trichoblasts and develop into root hairs. Five genes were identified that are specifically involved in the regulation of the differentiation in trichoblasts and atrichoblasts (Schneider *et al.*, 1997). *RHL1*, *RHL2* and *RHL3* are required to promote trichoblast development and corresponding mutants produce exclusively atrichoblasts. Downstream of these genes, *GL2* and *ERH3* act to maintain the atrichoblast cell fate and to prevent further root hair formation (Masucci *et al.*, 1996; Schneider *et al.*, 1997).

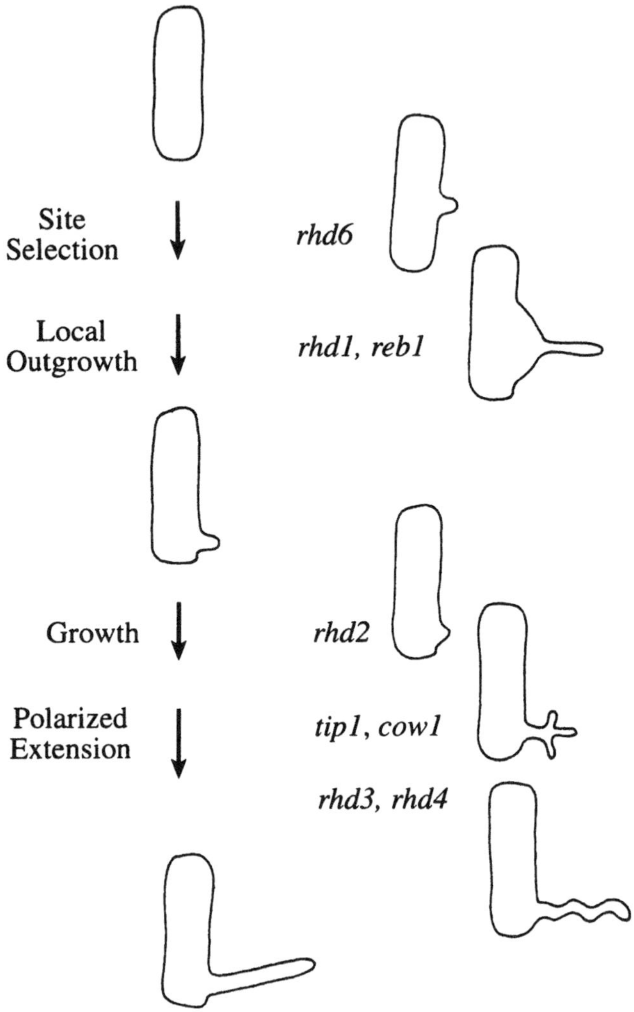

Fig. 5. Root hair differentiation and morphogenesis. Steps occurring during normal development are shown on the left. Mutants affecting these steps are shown on the right.

Incipient root hair cells expand extensively in the longitudinal direction. Then, a second growth axis is initiated at the base of the cell that will give rise to the root hair. The position of root hair initiation is impaired in the auxin-, ethylene-, and abscisic acid-resistant mutant *auxin resistant2* (*axr2*), the *ethylene resistant1* (*etr1*) mutant and *root hair defective6* (*rhd6*) mutants (Masucci and Schiefelbein, 1994). In these mutants root hair initiation is frequently shifted to a more apical position and, occasionally, more than one root hair per cell are observed. Furthermore, the inhibition of ethylene biosynthesis by drug treatment (i.e. aminoethoxy-vinylglycine) results in a similar phenotype. These data suggest that cell polarity and/or initiation of root hair outgrowth is regulated by plant hormones. The *RHD6* gene is likely to act in this hormone-dependent pathway as the addition of ethylene precursors can rescue the *rhd6* mutant phenotype (Masucci and Schiefelbein, 1994).

Tubular outgrowth at the base of the committed epidermal cell begins as a small bulge that continues to grow exclusively at the tip. The phenotype of two mutants, *root hair defective1* (*rhd1*) and *root epidermal blebbing1* (*reb1*) indicate that the localization of the initial outgrowth is a genetically controlled step in morphogenesis (Schiefelbein and Somerville, 1990; Baskin *et al.*, 1992). In these mutants, the initial bulge is at the correct position; further growth, however, is not restricted to a narrow point resulting in a large bulge. Further root hair growth is normal.

Six genes are known to act during root hair elongation, although the corresponding phenotypes suggest that they control different aspects. Once a small bulge has been formed, a phase of slow tip growth begins. *RHD2* is thought to be required at this point. Mutations in the *RHD2* gene result in the formation of bulges that do not elongate (Schiefelbein and Somerville, 1990). Further root hair growth is driven by rapid tip growth. Mutations in the *SINGED* (*SNE*) gene result in shortened root hairs suggesting a role of *SNE* in root hair elongation. Interestingly, mutations in the *SNE* gene also interfere with the directionality of trichome growth (cf. Section II.F). Two mutants, *tip1* and *cow1*, are defective in root hair elongation and frequently form branched root hairs (Schiefelbein *et al.*, 1993; Grierson *et al.*, 1997). In particular, the formation of branches suggests that the control of tip growth initiation seems to be less stringent in these mutants. By contrast, *rhd3* and *rhd4* mutants predominantly affect the extent and the directionality of tip growth. *rhd3* mutant root hairs are shorter and usually distorted. *rhd4* root hairs grow irregularly with bulges and constrictions along the growth axis; occasionally, branches are formed (Schiefelbein and Somerville, 1990).

The analysis of various double mutant combinations leads to a genetic model, which predicts that the known genes act in a sequence of steps that generally reflect the sequence of morphologically distinct events. In this genetic model, *COW1* acts in a separate pathway to *RHD3* and *TIP1*, and all three act upstream of *RHD4* (Grierson *et al.*, 1997).

C. MOLECULAR AND CELLULAR ANALYSIS OF ROOT HAIR MORPHOGENESIS

Tubular outgrowth of the root hair cell is exclusively governed by tip growth, i.e. new plasma membrane and cell wall components are incorporated only in a small region at the tip, whereas the remaining cell body does not grow. This growth mode is found in a variety of cell types, including pollen tubes, fungal hyphae and moss protonemata (Schnepf, 1986). Tip growth is correlated with a characteristic intracellular zonation. Large amounts of secretory vesicles are found at the extreme tip of growing root hairs. The subapical region contains large amounts of organelles that are involved in the biosynthesis and transport of membrane and cell wall components. The central vacuole is found at the base of the extending root hair.

The actin/tubulin cytoskeleton and the distribution of free cytoplasmic Ca^{2+} levels are important key regulators of the directionality and maintenance of tip growth. Cytoplasmatic Ca^{2+} in root hairs is found as a tip-high gradient (Bibikova *et al.*, 1997; Felle and Hepler, 1997). Application of calcium buffers or channel blockers that result in a breakdown of the Ca^{2+} gradient causes an arrest of tip growth (Hermann and Felle, 1995; Wymer *et al.*, 1997). Conversely, localized application of calcium ionophores, which cause a spatially defined influx of calcium, is sufficient to redirect tip growth in a new direction (Bibikova *et al.*, 1999). This indicates that the spatial regulation of Ca^{2+} influx is a critical step for tip growth directionality. The actin cytoskeleton is also important for tip growth maintenance as drug treatments affecting the stability of the actin cytoskeleton inhibit tip growth effectively (Bibikova *et al.*, 1999). By contrast, microtubule depolymerizing or stabilizing drugs have no affect on the growth rate. However, under certain experimental conditions, directionality of growth is affected and a root hair waving phenotype is observed (Bibikova *et al.*, 1999). Stronger treatments cause the initiation of multiple tip growth initiation sites. In an elegant series of experiments, it has been shown that microtubules are important to stabilize the growth axis via the Ca^{2+} distribution (Bibikova *et al.*, 1999). On the one hand, the initiation of extra root hair branches in drug-treated root hairs was strictly correlated with a new tip-high Ca^{2+} gradient. On the other hand, localized elevation of Ca^{2+} by caged calcium-ionophore experiments allowed a new growth direction to be triggered in the apical region of the growing root hair but not in subapical regions. However, if cells were treated with microtubule inhibitors, the generation of a new Ca^{2+} gradient was sufficient to initiate a new growth site also at basal regions of the root hair. This demonstrates that microtubules are essential to restrict the growth machinery to the apex of the elongating root hair.

Although many genes have been identified that affect the morphogenesis of root hair cell formation, a molecular analysis is available only for the *RHD3* gene. The *RHD3* gene was cloned by T-DNA tagging and encodes a protein with sequence similarity to GTP-binding proteins (Wang *et al.*, 1997). The activity of GTP-binding proteins is usually regulated by the exchange of GDP and GTP, and is involved in various cellular processes including signal transduction, the regulation of the cytoskeleton and intracellular traffic. The *rhd3* phenotype is characterized by a smaller central vacuole and irregularities in growth directionality (Galway *et al.*, 1997). This suggests that the *RHD3* gene is involved either in directing vesicle transport to the tip or in the biogenesis of the vacuole.

IV. CONCLUSIONS

By all morphological criteria, the development of root hairs and aerial trichomes is fairly distinct. However, their genetic analysis has revealed a surprisingly high number of genes that exhibit a functional overlap in regulating both types of epidermal hairs. Striking functional overlaps include the genes involved in the regulation of cell differentiation, such as *RHL1*, *RHL2*, *RHL3*, *ERH3* and *GL2*. Moreover, some genes involved in cell morphogenesis, i.e. *SINGED* and *RHD3*, are required in both cell types (Schiefelbein and Somerville, 1990; Marks and Esch, 1992). Although such genetic overlaps probably represent only a glimpse on the complexity of the underlying molecular mechanisms, the genetic identification of key players provides a convenient entry point in a detailed molecular and biochemical analysis.

ACKNOWLEDGEMENTS

We would like to thank Hilmar Ilgenfritz, Swen Schellmann and Arp Schnittger for helpful suggestions and critical reading of the manuscript. The work in our laboratory on this subject is supported by a DFG grant (HU397/2-2) and by the Volkswagen-Stiftung to M.H.

REFERENCES

Aeschbacher, R., Schiefelbein, J. W. and Benfey, P. N. (1994). The genetic and molecular basis of root development. *Annual Review of Plant Physiology and Plant Molecular Biology* **45**, 25–45.

Baskin, T. I., Betzner, A. S., Hoggart, R., Cork, A. and Williamson, R. E. (1992). Root morphology mutants in Arabidopsis thaliana. *Australian Journal of Plant Physiology* **19**, 427–438.

Bibikova, T. N., Zhigilei, A. and Gilroy, S. (1997). Root hair growth is directed by calcium and endogenous polarity. *Planta* **203**, 495–505.

Bibikova, T. N., Blancaflor, E. B. and Gilroy, S. (1999). Microtubules regulate tip growth and orientation in root hairs in Arabidopsis thaliana. *Plant Journal* **17**, 657–665.

Bowser, J. and Reddy, A. S. N. (1997). Localization of a kinesin-like calmodulin-binding protein in dividing cells of Arabidopsis and tobacco. *Plant Journal* **12**, 1429–1437.

Chang, C., Kwok, S. F., Bleecker, A. B. and Meyerowitz, E. M. (1993). Arabidopsis ethylene-response gene ETR1: similarity of product to two-component regulators. *Science* **262**, 539–544.

Feenstra, W. J. (1978). Contiguity of linkage groups 1 and 4, as revealed by linkage relationships of two newly isolated markers dis-1 and dis-2. *Arabidopsis Information Service* **15**, 35–38.

Felle, H. and Hepler, P. K. (1997). The cytosolic Ca^{2+} concentration gradient of Sinapsis alba root hairs as revealed by Ca^{2+} selective microelectrode tests and Fura-dextran ratio imaging. *Plant Physiology* **114**, 39–45.

Folkers, U., Berger, J. and Hülskamp, M. (1997). Cell morphogenesis of trichomes in Arabidopsis: differential regulation of primary and secondary branching by branch initiation regulators and cell size. *Development* **124**, 3779–3786.

Galway, M. E., Heckman, J. and Schiefelbein, J. W. (1997). Growth and ultrastructure of Arabidopsis root hairs: the rhd3 mutation alters vacuole enlargement and tip growth. *Planta* **201**, 209–218.

Grierson, C. S., Roberts, K., Feldmann, K. and Dolan, L. (1997). The COW1 locus of Arabidopsis acts after RHD2, and in parallel with RHD3 and TIP1, to determine the shape, rate of elongation, and number of root hairs produced from each site of hair formation. *Plant Physiology* **115**, 981–990.

Haughn, G. W. and Somerville, C. R. (1988). Genetic control of morphogenesis in Arabidopsis. *Developmental Genetics* **9**, 73–89.

Hermann, A. and Felle, H. (1995). Tip growth in root hair cells of Sinapsis alba L.: significance of internal and external Ca pH. *New Phytology* **129**, 523–533.

Hirokawa, N. (1998). Kinesin and dynein superfamily proteins and the mechanism of organelle transport. *Science* **279**, 519–526.

Hülskamp, M., Misera, S. and Jürgens, G. (1994). Genetic dissection of trichome cell development in Arabidopsis. *Cell* **76**, 555–566.

Hülskamp, M., Folkers, U. and Grini, P. (1998). Cell morphogenesis in Arabidopsis. *BioEssays* **20**, 20–29.

Hülskamp, M., Folkers, U. and Schnittger, A. (1999). Trichome development in Arabidopsis thaliana. *International Review of Cytology* **186**, 147–178.

Jacobsen, S. E., Binkowski, K. A. and Olszewski, N. E. (1996). SPINDLY, a tetratricopeptide repeat protein involved in gibberellin signal transduction in Arabidopsis. *Proceedings of the National Academy of Sciences (USA)* **93**, 9292–9296.

Koornneef, M., Dellaert, L. W. M. and Veen, J. H. V. d. (1982). EMS- and radiation-induced mutation frequencies at individual loci in Arabidopsis thaliana. *Mutation Research* **93**, 109–123.

Larkin, J. C., Young, N., Prigge, M. and Marks, M. D. (1996). The control of trichome spacing and number in Arabidopsis. *Development* **122**, 997–1005.

Marks, M. D. (1997). Molecular genetic analysis of trichome development in Arabidopsis. *Annual Review of Plant Physiology and Plant Molecular Biology* **48**, 137–163.

Marks, M. D. and Esch, J. J. (1992). Trichome formation in Arabidopsis as a genetic model for studying cell expansion. *Current Topics in Plant Biochemistry and Physiology* **11**, 131–142.

Masucci, J. D. and Schiefelbein, J. W. (1994). The rhd6 mutation of Arabidopsis thaliana alters root-hair initiation through an auxin- and ethylene-associated process. *Plant Physiology* **106**, 1335–1346.

Masucci, J. D. and Schiefelbein, J. W. (1996). Hormones act downstream of TTG and GL2 to promote root hair outgrowth during epidermis development in the Arabidopsis root. *Plant Cell* **8**, 1505–1517.

Masucci, J. D., Rerie, W. G., Foreman, D. R., Zhang, M., Galway, M. E., Marks, M. D. and Schiefelbein, J. W. (1996). The homeobox gene GLABRA 2 is required for position-dependent cell differentiation in the root epidermis of Arabidopsis thaliana. *Development* **122**, 1253–1260.

Mayer, U., Ruiz, R. A. T., Berleth, T., Misera, S. and Jürgens, G. (1991). Mutations affecting body organization in the Arabidopsis embryo. *Nature* **353**, 402–407.

McClinton, R. S. and Sung, Z. R. (1997). Organization of cortical microtubules at the plasma membrane in Arabidopsis. *Planta* **201**, 252–260.

Oppenheimer, D. (1998). Genetics of plant cell shape. *Current Opinion in Plant Biology* **1**, 520–524.

Oppenheimer, D. G., Pollock, M. A., Vacik, J., Szymanski, D. B., Ericson, B., Feldmann, K. and Marks, M. D. (1997). Essential role of a kinesin-like protein in Arabidopsis trichome morphogenesis. *Proceedings of the National Academy of Sciences USA* **94**, 6261–6266.

Perazza, D., Herzog, M., Hülskamp, M., Brown, S., Dorne, A. and Bonneville, J. (1999). Trichome cell growth in Arabidopsis thaliana can be depressed by mutations in at least five genes. *Genetics* **152**, 461–476.

Potikha, T. and Delmer, D. (1995). A mutant of Arabidopsis thaliana displaying altered patterns of cellulose deposition. *Plant Journal* **7**, 453–460.

Reddy, A. S. N., Safadi, F., Narasimhulu, S. B., Golovkin, M. and Hu, X. (1996). A novel plant calmodulin-binding protein with a kinesin heavy chain motor domain. *Journal of Biological Chemistry* **271**, 7052–7060.

Rerie, W. G., Feldmann, K. A. and Marks, M. D. (1994). The glabra 2 gene encodes a homeodomain protein required for normal trichome development in Arabidopsis. *Genes and Development* **8**, 1388–1399.

Schiefelbein, J. and Somerville, C. (1990). Genetic control of root hair development in Arabidopsis thaliana. *Plant Cell* **2**, 235–243.

Schiefelbein, J., Galway, M., Masucci, J. and Ford, S. (1993). Pollen tube and root-hair tip growth is disrupted in a mutant of Arabidopsis thaliana. *Plant Physiology* **103**, 979–985.

Schneider, K., Wells, B., Dolan, L. and Roberts, K. (1997). Structural and genetic analysis of epidermal cell differentiation in Arabidopsis primary roots. *Development* **124**, 1789–1798.

Schnepf, E. (1986). Cellular polarity. *Annual Review of Plant Physiology* **37**, 23–47.

Schnittger, A., Jürgens, G. and Hülskamp, M. (1998). Tissue layer and organ specificity of trichome formation are regulated by GLABRA1 and TRIPTYCHON in Arabidopsis. *Development* **125**, 2283–2289.

Song, H., Golovkin, M., Reddy, A. S. N. and Endow, S. A. (1997). In vitro motility of AtKCBP, a calmodulin-binding kinesin protein of Arabidopsis. *Proceedings of the National Academy of Sciences USA* **94**, 322–327.

Szymanski, D. B. and Marks, M. D. (1998). GLABROUS1 overexpression and TRIPTYCHON alter the cell cycle and trichome cell fate in Arabidopsis. *Plant Cell* **10**, 2047–2062.

Telfer, A., Bollman, K. M. and Poethig, R. S. (1997). Phase change and the regulation of trichome distribution in Arabidopsis thaliana. *Development* **124**, 645–654.

Torres-Ruiz, R. A. and Jürgens, G. (1994). Mutations in the FASS gene uncouple pattern formation and morphogenesis in Arabidopsis development. *Development* **120**, 2967–2978.

Traas, J., Bellini, C., Nacry, P., Kronenberger, J., Bouchez, D. and Caboche, M. (1995). Normal differentiation patterns in plants lacking microtubular preprophase bands. *Nature* **375**, 676–677.

Traas, J., Hülskamp, M., Gendreau, E. and Höfte, H. (1998). Endoreduplication and development: rule without dividing? *Current Opinion in Plant Biology* **1**, 498–503.

Tsuge, T., Tsukaya, H. and Uchimiya, H. (1996). Two independent and polarized processes of cell elongation regulate leaf blade expansion in Arabidopsis

thaliana (L.) Heynh. *Development* **122**, 1589–1600.
Wang, H., Lockwood, S. K., Hoeltzel, M. F. and Schiefelbein, J. W. (1997). The ROOT HAIR DEFECTIVE3 gene encodes an evolutionary conserved protein with GTP-binding motifs and is required for regulated cell enlargement in Arabidopsis. *Genes and Development* **11**, 799–811.
Wymer, C. L., Bibikova, T. N. and Gilroy, S. (1997). Cytoplasmic free calcium distribution during the development of root hairs of Arabidopsis. *Plant Journal* **12**, 427–439.

Trichome Plasmodesmata: A Model System for Cell-to-cell Movement

E. WAIGMANN[1] and P. ZAMBRYSKI[2]

[1]*Institute of Biochemistry, University of Vienna, Dr. Bohrg. 9, A-1030 Austria*
[2]*Department of Plant and Microbial Biology, 111 Koshland Hall, University of California, Berkeley, CA 94720, USA*

I. WHY ARE TRICHOMES A MODEL SYSTEM FOR CELL-TO-CELL MOVEMENT?

Cell-to-cell transport of molecules and communication between plant cells occur via plasmodesmata, cytoplasmic bridges that traverse the thick cell walls that surround plant cells (Lucas *et al.*, 1993; Zambryski, 1995; Ding, 1997; McLean *et al.*, 1997; Kragler *et al.*, 1998). Research on these complex

Advances in Botanical Research Vol. 31
ISBN 0-12-005931-2

structures, which have even been considered as "intercellular organelles" (Lucas and Wolf, 1993; Oparka, 1993; Lucas, 1995), has a long history. Plasmodesmata were first identified more than a hundred years ago. Originally termed "offene Communicationen" (Tangl, 1879), the name "plasmodesma" – a bond between cytoplasms – was coined by Strasburger in 1901. Even in early times, plasmodesmata fascinated many researchers. Numerous reports described their light and electron microscopic analyses in a range of different plants and tissues (for a comprehensive summary of citations, see Robards, 1976; Robards and Lucas, 1990). This wealth of structural information was accompanied by only minimal functional data. Notably, several studies were conducted in plant hairs and provided some of the first clues about cell–cell movement of low-molecular-weight fluorescent dyes and peptides (Schumacher, 1936; Bauer, 1953; Terry and Robards, 1987).

To study cell-to-cell movement, it was imperative to develop a method that allows the controlled introduction of substances into single cells, such as microinjection. Again, plant hairs were recognized early on as excellent target cells for microinjection. Janet Plowe, one of the pioneers of microinjection into plant cells, describes the advantages of hairs for the study of dye movement by microinjection (Plowe, 1931): "The wall is easily pierced. The absence of plastids, the transparency of the cytoplasm, the thinness of the walls, make internal details clearly visible. Injections into cytoplasm or vacuole can be made with no ambiguity as to location." In addition to these merits, many types of plant hairs are composed of a uniseriate succession of cells that facilitates observation and documentation of cell-to-cell movement of microinjected substances. For all these reasons, trichomes have evolved to become a model system for the study of intercellular molecular transport.

II. STRUCTURE OF PLASMODESMATA

Plasmodesmata are pores, lined with plasma-membrane, that are traversed in their centre by the desmotubule (Fig. 1) which is composed of membranes continuous with the endoplasmic reticulum (ER). Thus, two membrane systems form an unbroken link between plant cells (Fig. 1). Frequently, in micrographs showing cross-sections through plasmodesmata, the desmotubule is visualized as an electron-dense rod interpreted as appressed ER without a lumen (Fig. 4B,D; Lopez-Saez *et al.*, 1966; Gunning and Hughes, 1976; Olesen, 1979; Overall *et al.*, 1982; Gunning and Overall, 1983; Robinson-Beers and Evert, 1991; Ding *et al.*, 1992; Waigmann *et al.*, 1997), but in some ultrastructural studies a clear desmotubular lumen is observed (Fig. 4C,E; Eleftheriou and Hall, 1983b; Waigmann *et al.*, 1997). Proteinaceous particles may be embedded in the desmotubular membrane (Fig. 1; Ding *et al.*, 1992). For plasmodesmata of fern gametophytes, evidence

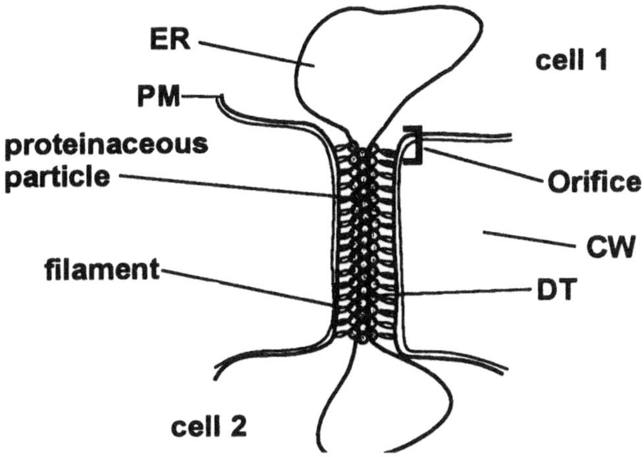

Fig. 1. Schematic representation of a longitudinal view through a plasmodesma. PM, plasma membrane; DT, desmotubule; CW, cell wall; ER, endoplasmic reticulum.

points to a mainly proteinaceous desmotubule (Tilney *et al.*, 1990). The space between the desmotubule and the plasma membrane is termed the cytoplasmic sleeve (Fig. 2) and is often segmented by filamentous connections between the desmotubule and the plasma membrane (Figs 1 and 4C,E; Thompson and Platt-Aloia, 1985; Ding *et al.*, 1992; Schulz, 1995; Waigmann *et al.*, 1997). Plasmodesmata exist either as simple pores (Fig. 1), which may be enlarged in the middle to generate a central cavity, or as branched entities, where multiple exits and entrances fuse into a common central cavity. The exit and entrance areas of plasmodesmata, termed orifices (Fig. 1), are considered to control molecular traffic. Orifices are occasionally surrounded by constricting electron-lucent collars ("neck constrictions"; Fig. 4C,E; Hughes and Gunning, 1980; Kronestedt *et al.*, 1986; Turner *et al.*, 1994; Waigmann *et al.*, 1997) or by a ring of external particles called the sphincter (Olesen, 1979; Badelt *et al.*, 1994) proposed to function as a valve regulating plasmodesmal permeability.

Both the neck constriction and the collar region have been suggested to be artifacts of chemical fixation in some tissues (Hughes and Gunning, 1980; Radford *et al.*, 1998). Chemical fixation by the most widely used fixative glutaraldehyde is a slow process that may require several minutes until completion (Mersey and McCully, 1978). During that time, cellular structures may be destroyed or altered due to stress responses (Mersey and McCully, 1978). For example, in some types of nectary trichomes, the presence of electron-lucent constricting collars around plasmodesmal orifices was observed and attributed to callose deposition, a wound response induced by glutaraldehyde fixation (Hughes and Gunning, 1980). When callose deposition was prevented by freeze substitution, plasmodesmata were evident as unconstricted channels throughout their length and not surrounded by collars (Hughes and Gunning, 1980). Similarly, neck constrictions of plasmodesmata in *Allium* roots were induced by glutaraldehyde fixation or tissue dissection (Radford *et al.*, 1998). Upon chemical inhibition of callose formation prior to physical wounding and fixation, plasmodesmata exhibited a funnel-shaped configuration suggesting that collars and neck constrictions were artifacts of sample preparation (Radford *et al.*, 1998). Alternatively, callose deposition could be a physiological mechanism to shut down plasmodesmal transport under certain conditions, such as wounding, or altered transport requirements due to changes in development and environment (Lucas *et al.*, 1993). Consequently, plasmodesmata with constricted necks may represent "closed" plasmodesmata, a status inherent to the dynamic nature of plasmodesmata. The latter interpretation is strengthened by observations in pea root tips, where there is an increased rate of phloem unloading correlated with widening of plasmodesmal orifices and a transient loss of neck constrictions in the majority of plasmodesmata. Glutaraldehyde was routinely included as a fixative in all experiments, suggesting that the presence or absence of neck constrictions was not dependent on the fixation protocol, but on the cell-to-cell transport activity of plasmodesmata (Schulz, 1995).

In principle, three routes for molecular transport through plasmodesmata are feasible. The route through the cytoplasmic sleeve is most commonly proposed (e.g. Olesen, 1975; Gunning and Robards, 1976; Lucas *et al.*, 1993). In addition, alternative routes either by lateral diffusion within the desmotubular membrane system suggested for hydrophobic molecules (Grabski *et al.*, 1993) or through the lumen of the desmotubule (Robinson-Beers and Evert, 1991; Gamalei *et al.*, 1994; Glockmann and Kollmann, 1996; Waigmann *et al.*, 1997) have been considered.

In contrast to numerous ultrastructural studies, there exists very little information on the molecular composition of plasmodesmata. So far, only actin (White *et al.*, 1994; Blackman and Overall, 1998), myosin (Blackman and Overall, 1998; Radford and White, 1998), ubiquitin (Ehlers *et al.*, 1996) and a 41 kDa protein isolated from maize mesocotyl cell walls (Epel *et al.*, 1996) have been identified as plasmodesmal components by immunoelectron microscopy. Actin also was functionally implicated in the regulation of plasmodesmal permeability (Ding *et al.*, 1996).

III. GLANDULAR TRICHOME PLASMODESMATA: A SYSTEM OPTIMIZED FOR HIGH TRANSPORT ACTIVITY

Many trichomes specialize in the secretion of a broad range of natural products, such as nectar, terpenoids, cannabinoids, sucrose esters and phenolic compounds (reviewed in Kelsey *et al.*, 1984). Such trichomes are termed glandular trichomes. They are generally composed of basal cells, which are embedded in the leaf tissue, and one or several stalk cells that coalesce into a tip region consisting of one or several secretory head cells (Fig. 2A,C and Fig. 4A). Often, synthesis of the secreted substances takes place in the secretory head cells themselves; in fact, isolated secretory cell clusters of glandular trichomes can synthesize secondary metabolites from basic precursors. For example, isolated peppermint trichome heads synthesize monoterpenes from sucrose (McCaskill *et al.*, 1992), or isolated tobacco trichome heads synthesize sucrose esters from sucrose and aspartate (Kandra *et al.*, 1990). Glandular trichomes rely on the import of carbon and energy sources from the underlying leaf tissue, since they are only minimally photosynthetic. In addition, their stalk cells are surrounded by a heavily cutinized cell wall that forms a barrier to apoplastic transport from the leaf (Findley and Mercer, 1971; Schnepf, 1974; Gunning and Hughes, 1976). Consequently, symplastic cell–cell transport via plasmodesmata is the major route for molecular traffic into trichome cells, and plasmodesmata of glandular trichomes are likely to be optimized for high transport activity.

A. *ABUTILON* NECTARY TRICHOMES

Nectary trichomes are an extremely active type of gland and hence are extensively studied. These glands expel large amounts of nectar through transient pores in the cuticle at the tip of the trichome. *Abutilon* nectary trichomes (Fig. 2A) excrete a droplet of 500 μm^3 every 6–20 seconds (Findley and Mercer, 1971; Gunning and Hughes, 1976). Depending on the type of nectary trichome, different mechanisms for nectar secretion from the hair cells have been suggested (Eleftheriou and Hall, 1983a; Robards and Stark, 1988; Sawidis *et al.*, 1989). Regardless of the mechanism of secretion, a flow of pre-nectar, equivalent to the volume of nectar expelled at the apex of the trichome, must pass via plasmodesmata from the phloem into the most proximal stalk cell (Gunning and Hughes, 1976; Fahn, 1979).

A model for the flow of pre-nectar and nectar secretion has been developed for *Abutilon* nectaries (Robards and Stark, 1988) based on several structural observations. First, an extensive endomembrane system termed the sarcoplasmic reticulum, which is in close association with the plasma membrane (Fig. 2B), develops in all trichome cells except the stalk cell, prior to the onset of secretion (Kronestedt *et al.*, 1986; Robards and Stark, 1988). Secondly, adjacent to nearly all cells, there is a free extracytoplasmic space between the plasma membrane and the cuticle surrounding the trichome (Fig. 2B; Robards and Stark, 1988). Thirdly, plasmodesmal frequency is highest in the walls of the stalk cell (Table I; 12.6 and 10.3 plasmodesmata per μm^2 in the distal and proximal wall, respectively) and decreases gradually to about 4–5 per μm^2 in intermediate cells close to the tip (Table I;

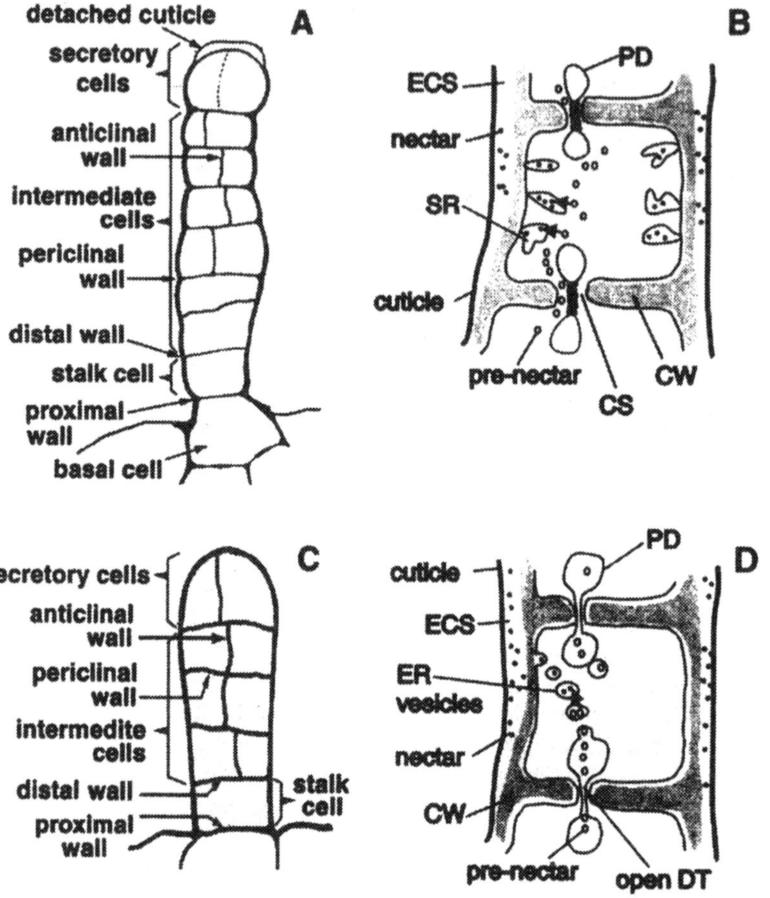

Fig. 2. (A,C) Schematic drawing of nectary trichomes. (A) *Abutilon* nectary trichome. (C) Extrafloral nectary trichome of *Gossypium hirsutum* (cotton). (B,D) Models linking symplastic transport of pre-nectar through plasmodesmata with secretion of nectar. (B) *Abutilon* nectary trichome. (D) Extrafloral nectary trichome of *Gossypium hirsutum* (cotton). Arrows indicate processing of pre-nectar into nectar. PD, plasmodesma; DT, desmotubule; CW, cell wall; SR, sarcoplasmic reticulum; ER, endoplasmic reticulum; CS, cytoplasmic sleeve; ECS, extracytoplasmic space. Modified from Eleftheriou and Hall (1983b), Kronestedt *et al.* (1986), and Robards and Stark (1988).

Kronestedt *et al.*, 1986). The model suggests that, in all trichome cells except the stalk cell, the sarcoplasmic reticulum is loaded with pre-nectar from the cytoplasm (Fig. 2B). Either upon loading or within the sarcoplasmic reticulum, pre-nectar is processed into nectar (Fig. 2B, arrows). The pressure in the sarcoplasmic reticulum builds up until the cell expels a small pulse of nectar into the extracytoplasmic space (Fig. 2B). Nectar from all cells accumulates in this compartment until, by force of pressure, pores in the cuticle at the tip open and release a drop of nectar to the exterior.

The model predicts that the amount of pre-nectar to be symplastically transported between trichome cells should be highest in the stalk cell and gradually decrease towards the tip cell. This prediction is supported by the marked decrease in plasmodesmal frequency from the stalk cell up to the tip cell. This further suggests that plasmodesmata cope with the enormous amount of solute to be transported by sheer force of numbers. Indeed, ultrastructural studies do not reveal unusual features in plasmodesmal substructure (Gunning and Hughes, 1976). The majority of plasmodesmata in the stalk cell contain an electron-dense desmotubule; only sometimes is an open lumen observed. Twenty per cent of plasmodesmata show neck constrictions, whereas, in the remaining 80%, desmotubule and plasma membrane appear parallel over the whole length. The diameter of plasmodesmata compartments is rather small – 43 nm for the outer leaflet of the plasma membrane and 16 nm for the desmotubule (Table I). Stalk cell plasmodesmata are notably short (87 nm) owing to the extremely thin cell wall of the stalk cell (Gunning and Hughes, 1976). In all likelihood, the short transport pathway through plasmodesmata is crucial to maintain high rates of solute flux.

Functional studies on the conductivity of *Abutilon* nectary trichome plasmodesmata were key to our understanding of diffusion-driven movement through plasmodesmata. Low-molecular-weight fluorescent dyes, fluorescently labelled amino acids and fluorescently labelled peptides were introduced into the tip cell by microinjection and movement into the adjacent cell scored after 1 minute (Terry and Robards, 1987). This study revealed that the effective size of the transported molecule, as measured by its Stokes' radius, is the major determinant of movement. In simple terms, if the Stokes' radius is smaller than the radius of the transport channel, then a molecule will be able to move from cell to cell. The closer the radius of the transported molecule approaches that of the transport channel, the slower the movement. *Abutilon* nectary plasmodesmata are functionally distinguished from other types of plasmodesmata by their ability to transport slightly larger molecules than those reported from other systems (Tucker, 1982; Goodwin, 1983). Also, aromatic amino acids that can reduce or abolish movement (Tucker, 1982; Erwee and Goodwin, 1984) had no negative effect on movement through *Abutilon* nectary plasmodesmata (Terry and Robards, 1987). Based on this latter study, it was concluded that channels through the cytoplasmic sleeve are the most likely route for diffusion of probes and that the width of individual channels is close to 3 nm. The size exclusion limit (SEL) for molecules diffusing through nectary trichome plasmodesmata was placed at a Stokes' radius of about 0.9 nm based on the mobility of the fluorescein isothiocyanate (FITC)-labelled dipeptide Trp-Phe. The largest molecule that moved between cells was FITC-$(Gly)_{12}$ with a molecular mass of 1090 Da (Table I; Terry and Robards, 1987).

TABLE I

Plasmodesmal characteristics in different types of trichomes and mesophyll cells

Plant species and tissue type	Size exclusion limit: Molecular mass (dalton)	Size exclusion limit: Stokes' radius (nm)	Type of wall	PD dimensions (nm): Outer diameter	PD dimensions (nm): DT diameter	PD dimensions (nm): Length	Ultrastructural appearance of PD	PD frequency (μm^{-2})	References
Abutilon nectary trichome	1090	0.9	Distal wall of stalk cell	43	16	87	Desmotubular lumen electron dense	12.6	Gunning and Hughes (1976) Terry and Robards (1987)
			Proximal wall of stalk cell	—[a]	—	—	—	10.3	Kronestedt *et al.* (1986)
			Periclinal walls in intermediate cells close to the tip	—	—	—	—	4–5	Kronestedt *et al.* (1986)
Cotton nectary papillae	—	—	Distal wall of stalk cell	42.4–55.4	19.3–21	143	Desmotubular lumen open	15.4	Eleftheriou and Hall (1983b)
			Proximal wall of stalk cell	40.9	—	235	Desmotubular lumen open	12.9	Eleftheriou and Hall (1983b)
			Periclinal walls of intermediate cells	—	—	—	—	9.1	Eleftheriou and Hall (1983b)
			Anticlinal walls of intermediate cells	—	—	—	—	5.7	Eleftheriou and Hall (1983b)
Hibiscus nectary trichome	—	—	Distal wall of stalk cell	—	—	715	—	11.8	Sawidis *et al.* (1987)
			Proximal wall of stalk cell	—	—	510	—	20.9	Sawidis *et al.* (1987)
			Periclinal walls of intermediate cells	—	—	—	—	10.3	Sawidis *et al.* (1987)
			Anticlinal walls of intermediate cells	—	—	—	—	7.2	Sawidis *et al.* (1987)
Nicotiana clevelandii leaf trichome	7000	2.15	All walls	59	~20	529–1129	Desmotubular lumen open	—	Waigmann and Zambryski (1995) Waigmann *et al.* (1997)
Mesophyll cells	<1000	0.73	All walls	60	25	823	Desmotubular lumen electron dense	—	Waigmann *et al.* (1997) Lucas *et al.* (1993) Wolf *et al.* (1989)

[a]No data available. PD, plasmodesma; DT, desmotubule.

For several other types of plasmodesmata, such as tobacco mesophyll (Wolf *et al.*, 1989), *Setcreasea purpurea* stamen hair (Tucker, 1982) and *Elodea* plasmodesmata (Goodwin, 1983), the molecular SEL ranges between 700 and 800 Da. In contrast, *Nicotiana clevelandii* leaf trichomes transport molecules with a molecular mass up to 7 kDa (Waigmann and Zambryski, 1995) corresponding to a Stokes' radius of approximately 2.15 nm (Table I). It should be noted that in the latter case FITC-labelled dextrans were used to define the SEL, whereas in previous studies predominantly FITC-labelled peptides were employed (Tucker, 1982; Terry and Robards, 1987). The different chemical compositions of the tracer molecules could influence their mobility independently of their Stokes' radii.

B. EXTRAFLORAL NECTARIES OF COTTON: MORE OF THE SAME?

Extrafloral nectary trichomes of cotton (also called papillae) display both similarities and differences compared to *Abutilon* nectary trichomes. Morphologically they consist of a single stalk cell, several intermediate cells and a head of up to 12 secretory cells (Fig. 2C; Eleftheriou and Hall, 1983b). Unlike *Abutilon* nectary trichomes, the mechanism for nectar secretion from secretory cells of cotton papillae seems to take place by reverse pinocytosis via vesicles that fuse with the plasma membrane (Fig. 3D; Eleftheriou and Hall, 1983a). The vesicles may originate from the ER, since the ER is prominent in secretory cells, and in addition to the multiple vesicles close to and in contact with the plasmalemma, occasionally even a whole ER cisterna appressed to the plasma membrane is observed (Eleftheriou and Hall, 1983b). Thus, nectar secretion in cotton nectaries is likely to occur via successive fusion of ER-derived vesicles with the plasma membrane. Like *Abutilon* nectary trichomes, transport of pre-nectar seems to take place symplastically via plasmodesmata. Again, plasmodesmal frequencies have been determined for various cells of the papillae (Fig. 2C; Eleftheriou and Hall, 1983b). Frequencies are highest in the walls of the stalk cell (Table I; 12.9 plasmodesmata per μm^2 in the proximal wall, 15.4 plasmodesmata per μm^2 in the distal wall) followed by a somewhat lower, but constant, number in the periclinal walls of all intermediate cells (Table I; 9.1 plasmodesmata per μm^2), and are lowest in anticlinal walls of intermediate cells and secretory cells (Table I; 5.7 plasmodesmata per μm^2). This suggests directional symplastic flow towards the tip cells.

An ultrastructural analysis reveals the dimensions of plasmodesmata of cotton papillae stalk cells (Eleftheriou and Hall, 1983b). The outer diameter at the orifice ranges from 40.9 to 55.4 nm, comparable to that of *Abutilon* plasmodesmata; the diameter of the desmotubule is between 19.3 and 21 nm, slightly larger than in *Abutilon* nectary plasmodesmata (Table I). Plasmodesmal length varies from 143 nm in the distal wall to 235 nm in the proximal wall of the stalk cell, 2–3 times the length of *Abutilon* nectary plasmodesmata

(Table I). In addition, plasmodesmata of cotton papillae display an unusual substructure, an "open" electron-lucent desmotubular lumen. Unfortunately, there are no functional data available about cell-to-cell transport in cotton papillae to complement the ultrastructural information.

The difference in ultrastructure of the desmotubule between *Abutilon* nectary plasmodesmata and cotton nectary plasmodesmata again raises the extensively discussed question of the exact pathway of movement through plasmodesmata. In *Abutilon* nectaries, functional studies involving microinjection of fluorescent dyes clearly show that only cytoplasmic injections result in cell-to-cell movement (Terry and Robards, 1987). Also, the *Abutilon* desmotubule is visualized as a solid rod in electron micrographs. Consequently, it was proposed that molecular transport most likely occurs via the cytoplasmic sleeve (Terry and Robards, 1987). There is no clear indication about the exact passageway through plasmodesmata in cotton papillae. Based on ultrastructural analysis, the unusual open lumen of the desmotubule may suggest involvement of the desmotubule in transport.

These two types of nectaries secrete nectar by two very different mechanisms; thus, a model linking the differences in plasmodesmal ultrastructure with the different secretion pathways could be envisaged (Fig. 2B,D): *Abutilon* nectary trichomes secrete nectar via the sarcoplasmic reticulum, an endomembrane system contacting the plasma membrane (Fig. 2B; Robards and Stark, 1988). Nectar production occurs by partial hydrolysis of pre-nectar sucrose into fructose and glucose, either during or after loading of the sarcoplasmic reticulum with pre-nectar (Fig. 2B, arrows; Robards and Stark, 1988). Pre-nectar is routed towards the sarcoplasmic reticulum via the cytoplasm (Robards and Stark, 1988). Thus, cell-to-cell transport of pre-nectar via the cytoplasmic sleeve of plasmodesmata (Fig. 2B) as suggested by ultrastructural evidence, positions pre-nectar into the right cellular compartment for eventual secretion. The situation differs in cotton papillae. There, nectar secretion occurs via reverse pinocytosis with vesicles that are likely to originate from the ER (Fig. 2D; Eleftheriou and Hall, 1983b) and nectar production from pre-nectar may take place in those vesicles (Fig. 2D). Symplastic cell–cell transport of pre-nectar through the "open" lumen of the desmotubule (Eleftheriou and Hall, 1983b) into the connected ER and subsequent budding of pre-nectar-filled vesicles (Fig. 2D) may be the most advantageous route to connect symplastic transport of pre-nectar with secretion of nectar.

C. OTHER TYPES OF NECTARY TRICHOMES

In *Hibiscus* nectary trichomes, plasmodesmal frequency has been evaluated and compared to that of other nectary trichomes (Sawidis *et al.*, 1987). The proximal wall of the stalk cell is traversed by 20.9 plasmodesmata per μm^2, an even larger number of plasmodesmata than in the corresponding wall in cotton papillae and *Abutilon* nectaries (Table I; Eleftheriou and Hall, 1983b). For the distal wall of the *Hibiscus* nectary stalk cell, plasmodesmal frequency was placed at 11.8 plasmodesmata per μm^2, a value close to that for *Abutilon* nectaries and somewhat less than in cotton papillae (Table I). The intermediate cells in *Hibiscus* nectary trichomes may be involved in nectar secretion (Sawidis *et al.*, 1989). Although there is a drop in plasmodesmal frequency from stalk cell to intermediate cells, it is unknown whether this decrease continues gradually towards the tip cells as in *Abutilon* nectaries. Based on ultrastructural data about ER form and abundance, direct involvement of the ER compartment of plasmodesmata in the transport of pre-nectar was proposed (Sawidis *et al.*, 1989). However, there are no data on the substructure of plasmodesmata to provide structural support for this hypothesis. Other nectaries that have been studied include *Chrysanthemum morifolium* cv. Dramatic (Vermeer and Peterson, 1979), *Ricinus* (Nichol and Hall, 1988), *Vigna* (Kuo and Pate, 1985; Pate *et al.*, 1985), *Acacia* (Marginson *et al.*, 1985) and *Digitalis purpurea* (Gaffal *et al.*, 1998). None of these analyses determines plasmodesmata frequency, their substructure or function.

D. *NICOTIANA* LEAF TRICHOMES: A MODEL SYSTEM TO STUDY MACROMOLECULAR TRAFFIC

In the last decade, plasmodesmata have emerged as a major communication system whereby plant cells actively transport proteins and nucleic acids. Plant viruses pirate this pathway to move from infected into neighbouring healthy cells. Viral spread depends essentially on a group of virally encoded proteins, the movement proteins (MP), which interact with plasmodesmata and mediate transfer of the viral genome. Viral MPs are often used as probes to track macromolecular traffic through plasmodesmata. MP function and macromolecular transport are commonly studied in mesophyll cells by microinjection of fluorescently labelled compounds and subsequent observation of their movement pattern. However, mesophyll cells have a major drawback: they are highly interconnected and form a complex three-dimensional network, such that only the surface cells can be clearly monitored with the light microscope. It is therefore not possible to observe or predict the exact pathway of movement. Studies on plasmodesmal transport are facilitated if the microinjected cell is part of a linear array. Leaf trichomes of *Nicotiana* species consisting of 4–8 cells topped by a secretory tip cell (Fig. 4A) provide such a system. Moreover, trichomes can be invaded by viruses. Thus, *Nicotiana* leaf trichomes represent an advantageous biological system to study viral spread and MP-mediated plasmodesmal transport.

1. Interaction between Viruses and Plasmodesmata

Historically, plasmodesmata have been described as simple pores that function in the cell-to-cell transport of ions and metabolites with a molecular mass <1 kDa. Several plant viruses also utilize plasmodesmata to move from an infected cell into neighbouring cells (for reviews, see Citovsky and Zambryski, 1993; Lucas and Gilbertson, 1994; Carrington *et al.*, 1996; Goshroy *et al.*, 1997), which might be viewed as a contradiction, since the dimensions of viral particles and even free viral nucleic acid are larger than the effective channel size of plasmodesmata. The key to this dilemma resides in the function of virally encoded proteins, the movement proteins. Tobacco mosaic virus (TMV), a simple, well-studied virus with an RNA genome, encodes one movement protein (TMV-MP). To date, three biological activities are assigned to TMV-MP: it increases ("gates") the SEL of plasmodesmata from the basal limit of 1 kDa to permit passage of 10–20 kDa dextrans (Wolf *et al.*, 1989; Waigmann *et al.*, 1994), it coöperatively binds to single-stranded nucleic acid (Citovsky *et al.*, 1990) and it interacts with cytoskeletal elements (Heinlein *et al.*, 1995; McLean *et al.*, 1995). Based on these three functions, a model for TMV-MP-mediated cell-to-cell movement of viral RNA was developed (Citovsky and Zambryski, 1991; Zambryski, 1995): TMV-MP binds and coats viral RNA to form a high-molecular-weight complex containing RNA in an extended form (Citovsky *et al.*, 1992). The MP:RNA complex then interacts with host-cell cytoskeletal elements and migrates towards plasmodesmata. There, MP gates plasmodesmata and mediates translocation of the complex into the next cell. In addition to TMV-MP, MPs from other viruses display similar functions, implying that the above model may hold true for several plant viruses (Fujiwara *et al.*, 1993; Poirson *et al.*, 1993; Noueiry *et al.*, 1994; Ding *et al.*, 1995).

Microinjection of movement proteins purified from *Escherichia coli* into mesophyll cells of various plants illustrated the versatility and dynamics of plasmodesmal function. Movement proteins not only modulate diffusion-driven transport by gating, but also move themselves between cells and mediate transport of nucleic acids (Fujiwara *et al.*, 1993; Noueiry *et al.*, 1994; Waigmann *et al.*, 1994; Ding *et al.*, 1995; Nguyen *et al.*, 1996). Intercellular traffic occurs within seconds to minutes after microinjection, strongly suggesting that movement proteins operate a pre-existing endogenous pathway for cell–cell transport (Waigmann and Zambryski, 1994).

2. Function and Ultrastructure: a Comparison between Trichome and Mesophyll Plasmodesmata

The development of microinjection as a method to introduce solutes into *N. clevelandii* leaf trichomes (Oparka *et al.*, 1991) allowed subsequent functional transport studies in this system. Microinjection into trichome cells is technically challenging. The high internal turgor pressure, 413 700 Pa (60 psi), complicates impalement and injection of solutes as well as withdrawing of the injection needle from the cell (Oparka *et al.*, 1991; E. Waigmann and P. Zambryski, personal observations). Loss of turgor in a trichome cell results in closure of plasmodesmata linking the cell closer to the leaf (Oparka and Prior, 1992). Cytoplasmic injections can be easily distinguished visually from vacuolar injections. Only cytoplasmic injections, where the thin cytoplasmic seam below the cell wall (Waigmann *et al.*, 1997) is precisely targeted, result in cell-to-cell movement of dyes.

The basal trichome SEL and kinetics of movement of low-molecular-weight dyes and fluorescently labelled dextrans of various size were assessed by microinjection. Since commercially available dextrans are polydisperse, they were fractionated and the molecular mass of individual fractions determined prior to use (Waigmann and Zambryski, 1995). Movement of low-molecular-weight dyes and dextrans is in principle bidirectional but often displays a preference in directionality towards the tip cell (Waigmann and Zambryski, 1995). As observed in *Abutilon* nectary trichomes, there is no dye movement from trichome cells into the leaf tissue (Terry and Robards, 1987; Waigmann and Zambryski, 1995). This may reflect inherent properties of the trichome plasmodesmata that routinely transport compounds from the leaf towards the glandular tip cell to support its synthetic and secretory activity. Dye movement between trichome cells is slow (1–2 minutes to spread to neighbouring cells, 5 minutes to spread throughout a trichome; Oparka and Prior, 1992; Waigmann and Zambryski, 1995; Angell *et al.*, 1996) and injected cells retain substantial amounts of fluorescence, whereas movement in mesophyll cells occurs in seconds and injected cells are virtually drained of dye (Wolf *et al.*, 1989). This difference is likely to be due to differences in the number of cells available for intercellular transport in the two tissues: mesophyll cells are a highly interconnected network such that dye can spread to many cells, while only 4–8 cells are connected in a *N. clevelandii* leaf trichome (Fig. 4A).

Surprisingly, the basal SEL of *N. clevelandii* trichome plasmodesmata is as high as 7 kDa based on movement of purified dextran fractions (Waigmann and Zambryski, 1995). This is considerably higher than the generally accepted value of <1 kDa for diffusion-driven transport in mesophyll cells and other tissue types (Table I; Tucker, 1982; Goodwin, 1983; Terry and Robards, 1987; Wolf *et al.*, 1989; Derrick *et al.*, 1990). Earlier studies concluded that the basal SEL for dextrans in trichome plasmodesmata is <4.4 kDa (Derrick *et al.*, 1992; Vaquero *et al.*, 1994). However, movement was scored after a considerably shorter observation time. In addition, unfractionated polydisperse dextrans were used, which complicates assessment of SEL. Nonetheless, Vaquero *et al.* (1994) report cell-to-cell movement of 4.4 kDa dextrans in 12% of microinjections into *N. tabacum* leaf trichomes, supporting that at least a subset of *N. tabacum* trichomes has a higher basal SEL. Similarly, Angell *et al.* (1996) score cell-to-cell movement of 4.4 kDa dextrans in 33% of microinjections into *N. clevelandii* trichomes, thereby confirming their high plasmodesmal SEL. A high basal SEL comparable to that of *N. clevelandii* trichomes occurs between companion cells and sieve elements (Kempers *et al.*, 1993; Kempers and van Bel, 1997). The enucleate sieve elements rely on companion cells for maintenance, requiring a continuous supply of proteins. Just as in trichomes, companion cell plasmodesmata are likely to be adapted to intense transport activity and the high basal SEL may reflect this adaptation.

Several viruses move through *Nicotiana* trichome plasmodesmata. Microinjection of tobacco rattle virus (TRV) particles into individual trichome cells resulted in the development of necrotic areas at the periphery of the leaf near the injected trichome (Derrick *et al.*, 1992), suggesting viral spread through the trichome into leaf tissue. Viruses tagged with green fluorescent protein (GFP; Chalfie *et al.*, 1994) provided a direct assay for viral replication and movement through trichomes. GFP-tagged potato virus X (PVX; Baulcombe *et al.*, 1995; Oparka *et al.*, 1995) and GFP-tagged TMV (Fig. 3; Casper and Curtis, 1996) replicate to a high level and move cell-to-cell in *N. clevelandii* and *N. benthamiana* trichomes, respectively. Thus, *Nicotiana* trichomes constitute a valid biological system to study various aspects of viral movement. Indeed, TMV-MP interacts with trichome plasmodesmata to move efficiently from a single injected cell throughout the

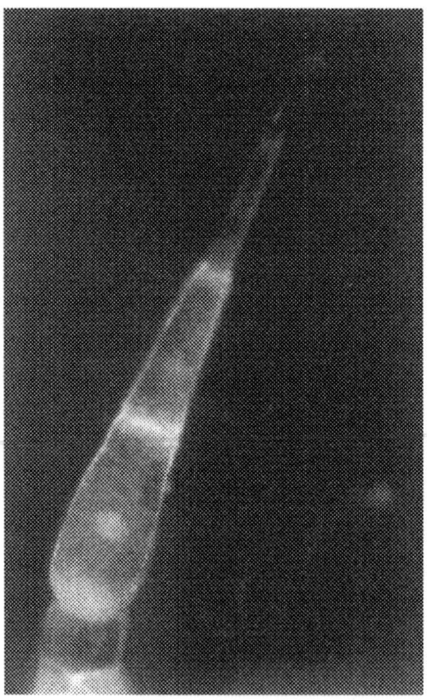

Fig. 3. Fluorescent image of *N. benthamiana* trichome infected with tobacco mosaic virus producing GFP as a reporter gene. GFP fluorescence is represented by different grey scales ranging from white (highest intensity) to black (background), and indicates that tobacco mosaic virus replication and translation took place in the respective cell.

whole trichome in less than 20 minutes (Waigmann and Zambryski, 1995). The presence of TMV-MP had no effect on either the SEL of trichome plasmodesmata nor the time course of dextran movement (Waigmann and Zambryski, 1995), whereas in mesophyll cells, TMV-MP induces a dramatic increase of SEL from the basal value of 1 kDa to 20 kDa (Waigmann *et al.*, 1994). Thus, not only their basal functional properties but also their interaction with TMV-MP is different in these two plasmodesmata types. In contrast to our observations, cell-to-cell movement of PVX particles in *N. clevelandii* trichomes is associated with an increase in SEL to permit passage of 10 kDa dextrans (Angell *et al.*, 1996). This difference may be due to the presence of additional viral proteins or to the different movement mechanism proposed for PVX involving transport of filamentous virions (Santa Cruz *et al.*, 1998).

Nicotiana clevelandii trichome plasmodesmata constitute an efficient protein transport system. Not only TMV-MP itself moves between trichome cells but TMV-MP mediates movement of a β-glucuronidase (GUS): TMV-MP fusion protein with a molecular mass of 90 kDa. GUS by itself with a smaller molecular mass of 60 kDa cannot move between cells (Waigmann and Zambryski, 1995). In trichomes, protein movement is not a simple diffusion process with the Stokes' radius of the transported molecule as the major selection criterion for movement, as assumed for dextrans and small compounds (Goodwin, 1983; Terry and Robards, 1987). Most likely, protein transport through trichome plasmodesmata is a chaperoned event that requires the presence of a plasmodesmal-targeting signal within the transported protein. This idea is consistent with most known examples for protein transport and targeting, such as transport through nuclear pores (Goldfarb and Michnaud, 1991).

Are functional differences between mesophyll and trichome plasmodesmata reflected by ultrastructural differences? This question prompted a comparative study of the ultrastructure of *N. clevelandii* leaf trichome plasmodesmata with mesophyll plasmodesmata. Fixation and embedding protocols were adapted to preserve the mechanically and osmotically labile morphology of *N. clevelandii* trichomes. Results show that plasmodesmata increase in complexity from the tip towards the base cell. Plasmodesmata in the cross-wall between the tip cell/cell 1 (for numbering of cells, see Fig. 4A) manifested as primarily simple, unbranched pores, while plasmodesmata connecting cells closer to the leaf were predominantly complex, branched structures (Waigmann *et al.*, 1997). For all types of *N. clevelandii* trichome plasmodesmata, the outer diameter at orifices was 59 nm (Table I), very close to the published value of 60 nm for mesophyll plasmodesmata (Lucas *et al.*, 1993; Waigmann *et al.*, 1997); their average length ranged from 529 nm in cross-walls connecting tip cell/cell 1 to 1129 nm in cross-walls connecting cell 3/cell 4 (Table I) compared to 823 nm for mesophyll plasmodesmata (Waigmann *et al.*, 1997). Special attention was paid to the entrance areas of

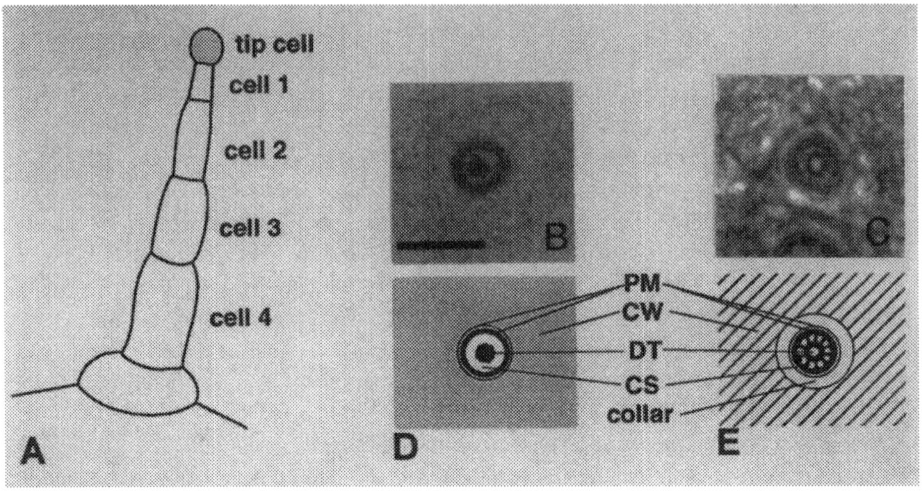

Fig. 4. Ultrastructural comparison of cross-sections through orifices of mesophyll and trichome plasmodesmata. (A) Schematic drawing of a *N. clevelandii* leaf trichome indicating the numbering of trichome cells. (B) Transmission electron microscopic (TEM) image of mesophyll plasmodesma. (C) TEM image of trichome plasmodesma. (D,E) Schematic representation of images in (B) and (C). PM, plasma membrane; DT, desmotubule; CW, cell wall; CS, cytoplasmic sleeve. Scale bar = 100 nm. Reproduced, with permission, from Waigmann *et al.* (1997).

plasmodesmata, the orifice areas (Fig. 1), since these are likely to be sites for control of plasmodesmal traffic. Despite the variation in overall complexity, orifice areas of all types of trichome plasmodesmata appeared structurally similar to each other but clearly distinct from mesophyll plasmodesmata. The most striking difference is the "open" desmotubule of trichome plasmodesmata (Fig. 4C,E) compared to the electron-dense desmotubule in mesophyll plasmodesmata (Fig. 4B,D). In addition, trichome plasmodesmata are surrounded by a structurally distinct area within the cell wall that has been termed the collar in other systems (Turner *et al.*, 1994), whereas mesophyll plasmodesmata are not. The desmotubule of trichome plasmodesmata appears connected to the inner leaflet of the plasma membrane by multiple spokes or filaments (Fig. 4C,E), while such connections are not perceptible in mesophyll plasmodesmata (Fig. 4B,D). These differences in ultrastructure are likely to reflect a difference in architecture between mesophyll and trichome plasmodesmata.

What are the functional implications of an open desmotubular lumen in *N. clevelandii* trichome plasmodesmata compared to a solid rod in mesophyll plasmodesmata? It is intriguing to speculate that the open lumen in trichome plasmodesmata accounts for their larger basal SEL. Possibly, the open desmotubular lumen could be interpreted as a channel serving as a passageway for large molecules. Movement of sugars via an open desmotubular lumen has been previously suggested for photosynthate transport into the phloem (Gamalei *et al.*, 1994). In a second scenario, the open desmotubular lumen would provide space for the adjacent cytoplasmic sleeve to expand into when engaged in transport (Waigmann *et al.*, 1997). The latter hypothesis is consistent with the currently held view that plasmodesmal transport proceeds through the cytoplasmic sleeve.

E. OTHER TYPES OF LEAF AND STEM TRICHOMES

A number of other glandular leaf and stem trichomes, such as petiolar hairs of *Lycopersicum esculentum* (tomato) and leaf and stem trichomes of *Cicer arietinum* (chickpea), have been assessed for their cell-to-cell transport mechanisms.

In tomato trichomes, the contribution of intracellular movement to the kinetics of cell-to-cell transport was studied. It was concluded that movement of low-molecular-weight dyes is not affected by inhibition of cytoplasmic streaming and that cell–cell transport progresses by diffusion (Barclay *et al.*, 1982).

Chickpea trichomes consist of one basal cell, three stalk cells and a head of 14 secretory cells (Lazzaro and Thomson, 1989). All cells of the trichome are interconnected by numerous plasmodesmata, indicating a symplastic pathway for movement of solutes throughout the trichome (Lazzaro and Thomson, 1989). Plasmodesmata connecting the head cells to each other or to the uppermost stalk cell are simple and unbranched, those connecting the lower stalk cells and the basal cell are mainly branched with a central cavity. This basipetal increase in complexity is also observed in *N. clevelandii* trichomes (Waigmann *et al.*, 1997). Interestingly, chickpea trichomes are distinct from nectary trichomes as they lack plasmodesmata between the basal cell and the surrounding mesophyll tissue. Since chickpea trichomes contain only few chloroplasts, they must be supplied with substrate from the underlying leaf tissue by a non-symplastic route (Lazzaro and Thomson, 1989). No detailed information on the substructure of plasmodesmata nor functional data obtained by microinjection are available.

IV. SUMMARY AND CONCLUSIONS

In the last two decades, research on cell–cell transport in trichomes has focused on two major biological systems, nectary trichomes and *Nicotiana* leaf trichomes. The favourable cellular architecture of trichomes prompted analyses of general aspects of cell–cell transport, and both systems contributed important insights to the understanding of molecular transport through plasmodesmata. Trichomes are a highly specialized array of cells evolved to transport, synthesize and secrete a defined set of molecules in large amounts. Consequently, trichome plasmodesmata are likely to have been optimally adjusted in structure and function to their respective transport role. Not surprisingly, specific aspects of plasmodesmal transport distinguish trichome plasmodesmata from plasmodesmata in other tissue types.

In the 1980s, it was predominantly nectary trichomes, with their abundant intercellular transport and secretion, that fascinated many scientists. Nectary trichomes route a high flux of low-molecular-weight compounds, primarily sucrose, from the base to the tip. Their transport requirement was solved by evolving a large number of plasmodesmata interconnecting individual cells. In the 1990s, research on intercellular traffic gained an enormous impetus from the availability of plant viral movement proteins as a molecular tool to probe plasmodesmal function. Consequently, trichomes that support movement of plant viruses such as *Nicotiana* leaf trichomes were studied. The functional differences distinguishing leaf hair plasmodesmata from mesophyll plasmodesmata expanded our knowledge on the diversity of transport mechanisms through plasmodesmata.

In spite of significant advances in understanding molecular traffic through plasmodesmata, major questions, such as the exact transport pathway within plasmodesmata or the molecular composition of these complex organelles, remain unsolved. To address these issues, diverse experimental tools, such as microinjection, electron microscopy and immunolocalization, combined with genetic approaches and molecular biology, will be required. There is no doubt that trichomes with their unique morphological and functional features will continue to be a valuable system.

ACKNOWLEDGEMENTS

E.W. was supported by fellowship APART 441 from the Austrian Academy of Sciences and grant P12614-MOB from the "Fonds zur Foerderung der wissenschaftlichen Forschung". P.Z. was supported by NIH grant GM45244.

REFERENCES

Angell, S. M., Davies, C. and Baulcombe, D. C. (1996). Cell-to-cell movement of potato virus X is associated with a change in the size-exclusion limit of plasmodesmata in trichome cells of *Nicotiana clevelandii*. *Virology* **216**, 197–201.

Badelt, K., White, R. G., Overall, R. L. and Vesk, M. (1994). Ultrastructural specializations of the cell wall sleeve around plasmodesmata. *American Journal of Botany* **81**, 1422–1427.

Barclay, G. F., Peterson, C. A. and Tyree, M. T. (1982). Transport of fluorescein in trichomes of *Lycopersicon esculentum*. *Canadian Journal of Botany* **60**, 397–402.
Bauer, L. (1953). Zur Frage der Stoffbewegung in der Pflanze mit besonderer Berücksichtigung der Wanderung von Fluorochromen. *Planta* **42**, 367–451.
Baulcombe, D., Chapman, S. and Santa Cruz, S. (1995). Jellyfish green fluorescent protein as a reporter for virus infections. *Plant Journal* **7**, 1045–1053.
Blackman, L. M. and Overall, R. L. (1998). Immunolocalisation of the cytoskeleton to plasmodesmata of *Chara corallina*. *Plant Journal* **14**, 733–741.
Carrington, J. C., Kasschau, K. D., Mahajan, S. K. and Schaad, M. C. (1996). Cell-to-cell and long-distance transport of viruses in plants. *Plant Cell* **8**, 1669–1681.
Casper, S. J. and Curtis, A. H. (1996). Expression of the green fluorescent protein-encoding gene from a tobacco mosaic virus-based vector. *Gene* **173**, 69–73.
Chalfie, M., Tu, Y., Euskirchen, G., Ward, W. W. and Prasher, D. C. (1994). Green fluorescent protein as a marker for gene expression. *Science* **263**, 802–805.
Citovsky, V. and Zambryski, P. (1991). How do plant virus nucleic acids move through intercellular connections? *Bioessays* **13**, 373–379.
Citovsky, V. and Zambryski, P. (1993). Transport of nucleic acids through membrane channels: snaking through small holes. *Annual Review of Microbiology* **47**, 167–197.
Citovsky, V., Knorr, D., Schuster, G. and Zambryski, P. (1990). The P30 movement protein of tobacco mosaic virus is a single strand nucleic acid binding protein. *Cell* **60**, 637–647.
Citovsky, V., Wong, M. L., Shaw, A., Prasad, B. V. V. and Zambryski, P. (1992). Visualization and characterization of tobacco mosaic virus movement protein binding to single-stranded nucleic acids. *Plant Cell* **4**, 397–411.
Derrick, P. M., Barker, H. and Oparka, K. J. (1990). Effect of virus infection on symplastic transport of fluorescent tracers in *Nicotiana clevelandii* leaf epidermis. *Planta* **181**, 555–559.
Derrick, P. M., Barker, H. and Oparka, K. J. (1992). Increase in plasmodesmatal permeability during cell-to-cell spread of tobacco rattle tobravirus from individually inoculated cells. *Plant Cell* **4**, 1405–1412.
Ding, B. (1997). Cell-to-cell transport of macromolecules through plasmodesmata: a novel signalling pathway in plants. *Trends in Cell Biology* **7**, 5–9.
Ding, B., Turgeon, R. and Parthasarathy, M. V. (1992). Substructure of freeze-substituted plasmodesmata. *Protoplasma* **169**, 28–41.
Ding, B., Li, Q., Nguyen, L., Palukaitis, P. and Lucas, W. J. (1995). Cucumber mosaic virus 3a protein potentiates cell-to-cell trafficking of CMV RNA in tobacco plants. *Virology* **207**, 345–353.
Ding, B., Kwan, M. O. and Warnberg, L. (1996). Evidence that actin filaments are involved in controlling the permeability of plasmodesmata in tobacco mesophyll. *Plant Journal* **10**, 157–164.
Ehlers, K., Schulz, M. and Kollmann, R. (1996). Subcellular localization of ubiquitin in plant protoplasts and the function of ubiquitin in selective degradation of outer-wall plasmodesmata in regenerating protoplasts. *Planta* **199**, 139–151.
Eleftheriou, E. P. and Hall, J. L. (1983a). The extrafloral nectaries of cotton. II. Cytochemical localization of ATPase activity and Ca^{2+}-binding sites, and selective osmium impregnation. *Journal of Experimental Botany* **34**, 1066–1079.
Eleftheriou, E. P. and Hall, J. L. (1983b). The extrafloral nectaries of cotton. I. Fine

structure of the secretory papillae. *Journal of Experimental Botany* **34**, 103–119.

Epel, B. L., van Lent, J. W. M., Cohen, L., Kotlizky, G., Katz, A. and Yahalom, A. (1996). A 41 kDa protein isolated from maize mesocotyl cell walls immunolocalizes to plasmodesmata. *Protoplasma* **191**, 70–78.

Erwee, M. G. and Goodwin, P. B. (1984). Characterisation of the *Egeria densa* leaf symplast: response to plasmolysis, deplasmolysis and to aromatic amino acids. *Protoplasma* **122**, 162–168.

Fahn, A. (1979). Ultrastructure of nectaries in relation to nectar secretion. *American Journal of Botany* **64**, 977–985.

Findley, N. and Mercer, F. V. (1971). Nectar production in Abutilon. I. Movement of nectar through the cuticle. *Australian Journal of Biological Science* **24**, 647–656.

Fujiwara, T., Giesman-Cookmeyer, D., Ding, B., Lommel, S. A. and Lucas, W. J. (1993). Cell-to-cell trafficking of macromolecules through plasmodesmata potentiated by the red clover necrotic mosaic virus movement protein. *Plant Cell* **5**, 1783–1794.

Gaffal, K. P., Heimler, W. and ElGammal, S. (1998). The floral nectary of *Digitalis purpurea* L., structure and nectar secretion. *Annals of Botany* **81**, 251–262.

Gamalei, Y. V., van Bel, A. J. E., Pakhomova, V. M. and Sjutkina, V. A. (1994). Effects of temperature on the conformation of the endoplasmic reticulum and on starch accumulation in leaves with the symplasmic minor-vein configuration. *Planta* **194**, 443–453.

Glockmann, C. and Kollmann, R. (1996). Structure and development of cell connections in the phloem of *Metasequoia glyptostroboides* needles I. Ultrastructural aspects of modified primary plasmodesmata in Strasburger cells. *Protoplasma* **193**, 191–203.

Goldfarb, D. and Michnaud, N. (1991). Pathways for the nuclear transport of proteins and RNAs. *Trends in Cell Biology* **1**, 20–24.

Goodwin, P. B. (1983). Molecular size limit for movement in the symplast of the *Elodea* leaf. *Planta* **157**, 124–130.

Goshroy, S., Lartey, R., Sheng, J. and Citovsky, V. (1997). Transport of proteins and nucleic acids through plasmodesmata. *Annual Review of Plant Physiology and Plant Molecular Biology* **48**, 27–50.

Grabski, S., de Feijter, A. W. and Schindler, M. (1993). Endoplasmic reticulum forms a dynamic continuum for lipid diffusion between contiguous soybean root cells. *Plant Cell* **5**, 25–38.

Gunning, B. E. S. and Hughes, J. E. (1976). Quantitative assessment of symplastic transport of pre-nectar into the trichomes of *Abutilon* nectaries. *Australian Journal of Plant Physiology* **3**, 619–637.

Gunning, B. E. S. and Overall, R. L. (1983). Plasmodesmata and cell-to-cell transport in plants. *BioScience* **33**, 260–265.

Gunning, B. E. S. and Robards, A. W. (1976). Plasmodesmata: current knowledge and outstanding problems. *In* "Intercellular Communication in Plants: Studies on Plasmodesmata" (B. S. E. Gunning and A. W. Robards, eds), pp. 297–311. Springer Verlag, Heidelberg.

Heinlein, M., Epel, B. L., Padgett, S. H. and Beachy, R. N. (1995). Interaction of tobamovirus movement proteins with the plant cytoskeleton. *Science* **270**, 1983–1985.

Hughes, J. E. and Gunning, B. E. S. (1980). Glutaraldehyde-induced deposition of callose. *Canadian Journal of Botany* **58**, 250–258.

Kandra, L., Severson, R. and Wagner, G. J. (1990). Modified branched-chain amino acid pathways give rise to acyl acids of sucrose esters exuded from tobacco leaf trichomes. *European Journal of Biochemistry* **188**, 385–391.

Kelsey, R. G., Reynolds, G. W. and Rodriguez, E. (1984). The chemistry of biologically active constituents secreted and stored in plant glandular trichomes. *In* "Biology and Chemistry of Plant Trichomes" (E. Rodriguez, P. L. Healey and I. Mehta, eds), pp. 187–241. Plenum Press, New York.

Kempers, R. and van Bel, A. J. E. (1997). Symplastic connections between sieve element and companion cell in the stem phloem of *Vicia faba* L. have a molecular exclusion limit of at least 10 kDa. *Planta* **201**, 195–201.

Kempers, R., Prior, D. A. M., Van Bel, A. J. E. and Oparka, K. J. (1993). Plasmodesmata between sieve element and companion cell of extrafascicular stem phloem of *Cucurbita maxima* permit passage of 3 kDa fluorescent probes. *Plant Journal* **4**, 567–575.

Kragler, F., Lucas, W. J. and Monzer, J. (1998). Plasmodesmata: dynamics, domains and patterning. *Annals of Botany* **81**, 1–10.

Kronestedt, E. C., Stark, M. and Robards, A. W. (1986). Development of trichomes in the Abutilon nectary gland. *Nordic Journal of Botany* **6**, 627–639.

Kuo, J. and Pate, J. S. (1985). The extrafloral nectaries of cowpea (Vigna unguiculata (L.) Walp.): I. Morphology, anatomy and fine structure. *Planta* **166**, 15–27.

Lazzaro, M. D. and Thomson, W. W. (1989). Ultrastructure of organic acid secreting trichomes of chickpea (*Cicer arietinum*). *Canadian Journal of Botany* **67**, 2669–2677.

Lopez-Saez, J. F., Gimenez-Martin, G. and Risueno, M. C. (1966). Fine structure of the plasmodesm. *Protoplasma* **61**, 81–84.

Lucas, W. J. (1995). Plasmodesmata: intercellular channels for macromolecular transport in plants. *Current Opinion in Cell Biology* **7**, 673–680.

Lucas, W. J. and Gilbertson, R. L. (1994). Plasmodesmata in relation to viral movement within leaf tissues. *Annual Review of Phytopathology* **32**, 387–411.

Lucas, W. J. and Wolf, S. (1993). Plasmodesmata: the intercellular organelles of green plants. *Trends in Cell Biology* **3**, 308–315.

Lucas, W., Ding, B. and Van der Schoot, C. (1993). Plasmodesmata and the supracellular nature of plants. *New Phytologist* **125**, 435–476.

Marginson, R., Sedgley, M. and Knox, R. B. (1985). Structure and histochemistry of the extrafloral nectary of *Acacia terminalis* (Leguminosae, Mimosidae). *Protoplasma* **127**, 21–30.

McCaskill, D., Gershenzon, J. and Croteau, R. (1992). Morphology and monoterpene biosynthetic capabilities of secretory cell clusters isolated from glandular trichomes of peppermint (*Mentha piperita* L.). *Planta* **187**, 445–454.

McLean, B. G., Zupan, J. and Zambryski, P. C. (1995). TMV P30 movement protein associates with the cytoskeleton in tobacco cells. *Plant Cell* **7**, 2101–2114.

McLean, B. G., Hempel, F. D. and Zambryski, P. C. (1997). Plant intercellular communication via plasmodesmata. *Plant Cell* **9**, 1043–1054.

Mersey, B. and McCully, M. E. (1978). Monitoring of the course of fixation of plant cells. *Journal of Microscopy* **114**, 49–76.

Nguyen, L., Lucas, W. J., Ding, B. and Zaitlin, M. (1996). Viral RNA trafficking is inhibited in replicase-mediated resistant transgenic tobacco plants. *Proceedings of the National Academy of Sciences (USA)* **93**, 12 643–12 647.

Nichol, P. and Hall, J. L. (1988). Characteristics of nectar secretion by the extrafloral nectaries of *Ricinus communis*. *Journal of Experimental Botany* **39**, 573–586.

Noueiry, A. O., Lucas, W. J. and Gilbertson, R. L. (1994). Two proteins of a plant virus coordinate nuclear and plasmodesmal transport. *Cell* **76**, 925–932.

Olesen, P. (1975). Plasmodesmata between mesophyll and bundle sheath cells in

relation to the exchange in C_4-acids. *Planta* **123**, 199–202.
Olesen, P. (1979). The neck constriction in plasmodesmata: evidence for a peripheral sphincter-like structure revealed by fixation with tannic acid. *Planta* **144**, 349–358.
Oparka, K. J. (1993). Signalling via plasmodesmata – the neglected pathway. *Seminars in Cell Biology* **4**, 131–138.
Oparka, K. J. and Prior, D. A. M. (1992). Direct evidence for pressure-generated closure of plasmodesmata. *Plant Journal* **2**, 741–750.
Oparka, K. J., Murphy, R., Derrick, P. M., Prior, D. A. M. and Smith, J. A. C. (1991). Modification of the pressure-probe technique permits controlled intracellular microinjection of fluorescent probes. *Journal of Cell Science* **98**, 539–544.
Oparka, K. J., Roberts, A. G., Prior, D. A. M., Chapman, S., Baulcombe, D. and Santa Cruz, S. (1995). Imaging the green fluorescent protein in plants – viruses carry the torch. *Protoplasma* **189**, 133–141.
Overall, R. L., Wolfe, J. and Gunning, B. E. S. (1982). Intercellular communication in *Azolla* roots: I. Ultrastructure of plasmodesmata. *Protoplasma* **111**, 134–150.
Pate, J. S., Peoples, M. B., Storer, B. J. and Atkins, C. A. (1985). The extrafloral nectaries of cowpea (*Vigna unguiculata* (L.) Walp.). II. Nectar composition, origin of nectar solutes, and nectary functioning. *Planta* **166**, 28–38.
Plowe, J. Q. (1931). Membranes in the plant cell. II. Localization of differential permeability in the plant protoplast. *Protoplasma* **12**, 1221–1240.
Poirson, A., Turner, A. P., Giovane, C., Berna, A., Roberts, K. and Godefroy-Colburn, T. (1993). Effect of the alfalfa mosaic virus movement protein expressed in transgenic plants on the permeability of plasmodesmata. *Journal of General Virology* **74**, 2459–2461.
Radford, J. E., Vesk, M. and Overall, R. L. (1998). Callose deposition at plasmodesmata. *Protoplasma* **201**, 30–37.
Radford, J. E. and White, R. G. (1998). Localization of a myosin-like protein to plasmodesmata. *Plant Journal* **14**, 743–750.
Robards, A. W. (1976). Plasmodesmata in higher plants. *In* "Intercellular Communication in Plants: Studies on Plasmodesmata" (B. E. S. Gunning and A. W. Robards, eds), pp. 15–53. Springer-Verlag, Berlin, Heidelberg, New York.
Robards, A. W. and Lucas, W. J. (1990). Plasmodesmata. *Annual Review of Plant Physiology and Plant Molecular Biology* **41**, 369–419.
Robards, A. W. and Stark, M. (1988). Nectar secretion in *Abutilon*: a new model. *Protoplasma* **142**, 79–91.
Robinson-Beers, K. and Evert, R. F. (1991). Fine structure of plasmodesmata in mature leaves of sugar cane. *Planta* **184**, 307–318.
Santa Cruz, S., Roberts, A. G., Prior, A. M., Chapman, S. and Oparka, K. J. (1998). Cell-to-cell and phloem-mediated transport of potato virus X: the role of virions. *Plant Cell* **10**, 495–510.
Sawidis, T., Eleftheriou, E. and Tsekos, I. (1987). The floral nectaries of *Hibiscus rosa-sinensis* L. II. Plasmodesmal frequencies. *Phyton* **27**, 155–164.
Sawidis, T., Eleftheriou, E. P. and Tsekos, I. (1989). The floral nectaries of Hibiscus rosa-sinensis III. A morphometric and ultrastructural approach. *Nordic Journal of Botany* **9**, 63–71.
Schnepf, E. (1974). Gland cells. *In* "Dynamic Aspects of Plant Ultrastructure" (A. W. Robards, ed.), pp. 331–353. McGraw-Hill Book Co. Ltd, Maidenhead.
Schulz, A. (1995). Plasmodesmal widening accompanies the short-term increase in symplastic phloem unloading in pea root tips under osmotic stress. *Protoplasma* **188**, 22–37.
Schumacher, W. (1936). Untersuchungen über die Wanderung des Fluoreszein in den

Haaren von Cucurbita pepo. *Jahrbuch der wissenschaftlichen Botanik* **82**, 507–533.

Strasburger, E. (1901). Ueber Plasmaverbindungen pflanzlicher Zellen. *Jahrbuch der wissenschaftlichen Botanik* **36**, 493–610.

Tangl, E. (1879). Ueber offene Communicationen zwischen den Zellen des Endosperms einiger Samen. *Jahrbuch der wissenschaftlichen Botanik* **12**, 170–190.

Terry, B. R. and Robards, A. W. (1987). Hydrodynamic radius alone governs the mobility of molecules through plasmodesmata. *Planta* **171**, 145–157.

Thompson, W. W. and Platt-Aloia, K. (1985). The ultrastructure of the plasmodesmata of the salt glands of Tamarix as revealed by plasmolysis, detergent extraction, and protease digestion. *Protoplasma* **125**, 13–23.

Tilney, L. G., Cooke, T. J., Connelly, P. S. and Tilney, M. S. (1990). The distribution of plasmodesmata and its relationship to morphogenesis in fern gametophytes. *Development* **110**, 1209–1221.

Tucker, E. B. (1982). Translocation in the staminal hairs of *Setcreasea purpurea*. 1. A study of cell ultrastructure and cell-to-cell passage of molecular probes. *Protoplasma* **113**, 193–201.

Turner, A., Wells, B. and Roberts, K. (1994). Plasmodesmata of maize root tips: structure and composition. *Journal of Cell Science* **107**, 3351–3361.

Vaquero, C., Turner, P. A., Demangeat, G., Sanz, A., Serra, M. T., Roberts, K. and Garcia-Luque, I. (1994). The 3a protein from cucumber mosaic virus increases the gating capacity of plasmodesmata in transgenic tobacco plants. *Journal of General Virology* **75**, 3193–3197.

Vermeer, J. and Peterson, R. L. (1979). Glandular trichomes on the inflorescence of *Chrysanthemum morifolium* cv. Dramatic (Compositae). II. Ultrastructure and histochemistry. *Journal of Botany* **57**, 714–729.

Waigmann, E. and Zambryski, P. (1994). Plasmodesmata: gateways for rapid information transfer. *Current Biology* **4**, 713–716.

Waigmann, E. and Zambryski, P. (1995). Tobacco mosaic virus movement protein-mediated protein transport between trichome cells. *Plant Cell* **7**, 2069–2079.

Waigmann, E., Lucas, W., Citovsky, V. and Zambryski, P. (1994). Direct functional assay for tobacco mosaic virus cell-to-cell movement protein and identification of a domain involved in increasing plasmodesmal permeability. *Proceedings of the National Academy of Sciences (USA)* **91**, 1433–1437.

Waigmann, E., Turner, A., Paert, J., Roberts, K. and Zambryski, P. (1997). Ultrastructural analysis of trichome plasmodesmata reveals major differences from mesophyll plasmodesmata. *Planta* **203**, 75–84.

White, R. G., Badelt, K., Overall, R. L. and Vesk, M. (1994). Actin associated with plasmodesmata. *Protoplasma* **180**, 169–184.

Wolf, S., Deom, C. M., Beachy, R. N. and Lucas, W. J. (1989). Movement protein of tobacco mosaic virus modifies plasmodesmatal size exclusion limit. *Science* **246**, 377–379.

Zambryski, P. (1995). Plasmodesmata: plant channels for molecules on the move. *Science* **270**, 1943–1944.

AUTHOR INDEX

Numbers in *italic* refer to pages on which full references are listed

Q

R

Y

Z

SUBJECT INDEX

Q

R

U

V

W

X

Z

FSC
www.fsc.org
MIX
Papier | Fördert
gute Waldnutzung
FSC® C083411

Kolibri 360 GmbH
Ferdinand-Jühlke-Straße 7
99095 Erfurt, Deutschland
produktsicherheit@kolibri360.de